T0215831

Lecture Notes in Computer Science   9928

Commenced Publication in 1973
Founding and Former Series Editors:
Gerhard Goos, Juris Hartmanis, and Jan van Leeuwen

Advanced Research in Computing and Software Science
Subline of Lecture Notes in Computer Science

More information about this series at http://www.springer.com/series/7409

Martin Gairing · Rahul Savani (Eds.)

Algorithmic Game Theory

9th International Symposium, SAGT 2016
Liverpool, UK, September 19–21, 2016
Proceedings

 Springer

Editors
Martin Gairing
University of Liverpool
Liverpool
UK

Rahul Savani
University of Liverpool
Liverpool
UK

ISSN 0302-9743   ISSN 1611-3349   (electronic)
Lecture Notes in Computer Science
ISBN 978-3-662-53353-6   ISBN 978-3-662-53354-3   (eBook)
DOI 10.1007/978-3-662-53354-3

Library of Congress Control Number: 2016949566

LNCS Sublibrary: SL3 – Information Systems and Applications, incl. Internet/Web, and HCI

Printed on acid-free paper

This Springer imprint is published by Springer Nature
The registered company is Springer-Verlag GmbH Berlin Heidelberg

Preface

This volume contains the proceedings of the 9th International Symposium on Algorithmic Game Theory (SAGT), held in Liverpool, UK, in September 2016.

The program of SAGT 2016 consisted of 3 invited lectures and 28 presentations of refereed submissions. The invited speakers were Constantinos Daskalakis (MIT), Olivier Gossner (LSE and École Polytechnique), and Kurt Mehlhorn (Max-Planck-Institut für Informatik). After a careful reviewing process, the Program Committee selected 28 out of 62 submissions. To accommodate the publishing traditions of different fields, authors of accepted papers could ask that only a one-page abstract of the paper appears in the proceedings. Among the 28 accepted papers, the authors of two papers opted for publication as a one-page abstract. The accepted submissions cover various important aspects of algorithmic game theory such as computational aspects of games, congestion games and networks, matching and voting, auctions and markets, and mechanism design. This year, with financial support from Springer, we introduced a best paper award, which was given to "The Big Match in Small Space" by Hansen, Ibsen-Jensen, and Koucky.

We would like to thank all authors who submitted their research work and all Program Committee members and external reviewers for their effort in selecting the program for SAGT 2016. We thank ACM SIGecom, EATCS, Facebook, and Springer for their generous support. We thank Anna Kramer and Alfred Hofmann at Springer for helping with the proceedings. We are grateful for the use of the EasyChair paper management system.

July 2016

Martin Gairing
Rahul Savani

Organization

Program Committee

Yakov Babichenko	Technion, Israel
Umang Bhaskar	Tata Institute of Fundamental Research, India
Yang Cai	McGill University, Canada
Xi Chen	Columbia University, USA
Paul Duetting	ETH Zürich, Switzerland
Edith Elkind	University of Oxford, UK
Martin Gairing (Co-chair)	University of Liverpool, UK
Gagan Goel	Google Research, USA
Paul Goldberg	University of Oxford, UK
Tobias Harks	Augsburg University, Germany
Max Klimm	Technische Universität Berlin, Germany
Kostas Kollias	Google Research, USA
Sebastien Lahaie	Microsoft Research, USA
Pascal Lenzner	Hasso-Plattner-Institute Potsdam, Germany
Stefano Leonardi	University of Rome "La Sapienza", Italy
Troels Bjerre Lund	IT-University of Copenhagen, Denmark
David Manlove	University of Glasgow, UK
Vangelis Markakis	Athens University of Economics and Business, Greece
Ruta Mehta	University of Illinois at Urbana-Champaign, USA
Britta Peis	RWTH Aachen, Germany
Ron Peretz	Bar Ilan University, Israel
Georgios Piliouras	Singapore University of Technology and Design, Singapore
Maria Polukarov	University of Southampton, UK
Rahul Savani (Co-chair)	University of Liverpool, UK
Marco Scarsini	LUISS Rome, Italy
Alexander Skopalik	Heinz Nixdorf Institute and Paderborn University, Germany
Eric Sodomka	Facebook Research, USA
Orestis Telelis	University of Piraeus, Greece

Steering Committee

Elias Koutsoupias	University of Oxford, UK
Marios Mavronicolas	University of Cyprus, Cyprus
Dov Monderer	Technion, Israel
Burkhard Monien	Paderborn University, Germany

Christos Papadimitriou	UC Berkeley, USA
Giuseppe Persiano	University of Salerno, Italy
Paul Spirakis (Chair)	University of Liverpool, UK and CTI Patras, Greece

Organizing Team

Eleftherios Anastasiadis	University of Liverpool, UK
Giorgos Christodoulou (Chair)	University of Liverpool, UK
Argyrios Deligkas	University of Liverpool, UK
Tobenna Peter Igwe	University of Liverpool, UK
Grammateia Kotsialou	University of Liverpool, UK
Rebekah Martin	University of Liverpool, UK
Alkmini Sgouritsa	University of Liverpool, UK
Elaine Smith	University of Liverpool, UK
Lisa Smith	University of Liverpool, UK

Additional Reviewers

Abed, Fidaa
Amir, Nadav
Barman, Siddharth
Birmpas, Georgios
Bousquet, Nicolas
Branzei, Simina
Cheng, Yu
Colini Baldeschi, Riccardo
Cord-Landwehr, Andreas
Dall'Aglio, Marco
De Keijzer, Bart
Farczadi, Linda
Fearnley, John
Feldotto, Matthias
Ferraioli, Diodato
Filos-Ratsikas, Aris
Fu, Hu
Gkatzelis, Vasilis
Gonczarowski, Yannai

Hackfeld, Jan
Huang, Zhiyi
Igarashi, Ayumi
Kotsialou, Grammateia
Lang, Jérôme
Lewenberg, Yoad
Li, Shouwei
Liu, Qingmin
Lykouris, Thodoris
Marmolejo, Francisco
Matsui, Tomomi
Matuschke, Jannik
Mestre, Julian
Monnot, Barnabé
Moscardelli, Luca
Obraztsova, Svetlana
Ordonez, Fernando
Panageas, Ioannis
Paparas, Dimitris
Pasquale, Francesco

Rey, Anja
Schmand, Daniel
Segal-Halevi, Erel
Simon, Sunil Easaw
Skowron, Piotr
Staudigl, Mathias
Tzamos, Christos
Tönnis, Andreas
Wakayama, Takuma
Wako, Jun
Weil, Vera
Wierz, Andreas
Xefteris, Dimitrios
Yang, Ger
Yazdanbod, Sadra
Zhang, Qiang
Zhao, Mingfei
Ziani, Juba
Ziliotto, Bruno

Contents

Matching and Voting

Auctions and Markets

Mechanism Design

Abstracts

Computational Aspects of Games

Logarithmic Query Complexity for Approximate Nash Computation in Large Games

Paul W. Goldberg[1], Francisco J. Marmolejo Cossío[1($\boxtimes$)],
and Zhiwei Steven Wu[2]

[1] University of Oxford, Oxford, UK
{paul.goldberg,francisco.marmolejocossio}@cs.ox.ac.uk
[2] University of Pennsylvania, Philadelphia, USA
wuzhiwei@cis.upenn.edu

Abstract. We investigate the problem of equilibrium computation for "large" n-player games where each player has two pure strategies. Large games have a Lipschitz-type property that no single player's utility is greatly affected by any other individual player's actions. In this paper, we assume that a player can change another player's payoff by at most $\frac{1}{n}$ by changing her strategy. We study algorithms having query access to the game's payoff function, aiming to find ε-Nash equilibria. We seek algorithms that obtain ε as small as possible, in time polynomial in n.

Our main result is a randomised algorithm that achieves ε approaching $\frac{1}{8}$ in a *completely uncoupled* setting, where each player observes her own payoff to a query, and adjusts her behaviour independently of other players' payoffs/actions. $O(\log n)$ rounds/queries are required. We also show how to obtain a slight improvement over $\frac{1}{8}$, by introducing a small amount of communication between the players.

1 Introduction

In studying the computation of solutions of multi-player games, we have the well-known problem that a game's payoff function has description length exponential in the number of players. One approach is to assume that the game comes from a concisely-represented class (for example, graphical games, anonymous games, or congestion games), and another one is to consider algorithms that have query access to the game's payoff function.

In this paper, we study the computation of approximate Nash equilibria of multi-player games having the feature that if a player changes her behaviour, she only has a small effect on the payoffs that result to any other player. These games, sometimes called *large* games, or *Lipschitz* games, have recently been studied in the literature, since they model various real-world economic interactions; for example, an individual's choice of what items to buy may have a small effect on prices, where other individuals are not strongly affected. Note that these games do not have concisely-represented payoff functions, which makes them a natural class of games to consider from the query-complexity perspective.

© Springer-Verlag Berlin Heidelberg 2016
M. Gairing and R. Savani (Eds.): SAGT 2016, LNCS 9928, pp. 3–14, 2016.
DOI: 10.1007/978-3-662-53354-3_1

It is already known how to compute approximate *correlated equilibria* for unrestricted n-player games. Here we study the more demanding solution concept of approximate Nash equilibrium.

Large games (equivalently, small-influence games) are studied in Kalai [16] and Azrieli and Shmaya [1]. In both of these the existence of pure ε-Nash equilibria for $\varepsilon = \gamma\sqrt{8n\log(2mn)}$ is established, where γ is the largeness/Lipschitz parameter of the game. In particular, since we assume that $\gamma = \frac{1}{n}$ and $m = 2$ we notice that $\varepsilon = O(n^{-1/2})$ so that there exist arbitrarily accurate pure Nash equilibria in large games as the number of players increases. Kearns et al. [17] study this class of games from the mechanism design perspective of mediators who aim to achieve a good outcome to such a game via recommending actions to players. Babichenko [2] studies large binary-action *anonymous* games. Anonymity is exploited to create a randomised dynamic on pure strategy profiles that with high probability converges to a pure approximate equilibrium in $O(n\log n)$ steps.

Payoff query complexity has been recently studied as a measure of the difficulty of computing game-theoretic solutions, for various classes of games. Upper and lower bounds on query complexity have been obtained for bimatrix games [6,7], congestion games [7], and anonymous games [11]. For general n-player games (where the payoff function is exponential in n), the query complexity is exponential in n for exact Nash, also exact correlated equilibria [15]; likewise for approximate equilibria with deterministic algorithms (see also [4]). For randomised algorithms, query complexity is exponential for *well-supported* approximate equilibria [3], which has since been strengthened to any ε-Nash equilibria [5]. With randomised algorithms, the query complexity of approximate correlated equilibrium is $\Theta(\log n)$ for any positive ε [10].

Our main result applies in the setting of *completely uncoupled dynamics* in equilibria computation. These dynamics have been studied extensively: Hart and Mas-Colell [13] show that there exist finite-memory uncoupled strategies that lead to pure Nash equilibria in every game where they exist. Also, there exist finite memory uncoupled strategies that lead to ε-NE in every game. Young's interactive trial and error [18] outlines completely uncoupled strategies that lead to pure Nash equilibria with high probability when they exist. Regret testing from Foster and Young [8] and its n-player extension by Germano and Lugosi in [9] show that there exist completely uncoupled strategies that lead to an ε-Nash equilibrium with high probability. Randomisation is essential in all of these approaches, as Hart and Mas-Colell [14] show that it is impossible to achieve convergence to Nash equilibria for all games if one is restricted to deterministic uncoupled strategies. This prior work is not concerned with rate of convergence; by contrast here we obtain efficient bounds on runtime. Convergence in adaptive dynamics for exact Nash equilibria is also studied by Hart and Mansour in [12] where they provide exponential lower bounds via communication complexity results. Babichenko [3] also proves an exponential lower bound on the rate of convergence of adaptive dynamics to an approximate Nash equilibrium for general binary games. Specifically, he proves that there is no k-queries dynamic that converges to an ε-WSNE in $\frac{2^{\Omega(n)}}{k}$ steps with probability of at least $2^{-\Omega(n)}$

in all n-player binary games. Both of these results motivate the study of specific subclasses of these games, such as "large" games.

2 Preliminaries

We consider games with n players where each player has two actions $\mathcal{A} = \{0, 1\}$. Let $a = (a_i, a_{-i})$ denote an *action profile* in which player i plays action a_i and the remaining players play action profile a_{-i}. We also consider *mixed strategies*, which are defined by the probability distributions over the action set $\mathcal{A}$. We write $p = (p_i, p_{-i})$ to denote a *mixed-strategy profile* where player i plays action 1 with probability p_i and the remaining players play the profile p_{-i}. We will sometimes abuse notation to use p_i to denote i's mixed strategy, and write p_{i0} and p_{i1} to denote the probabilities that player i plays the action 0 and 1 respectively.

Each player i has a payoff function $u_i \colon \mathcal{A}^n \to [0, 1]$ mapping an action profile to some value in $[0, 1]$. We will sometimes write $u_i(p) = \mathbb{E}_{a \sim p}[u_i(a)]$ to denote the expected payoff of player i under mixed strategy p. An action a is player i's *best response* to mixed strategy profile p if $a \in \text{argmax}_{j \in \{0,1\}} u_i(j, p_{-i})$.

We assume our algorithms or the players have no other prior knowledge of the game but can access payoff information through querying a *payoff oracle* $\mathcal{Q}$. For each *payoff query* specified by an action profile $a \in \mathcal{A}^n$, the query oracle will return $(u_i(a))_{i=1}^n$ the n-dimensional vector of payoffs to each player. Our goal is to compute an *approximate Nash equilibrium* with a small number of queries. In the completely uncoupled setting, a query works as follows: each player i chooses her own action a_i independently of the other players, and learns her own payoff $u_i(a)$ but no other payoffs.

Definition 1 (Regret; (approximate) Nash equilibrium). *Let p be a mixed strategy profile, the* regret *for player i at p is*

$$reg(p, i) = \max_{k \in \{0,1\}} \mathbb{E}_{a_{-i} \sim p_{-i}}[u_i(k, a_{-i})] - \mathbb{E}_{a \sim p}[u_i(a)]$$

A mixed strategy profile p is an ε-approximate Nash equilibrium if for each player i, the regret satisfies $reg(p, i) \leq \varepsilon$.

Observation. *To find an exact Nash (or even, correlated) equilibrium of a large game, in the worst case it is necessary to query the game exhaustively, even with randomised algorithms. This uses a similar negative result for general games due to [15], and noting that we can obtain a strategically equivalent large game, by scaling down the payoffs into the interval $[0, \frac{1}{n}]$.*

We will also use the following useful notion of *discrepancy*.

Definition 2 (Discrepancy). *Letting p be a mixed strategy profile, the discrepancy for player i at p is*

$$disc(p, i) = \left| \mathbb{E}_{a_{-i} \sim p_{-i}}[u_i(0, a_{-i})] - \mathbb{E}_{a_{-i} \sim p_{-i}}[u_i(1, a_{-i})] \right|$$

We will assume the following *largeness* condition in our games. Informally, such largeness condition implies that no single player has a large influence on any other player's utility function.

Definition 3 (Large Games). *A game is* large *if for any two distinct players* $i \neq j$, *any two distinct actions* a_j *and* a'_j *for player* j, *and any tuple of actions* a_{-j} *for everyone else:*

$$|u_i(a_j, a_{-j}) - u_i(a'_j, a_{-j})| \leq \frac{1}{n}.$$

One immediate implication of the largeness assumption is the following Lipschitz property of the utility functions.

Lemma 1. *For any player* $i \in [n]$, *and any action* $j \in \{0,1\}$, *the utility* $u_i(j, p_{-i})$ *is a* $(\frac{1}{n})$-*Lipschitz function in the second coordinate* p_{-i} *w.r.t. the* ℓ_1 *norm, and the mixed strategy profile of all other players.*

Estimating payoffs for mixed profiles. We can approximate the expected payoffs for any mixed strategy profile by repeated calls to the oracle $\mathcal{Q}$. In particular, for any target accuracy parameter β and confidence parameter δ, consider the following procedure to implement an oracle $\mathcal{Q}_{\beta,\delta}$:

- For any input mixed strategy profile p, compute a new mixed strategy profile $p' = (1 - \frac{\beta}{2})p + (\frac{\beta}{2})\mathbf{1}$ such that each player i is playing uniform distribution with probability $\frac{\beta}{2}$ and playing distribution p_i with probability $1 - \frac{\beta}{2}$.
- Let $N = \frac{64}{\beta^3} \log(8n/\delta)$, and sample N payoff queries randomly from p, and call the oracle $\mathcal{Q}$ with each query as input to obtain a payoff vector.
- Let $\widehat{u}_{i,j}$ be the average sampled payoff to player i for playing action j.[1] Output the payoff vector $(\widehat{u}_{ij})_{i \in [n], j \in \{0,1\}}$.

Lemma 2. *For any* $\beta, \delta \in (0,1)$ *and any mixed strategy profile* p, *the oracle* $\mathcal{Q}_{\beta,\delta}$ *with probability at least* $1 - \delta$ *outputs a payoff vector* $(\widehat{u}_i)_{i \in [n], j \in \{0,1\}}$ *that has an additive error of at most* β, *that is for each player* i, *and each action* $j \in \{0,1\}$,

$$|u_i(j, p_{-i}) - \widehat{u}_{i,j}| \leq \beta.$$

The lemma follows from Proposition 1 of [10] and the largeness property.

Extension to Stochastic Utilities. We consider a generalisation where the utility to player i of any pure profile a may consist of a probability distribution $D_{a,i}$ over $[0,1]$, and if a is played, i receives a sample from $D_{a,i}$. The player wants to maximise her expected utility with respect to sampling from a (possibly mixed) profile, together with sampling from any $D_{a,i}$ that results from a being chosen. If we extend the definition of $\mathcal{Q}$ to output samples of the $D_{a,i}$ for any queried profile a, then $\mathcal{Q}_{\beta,\delta}$ can be defined in a similar way as before, and simulated as above using samples from $\mathcal{Q}$. Our algorithmic results extend to this setting.

---

[1] If the player i never plays an action j in any query, set $\widehat{u}_{i,j} = 0$.

3   Warm-Up: 0.25-Approximate Equilibrium

As a starting point, we will show that without any payoff queries, we could easily give a $\frac{1}{2}$-approximate Nash equilibrium.

Observation. *Consider the following "uniform" mixed strategy profile. Each player puts $\frac{1}{2}$ probability mass on each action: for all i, $p_i = \frac{1}{2}$. Such a mixed strategy profile is a $\frac{1}{2}$-approximate Nash equilibrium.*

In this section, we will present two simple and query-efficient algorithms that allows us to get a better approximation than $\frac{1}{2}$. Both algorithms could be regarded as a simple refinement of the above "uniform" mixed strategy. For simplicity of presentation, we will assume that we have access to a mixed strategy query oracle $\mathcal{Q}_M$ that returns exact expected payoff values for any input mixed strategy p. Our results continue to hold if we replace $\mathcal{Q}_M$ by $\mathcal{Q}_{\beta,\delta}$.[2]

Obtaining $\varepsilon = 0.272$. First, we show that having each player making small adjustment from the "uniform" strategy can improve ε from $\frac{1}{2}$ to around 0.27. We simply let players with large regret shift more probability weight towards their best responses. More formally, consider the following algorithm **OneStep** with two parameters $\alpha, \Delta \in [0,1]$:

- Let the players play the "uniform" mixed strategy. Call the oracle $\mathcal{Q}_M$ to obtain the payoff values of $u_i(0, p_{-i})$ and $u_i(1, p_{-i})$ for each player i.
- For each player i, if $u_i(0, p_{-i}) - u_i(1, p_{-i}) > \alpha$, then set $p_{i0} = \frac{1}{2} + \Delta$ and $p_{i1} = \frac{1}{2} - \Delta$; if $u_i(1, p_{-i}) - u_i(0, p_{-i}) > \alpha$, set $p_{i1} = \frac{1}{2} + \Delta$ and $p_{i0} = \frac{1}{2} - \Delta$; otherwise keep playing $p_i = \frac{1}{2}$.

Lemma 3. *If we set the parameters $\alpha = 2 - \sqrt{11/3}$ and $\Delta = \sqrt{11/48} - 1/4$ in the instantiation of the algorithm **OneStep**, then the resulting mixed strategy profile is an ε-approximate Nash equilibrium with $\varepsilon \le 0.272$.*

Obtaining $\varepsilon = 0.25$. We now give a slightly more sophisticated algorithm than the previous one. We will again have the players starting with the "uniform" mixed strategy, then let players shift more weights toward their best responses, and finally let some of the players switch back to the uniform strategy if their best responses change in the adjustment. Formally, the algorithm **TwoStep** proceeds as:

- Start with the "uniform" mixed strategy profile, and query the oracle $\mathcal{Q}_M$ for the payoff values. Let b_i be player i's best response.
- For each player i, set the probability of playing their best response b_i to be $\frac{3}{4}$. Call $\mathcal{Q}_M$ to obtain payoff values for this mixed strategy profile, and let b_i' be each player i's best response in the new profile.
- For each player i, if $b_i \ne b_i'$, then resume playing $p_{i0} = p_{i1} = \frac{1}{2}$. Otherwise maintain the same mixed strategy from the previous step.

Lemma 4. *The mixed strategy profile output by **TwoStep** is an ε-approximate Nash equilibrium with $\varepsilon \le 0.25$.*

---

[2] In particular, if we use $\mathcal{Q}_{\beta,\delta}$ for our query access, we will get $(\varepsilon + O(\beta))$-approximate equilibrium, with $\varepsilon = 0.272, 0.25$ with probability at least $1 - \delta$.

4 $\frac{1}{8}$-Approximate Equilibrium via Uncoupled Dynamics

In this section, we present our main algorithm that achieves approximate equilibria with $\varepsilon \approx \frac{1}{8}$ in a completely uncoupled setting. In order to arrive at this we first model game dynamics as an uncoupled continuous-time dynamical system where a player's strategy profile updates depend only on her own mixed strategy and payoffs. Afterwards we present a discrete-time approximation to these continuous dynamics to arrive at a query-based algorithm for computing $(\frac{1}{8} + \alpha)$-Nash equilibrium with logarithmic query complexity in the number of players. Finally, as mentioned in Sect. 2, we recall that these algorithms carry over to games with stochastic utilities, where we can show that our algorithm uses an essentially optimal number of queries.

Throughout the section, we will rely on the following notion of a *strategy-payoff state*, capturing the information available to a player at any moment of time.

Definition 4 (Strategy-payoff state). *For any player i, the* strategy-payoff state *for player i is defined as the ordered triple $s_i = (v_{i1}, v_{i0}, p_i) \in [0, 1]^3$, where v_{i1} and v_{i0} are the player's utilities for playing pure actions 1 and 0 respectively, and p_i denotes the player's probability of playing action 1. Furthermore, we denote the player's discrepancy by $D_i = |v_{i1} - v_{i0}|$ and we let p_i^* denote the probability mass on the best response, that is if $v_{i1} \geq v_{i0}$, $p_i^* = p_i$, otherwise $p_i^* = 1 - p_i$.*

4.1   Continuous-Time Dynamics

First, we will model game dynamics in continuous time, and assume that a player's strategy-payoff state (and thus all variables it contains) is a differentiable time-valued function. When we specify these values at a specific time t, we will write $s_i(t) = (v_{i1}(t), v_{i0}(t), p_i(t))$. Furthermore, for any time-differentiable function g, we denote its time derivative by $\dot{g} = \frac{d}{dt}g$. We will consider continuous game dynamics formally defined as follows.

Definition 5 (Continuous game dynamic). *A* continuous game dynamic *consists of an update function f that specifies a player's strategy update at time t. Furthermore, f depends only on $s_i(t)$ and $\dot{s}_i(t)$. In other words, $\dot{p}_i(t) = f(s_i(t), \dot{s}_i(t))$ for all t.*

Observation. *We note that in this framework, a specific player's updates do not depend on other players' strategy-payoff states nor their history of play. This will eventually lead us to uncoupled Nash equilibria computation in Sect. 4.2.*

A central object of interest in our continuous dynamic is a linear sub-space $\mathcal{P} \subset [0, 1]^3$ such that all strategy-payoff states in it incur a bounded regret. Formally, we will define $\mathcal{P}$ via its normal vector $\boldsymbol{n} = (-\frac{1}{2}, \frac{1}{2}, 1)$ so that $\mathcal{P} = \{s_i | \ s_i \cdot \boldsymbol{n} = \frac{1}{2}\}$. Equivalently, we could also write $\mathcal{P} = \{s_i \mid p_i^* = \frac{1}{2}(1 + D_i)\}$. (See Fig. 1 for a visualisation.) With this observation, it is straightforward to see that any player with strategy-payoff state in $\mathcal{P}$ has regret at most $\frac{1}{8}$.

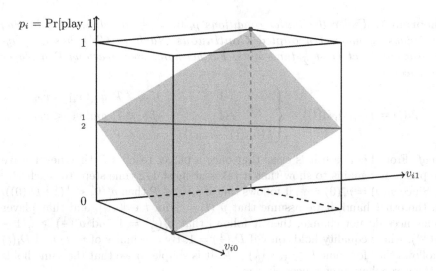

Fig. 1. Visualisation of $\mathcal{P}$; on the red line, $v_{i0} = v_{i1}$ so the player is indifferent and mixes with equal probabilities; at the red points the player has payoffs of 0 and 1, and makes a pure best response. (Color figure online)

Lemma 5. *Suppose that the player's strategy-payoff state satisfies $s_i \in \mathcal{P}$, then her regret is at most $\frac{1}{8}$.*

Proof. This follows from the fact that a player's regret can be expressed as $D_i(1 - p_i^*)$ and the fact that all points on $\mathcal{P}$ also satisfy $p_i^* = \frac{1}{2}(1 + D_i)$. In particular, the maximal regret of $\frac{1}{8}$ is achieved when $D_i = \frac{1}{2}$ and $p_i^* = \frac{3}{4}$.

Next, we want to show there exists a dynamic that allows all players to eventually reach $\mathcal{P}$ and remain on it over time. We notice that for a specific player, $\dot{v}_{i1}$, $\dot{v}_{i0}$ and subsequently $\dot{D}_i$ measure the cumulative effect of other players shifting their strategies. However, if we limit how much any individual player can change their mixed strategy over time by imposing $|\dot{p}_i| \leq 1$ for all i, Lemma 1 guarantees $|\dot{v}_{ij}| \leq 1$ for $j = 0, 1$ and consequently $|\dot{D}_i| \leq 2$. With these quantities bounded, we can consider an adversarial framework where we construct game dynamics by solely assuming that $|\dot{p}_i(t)| \leq 1$, $|\dot{v}_{ij}(t)| \leq 1$ for $j = 0, 1$ and $|\dot{D}_i(t)| \leq 2$ for all times $t \geq 0$.

Now assume an adversary controls $\dot{v}_{i0}$, $\dot{v}_{i1}$ and hence $\dot{D}_i$, one can show that if a player sets $\dot{p}_i(t) = \frac{1}{2}(\dot{v}_{i1}(t) - \dot{v}_{i0}(t))$, then she could stay on $\mathcal{P}$ whenever she reaches the subspace.

Lemma 6. *If $s_i(0) \in \mathcal{P}$, and $\dot{p}_i(t) = \frac{1}{2}(\dot{v}_{i1}(t) - \dot{v}_{i0}(t))$, then $s_i(t) \in \mathcal{P} \ \forall \ t \geq 0$.*

Theorem 1. *Under the initial conditions $p_i(0) = \frac{1}{2}$ for all i, the following continuous dynamic,* **Uncoupled Continuous Nash (UCN)**, *has all players reach $\mathcal{P}$ in at most $\frac{1}{2}$ time units. Furthermore, upon reaching $\mathcal{P}$ a player never leaves.*

$$\dot{p}_i(t) = f(s_i(t), \dot{s}_i(t)) = \begin{cases} 1 & \text{if } s_i \notin \mathcal{P} \text{ and } v_{i1} \geq v_{i0} \\ -1 & \text{if } s_i \notin \mathcal{P} \text{ and } v_{i1} < v_{i0} \\ \frac{1}{2}(\dot{v}_{i1}(t) - \dot{v}_{i0}(t)) & \text{if } s_i \in \mathcal{P} \end{cases}$$

Proof. From Lemma 6 it is clear that once a player reaches $\mathcal{P}$ they never leave the plane. It remains to show that it takes at most $1/2$ time steps to reach $\mathcal{P}$.

Since $p_i(0) = p_i^*(0) = \frac{1}{2}$, it follows that if $s_i(0) \notin \mathcal{P}$ then $\dot{p}_i^*(0) < \frac{1}{2}(1+D_i(0))$. On the other hand, if we assume that $\dot{p}_i^*(t) = 1$ for $t \in [0, \frac{1}{2}]$, and that player preferences do not change, then it follows that $p_i^*(\frac{1}{2}) = 1$ and $p_i^*(\frac{1}{2}) \geq \frac{1}{2}(1 + D_i(\frac{1}{2}))$, where equality holds only if $D_i(\frac{1}{2}) = 1$. By continuity of $p_i^*(t)$ and $D_i(t)$ it follows that for some $k \leq \frac{1}{2}$, $s_i(k) \in \mathcal{P}$. It is simple to see that the same holds in the case where preferences change.

4.2   Discrete Time-Step Approximation

The continuous-time dynamics of the previous section hinge on obtaining expected payoffs in mixed strategy profiles, thus we will approximate expected payoffs via $\mathcal{Q}_{\beta,\delta}$. Our algorithm will have each player adjusting their mixed strategy over rounds, and each round query $\mathcal{Q}_{\beta,\delta}$ to obtain the payoff values.

Since we are considering discrete approximations to **UCN**, the dynamics will no longer guarantee that strategy-payoff states stay on the plane $\mathcal{P}$. For this reason we define the following region around $\mathcal{P}$:

Definition 6. *Let $\mathcal{P}^\lambda = \{s_i \mid s_i \cdot \boldsymbol{n} \in [\frac{1}{2} - \lambda, \frac{1}{2} + \lambda]\}$, with normal vector $\boldsymbol{n} = (-\frac{1}{2}, \frac{1}{2}, 1)$. Equivalently, $\mathcal{P} = \{s_i \mid p_i^* = \frac{1}{2}(1 + D_i) + c, \ c \in [-\lambda, \lambda]\}$.*

Just as in the proof of Lemma 5, we can use the fact that a player's regret is $D_i(1 - p_i^*)$ to bound regret on $\mathcal{P}^\lambda$.

Lemma 7. *The worst case regret of any strategy-payoff state in $\mathcal{P}^\lambda$ is $\frac{1}{8}(1+2\lambda)^2$. This is attained on the boundary: $\partial \mathcal{P}^\lambda = \{s_i \mid s_i \cdot \boldsymbol{n} = \frac{1}{2} \pm \lambda\}$*

Corollary 1. *For a fixed $\alpha > 0$, if $\lambda = \frac{\sqrt{1+8\alpha}-1}{2}$, then $\mathcal{P}^\lambda$ attains a maximal regret of $\frac{1}{8} + \alpha$.*

We present an algorithm in the completely uncoupled setting, **UN**(α, η), that for any parameters $\alpha, \eta \in (0, 1]$ computes a $(\frac{1}{8} + \alpha)$-Nash equilibrium with probability at least $1 - \eta$.

Since $p_i(t) \in [0, 1]$ is the mixed strategy of the i-th player at round t we let $p(t) = (p_i(t))_{i=1}^n$ be the resulting mixed strategy profile of all players at round t. Furthermore, we use the mixed strategy oracle $\mathcal{Q}_{\beta,\delta}$ from Lemma 2 that for a given mixed strategy profile p returns the vector of expected payoffs for all players with an additive error of β and a correctness probability of $1 - \delta$.

The following lemma is used to prove the correctness of **UN**(α, η):

Lemma 8. *Suppose that $w \in \mathbb{R}^3$ with $\|w\|_\infty \le \lambda$ and let function $h(x) = x \cdot \boldsymbol{n}$, where $\boldsymbol{n}$ is the normal vector of $\mathcal{P}$. Then $h(x+w) - h(x) \in [-2\lambda, 2\lambda]$. Furthermore, if $w_3 = 0$, then $h(x+w) - h(x) \in [-\lambda, \lambda]$.*

Proof. The statement follows from the following expression:

$$h(x+w) - h(x) = w \cdot \boldsymbol{n} = \frac{1}{2}(w_2 - w_1) + w_3$$

We now give a formal description for $\mathbf{UN}(\alpha, \eta)$:

1. Set $\lambda = \frac{\sqrt{1+8\alpha}-1}{2}$, $\Delta = \frac{\lambda}{4}$, and $N = \lceil \frac{2}{\Delta} \rceil$
2. For each player i, let $p_i(0) = \frac{1}{2}$ and $\widehat{v}_{ij}(-1) = \left(\mathcal{Q}_{(\Delta, \frac{\eta}{N})}(p(0)) \right)_{i,j}$ for $j = 0, 1$
3. During $t \le N$ rounds, for each player i, calculate $\widehat{v}_{ij}(t) = \left(\mathcal{Q}_{(\Delta, \frac{\eta}{N})}(p(t)) \right)_{i,j}$
   and let $\Delta \widehat{v}_{ij}(t) = \widehat{v}_{ij}(t) - \widehat{v}_{ij}(t-1)$ for $j = 0, 1$.
4. if $\widehat{s}_i(t) = (\widehat{v}_{i1}(t), \widehat{v}_{i0}(t), p_i(t)) \notin \mathcal{P}^{\lambda/4}$, then $p_i^*(t+1) = p_i^*(t) + \Delta$, otherwise
   $p_i^*(t+1) = p_i^*(t) + \frac{1}{2}(\Delta \widehat{v}_{i1}(t) - \Delta \widehat{v}_{i0}(t))$
5. return $p(t)$

Theorem 2. *With probability $1 - \eta$, $\mathbf{UN}(\alpha, \eta)$ correctly returns a $(\frac{1}{8} + \alpha)$ approximate Nash equilibrium by using $O(\frac{1}{\alpha^4} \log\left(\frac{n}{\alpha\eta}\right))$ queries.*

Proof. By Lemma 2 and union bound, we can guarantee that with probability at least $1 - \eta$ all sample approximations to mixed payoff queries have an additive error of at most $\Delta = \frac{\lambda}{4}$. We will condition on this accuracy guarantee in the remainder of our argument. Now we can show that for each player there will be some round $k \le N$, such that at the beginning of the round their strategy-payoff state lies in $\mathcal{P}^{\lambda/2}$. Furthermore, at the beginning of all subsequent rounds $t \ge k$, it will also be the case that their strategy-payoff state lies in $\mathcal{P}^{\lambda/2}$.

The reason any player generally reaches $\mathcal{P}^{\lambda/2}$ follows from the fact that in the worst case, after increasing p^* by Δ for N rounds, $p^* = 1$, in which case a player is certainly in $\mathcal{P}^{\lambda/2}$. Furthermore, Lemma 8 guarantees that each time p^* is increased by Δ, the value of $\widehat{s}_i \cdot \boldsymbol{n}$ changes by at most $\frac{\lambda}{2}$ which is why $\widehat{s}_i$ are always steered towards $\mathcal{P}^{\lambda/4}$. Due to inherent noise in sampling, players may at times find that $\widehat{s}_i$ slightly exit $\mathcal{P}^{\lambda/4}$ but since additive errors are at most $\frac{\lambda}{4}$. We are still guaranteed that true s_i lie in $\mathcal{P}^{\lambda/2}$.

The second half of step 4 forces a player to remain in $\mathcal{P}^{\lambda/2}$ at the beginning of any subsequent round $t \ge k$. The argumentation for this is identical to that of Lemma 6 in the continuous case.

Finally, the reason that individual probability movements are restricted to $\Delta = \frac{\lambda}{4}$ is that at the end of the final round, players will move their probabilities and will not be able to respond to subsequent changes in their strategy-payoff states. From the second part of Lemma 8, we can see that in the worst case this can cause a strategy-payoff state to move from the boundary of $\mathcal{P}^{\lambda/2}$ to the boundary of $\mathcal{P}^{\frac{3\lambda}{4}} \subset \mathcal{P}^{\lambda}$. However, λ is chosen in such a way so that the worst-case regret within $\mathcal{P}^{\lambda}$ is at most $\frac{1}{8}+\alpha$, therefore it follows that $\mathbf{UN}(\alpha, \eta)$ returns a $\frac{1}{8} + \alpha$ approximate Nash equilibrium. Furthermore, the number of queries is

$$(N + 1) \left(\frac{1024}{\lambda^3} \log \left(\frac{8nN}{\eta} \right) \right) = \left(\frac{1}{\lambda} + 1 \right) \left(\frac{1024}{\lambda^3} \log \left(\frac{8n}{\lambda\eta} \right) \right).$$

It is not difficult to see that $\frac{1}{\lambda} = O(\frac{1}{\alpha})$ which implies that the number of queries made is $O\left(\frac{1}{\alpha^4} \log \left(\frac{n}{\alpha\eta} \right) \right)$ in the limit.

4.3   Logarithmic Lower Bound

As mentioned in the preliminaries section, all of our previous results extend to stochastic utilities. In particular, if we assume that G is a game with stochastic utilities where expected payoffs are large with parameter $\frac{1}{n}$, then we can apply $\mathbf{UN}(\alpha, \eta)$ with $O(\log(n))$ queries to obtain a mixed strategy profile where no player has more than $\frac{1}{8} + \alpha$ incentive to deviate. Most importantly, for $k > 2$, we can use the same methods as [10] to lower bound the query complexity of computing a mixed strategy profile where no player has more than $(\frac{1}{2} - \frac{1}{k})$ incentive to deviate.

Theorem 3. *If $k > 2$, the query complexity of computing a mixed strategy profile where no player has more than $(\frac{1}{2} - \frac{1}{k})$ incentive to deviate for stochastic utility games is $\Omega(\log_{k(k-1)}(n))$. Alongside Theorem 2 this implies the query complexity of computing mixed strategy profiles where no player has more than $\frac{1}{8}$ incentive to deviate in stochastic utility games is $\Theta(\log(n))$.*

5   Achieving $\varepsilon < \frac{1}{8}$ with Communication

We return to continuous dynamics to show that we can obtain a worst-case regret of slightly less than $\frac{1}{8}$ by using limited communication between players, thus breaking the uncoupled setting we have been studying until now.

First of all, let us suppose that initially $p_i(0) = \frac{1}{2}$ for each player i and that $\mathbf{UCN}$ is run for $\frac{1}{2}$ time units so that strategy-payoff states for each player lie on $\mathcal{P} = \{s_i \mid p_i^* = \frac{1}{2}(1 + D_i)\}$. We recall from Lemma 5 that the worst case regret of $\frac{1}{8}$ on this plane is achieved when $p_i^* = \frac{3}{4}$ and $D_i = \frac{1}{2}$. We say a player is *bad* if they achieve a regret of at least 0.12, which on $\mathcal{P}$ corresponds to having $p_i^* \in [0.7, 0.8]$. Similarly, all other players are *good*. We denote $\theta \in [0, 1]$ as the proportion of players that are bad. Furthermore, as the following lemma shows, we can in a certain sense assume that $\theta \leq \frac{1}{2}$.

Lemma 9. *If $\theta > \frac{1}{2}$, then for a period of 0.15 time units, we can allow each bad player to shift to their best response with unit speed, and have all good players update according to $\mathbf{UCN}$ to stay on $\mathcal{P}$. After this movement, at most $1 - \theta$ players are bad.*

Proof. If i is a bad player, in the worst case scenario, $\dot{D}_i = 2$, which keeps their strategy-payoff state, s_i, on $\mathcal{P}$. At the end of 0.15 time units however, $p_i^* > 0.85$, hence they will no longer be bad. On the other hand, good players stay on $\mathcal{P}$, so at worst, all of them become bad.

Observation. *After this movement, players who were bad are the only players possibly away from $\mathcal{P}$ and they have a discrepancy that is greater than 0.1. Furthermore, all players who become bad lie on $\mathcal{P}$.*

We can now outline a continuous-time dynamic that utilises Lemma 9 to obtain a $(\frac{1}{8} - \frac{1}{220})$ maximal regret.

1. Have all players begin with $p_i(0) = \frac{1}{2}$
2. Run **UCN** for $\frac{1}{2}$ time units.
3. Measure, θ, the proportion of bad players. If $\theta > \frac{1}{2}$ apply the dynamics of Lemma 9.
4. Let all bad players use $\dot{p}_i^* = 1$ for $\Delta = \frac{1}{220}$ time units.

Theorem 4. *If all players follow the aforementioned dynamic, no single player will have a regret greater than $\frac{1}{8} - \frac{1}{220}$.*

Proof. Technical details of this proof can be found in the full paper, but in essence one shows that if Δ is a small enough time interval (less than 0.1 to be exact), then all bad players will unilaterally decrease their regret by at least 0.1Δ and good players won't increase their regret by more than Δ. The time step $\Delta = \frac{1}{220}$ is thus chosen optimally.

As a final note, we see that this process requires one round of communication in being able to perform the operations in Lemma 9, that is we need to know if $\theta > \frac{1}{2}$ or not to balance player profiles so that there are at most the same number of bad players to good players. Furthermore, in exactly the same fashion as $\mathbf{UN}(\alpha, \eta)$, we can discretise the above process to obtain a query-based algorithm that obtains a regret of $\frac{1}{8} - \frac{1}{220} + \alpha < \frac{1}{8}$ for arbitrary α.

6   Conclusion and Further Research

We have assumed a largeness parameter of $\gamma = \frac{1}{n}$, but in the full paper we extend our techniques to $\gamma = \frac{c}{n}$ for constant c. We can obtain approximate equilibria approaching $\varepsilon = \frac{c}{8}$ for $c \leq 2$ and $\varepsilon = \frac{1}{2} - \frac{1}{2c}$ for $c > 2$. In the full paper, we also extend our techniques to games where players have k strategies.

An obvious question raised by our results is the possible improvement in the additive approximation obtainable since *pure* approximate equilibria are known to exist for these games. A slightly weaker objective than this would be the search for *well-supported* approximate equilibria. It would also be interesting to investigate lower bounds in the completely uncoupled setting. Finally, since our algorithms are randomised, it would be interesting to see what can be achieved using deterministic algorithms.

References

1. Azrieli, Y., Shmaya, E.: Lipschitz games. Math. Oper. Res. **38**(2), 350–357 (2013)
2. Babichenko, Y.: Best-reply dynamics in large binary-choice anonymous games. Games Econ. Behav. **81**, 130–144 (2013)
3. Babichenko, Y.: Query complexity of approximate Nash equilibria. In: Proceedings of 46th STOC, pp. 535–544 (2014)
4. Babichenko, Y., Barman, S.: Query complexity of correlated equilibrium. ACM. Trans. Econ. Comput. **4**(3), 1–35 (2015)
5. Chen, X., Cheng, Y., Tang, B.: Well-supported versus approximate Nash equilibria: Query complexity of large games (2015). ArXiv rept. 1511.00785
6. Fearnley, J., Savani, R.: Finding approximate Nash equilibria of bimatrix games via payoff queries. In: Proceedings of 15th ACM EC, pp. 657–674 (2014)
7. Fearnley, J., Gairing, M., Goldberg, P.W., Savani, R.: Learning equilibria of games via payoff queries. J. Mach. Learn. Res. **16**, 1305–1344 (2015)
8. Foster, D.P., Young, H.P.: Regret testing: learning to play Nash equilibrium without knowing you have an opponent. Theor. Econ. **1**(3), 341–367 (2006)
9. Germano, F., Lugosi, G.: Global Nash convergence of Foster and Young's regret testing (2005). http://www.econ.upf.edu/lugosi/nash.pdf
10. Goldberg, P.W., Roth, A.: Bounds for the query complexity of approximate equilibria. In: Proceedings of the 15th ACM-EC Conference, pp. 639–656 (2014)
11. Goldberg, P.W., Turchetta, S.: Query complexity of approximate equilibria in anonymous games. In: Markakis, E., et al. (eds.) WINE 2015. LNCS, vol. 9470, pp. 357–369. Springer, Heidelberg (2015). doi:10.1007/978-3-662-48995-6_26
12. Hart, S., Mansour, Y.: How long to equilibrium? the communication complexity of uncoupled equilibrium procedures. Games Econ. Behav. **69**(1), 107–126 (2010)
13. Hart, S., Mas-Colell, A.: A simple adaptive procedure leading to correlated equilibrium. Econometrica **68**(5), 1127–1150 (2000)
14. Hart, S., Mas-Colell, A.: Uncoupled dynamics do not lead to Nash equilibrium. Am. Econ. Rev. **93**(5), 1830–1836 (2003)
15. Hart, S., Nisan, N.: The query complexity of correlated equilibria (2013). ArXiv tech rept. 1305.4874
16. Kalai, E.: Large robust games. Econometrica **72**(6), 1631–1665 (2004)
17. Kearns, M., Pai, M.M., Roth, A., Ullman, J.: Mechanism design in large games: Incentives and privacy. Am. Econ. Rev. **104**(5), 431–435 (2014). doi:10.1257/aer.104.5.431
18. Young, H.P.: Learning by trial and error (2009). http://www.econ2.jhu.edu/people/young/Learning5June08.pdf

Lipschitz Continuity and Approximate Equilibria

Argyrios Deligkas[1(✉)], John Fearnley[1], and Paul Spirakis[1,2]

[1] University of Liverpool, Liverpool, UK
a.deligkas@liverpool.ac.uk
[2] Computer Technology Institute (CTI), Patras, Greece

Abstract. In this paper, we study games with continuous action spaces and non-linear payoff functions. Our key insight is that Lipschitz continuity of the payoff function allows us to provide algorithms for finding approximate equilibria in these games. We begin by studying Lipschitz games, which encompass, for example, all concave games with Lipschitz continuous payoff functions. We provide an efficient algorithm for computing approximate equilibria in these games. Then we turn our attention to penalty games, which encompass biased games and games in which players take risk into account. Here we show that if the penalty function is Lipschitz continuous, then we can provide a quasi-polynomial time approximation scheme. Finally, we study distance biased games, where we present simple strongly polynomial time algorithms for finding best responses in L_1, L_2^2, and L_∞ biased games, and then use these algorithms to provide strongly polynomial algorithms that find 2/3, 5/7, and 2/3 approximations for these norms, respectively.

1 Introduction

Nash equilibria [18] are the central solution concept in game theory. However, recent advances have shown that computing an *exact* Nash equilibrium is PPAD-complete [8,9], and so there are unlikely to be polynomial time algorithms for this problem. The hardness of computing exact equilibria has lead to the study of *approximate* equilibria: while an exact equilibrium requires that all players have no incentive to deviate from their current strategy, an ϵ-approximate equilibrium requires only that their incentive to deviate is less than ϵ.

A fruitful line of work has developed studying the best approximations that can be found in polynomial-time for *bimatrix games*, which are two-player strategic form games. There, after a number of papers [5,10,11], the best known algorithm was given by Tsaknakis and Spirakis [21], who provide a polynomial time algorithm that finds a 0.3393-equilibrium. The existence of an FPTAS was ruled out by Chen, Deng, and Teng [8] unless PPAD = P. Recently, Rubinstein [20] proved that there is no PTAS for the problem, assuming the Exponential Time Hypothesis for PPAD. However, there is a *quasi-polynomial* approximation scheme given by Lipton, Markakis, and Mehta [16].

In a strategic form game, the game is specified by giving each player a finite number of strategies, and then specifying a table of payoffs that contains one

© Springer-Verlag Berlin Heidelberg 2016
M. Gairing and R. Savani (Eds.): SAGT 2016, LNCS 9928, pp. 15–26, 2016.
DOI: 10.1007/978-3-662-53354-3_2

entry for every possible combination of strategies that the players might pick. The players are allowed to use mixed strategies, and so ultimately the payoff function is a convex combination of the payoffs given in the table. However, some games can only be modelled in a more general setting where the action spaces are continuous, or the payoff functions are non-linear.

For example, Rosen's seminal work [19] considered *concave games*, where each player picks a vector from a convex set. The payoff to each player is specified by a function that satisfies the following condition: if every other player's strategy is fixed, then the payoff to a player is a concave function over his strategy space. Rosen proved that concave games always have an equilibrium. A natural subclass of concave games, studied by Caragiannis, Kurokawa, and Procaccia [6], is the class of biased games. A biased game is defined by a strategic form game, a *base strategy* and a *penalty function*. The players play the strategic form game as normal, but they all suffer a penalty for deviating from their base strategy. This penalty can be a non-linear function, such as the L_2^2 norm.

In this paper, we study the computation of approximate equilibria in such games. Our main observation is that Lipschitz continuity of the players' payoff functions (with respect to changes in the strategy space) allows us to provide algorithms that find approximate equilibria. Several papers have studied how the Lipschitz continuity of the players' payoff functions affects the existence, the quality, and the complexity of the equilibria of the underlying game. Azrieli and Shmaya [1] studied many player games and derived bounds for the Lipschitz constant of the utility functions for the players that guarantees the existence of pure approximate equilibria for the game. We have to note though, that the games Azrieli and Shmaya study are significantly different from our games. In [1] the Lipschitz coefficient refers to the payoff function of player i as a function of $\mathbf{x}_{-i}$, i.e. when x_i is fixed. In this paper, the Lipschitz coefficient refers to the payoff function of player i as a function of x_i when the $\mathbf{x}_{-i}$ is fixed. We used this definition of the Lipschitz continuity in order to follow Rosen's definition of concave games that requires the payoff function of player i to be concave for every fixed strategy profile for the rest of the players. Daskalakis and Papadimitriou [12] proved that anonymous games posses pure approximate equilibria whose quality depends on the Lipschitz constant of the payoff functions and the number of pure strategies the players have and proved that these approximate equilibria can be computed in polynomial time. Furthermore, they gave a polynomial-time approximation scheme for anonymous games with many players and constant number of pure strategies. Babichenko [2] presented a best-reply dynamic for n-players Lipschitz anonymous games with two strategies which reaches an approximate pure equilibrium in $O(n \log n)$ steps. Deb and Kalai [13] studied how some variants of the Lipschitz continuity of the utility functions are sufficient to guarantee hindsight stability of equilibria.

1.1   Our Contribution

Lipschitz Games. We begin by studying a very general class of games, where each player's strategy space is continuous, and represented by a convex set of vectors, and where the only restriction is that the payoff function is Lipschitz continuous. This class is so general that exact equilibria, and even approximate equilibria may not exist. Nevertheless, we give an efficient algorithm that either outputs an ϵ-equilibrium, or determines that the game has no exact equilibria. More precisely, for M player games with a strategy space defined as the convex hull of n vectors, that have λ-Lipschitz continuous payoff functions in the L_p norm, for $p \geq 2$, and where $\gamma = \max \|\mathbf{x}\|_p$ over all $\mathbf{x}$ in the strategy space, we either compute an ϵ-equilibrium or determine that no exact equilibrium exists in time $O\left(Mn^{Mk+l}\right)$, where $k = O\left(\frac{\lambda^2 M p \gamma^2}{\epsilon^2}\right)$ and $l = O\left(\frac{\lambda^2 p \gamma^2}{\epsilon^2}\right)$. Observe that this is a polynomial time algorithm when λ, p, γ, M, and ϵ are constant.

To prove this result, we utilize a recent result of Barman [4], which states that for every vector in a convex set, there is another vector that is ϵ close to the original in the L_p norm, and is a convex combination of b points on the convex hull, where b depends on p and ϵ, but does not depend on the dimension. Using this result, and the Lipschitz continuity of the payoffs, allows us to reduce the task of finding an ϵ-equilibrium to checking only a small number of strategy profiles, and thus we get a brute-force algorithm that is reminiscent of the QPTAS given by Lipton, Markakis, and Mehta for bimatrix games [16] and by the QPTAS of Babichenko, Barman, and Peretz [3] for many player games.

However, life is not so simple for us. Since we study a very general class of games, verifying whether a given strategy profile is an ϵ-equilibrium is a nontrivial task. It requires us to compute a *regret* for each player, which is the difference between the player's best response payoff and their actual payoff. Computing a best response in a bimatrix game is trivial, but for Lipschitz games, it may be a hard problem. We get around this problem by instead giving an algorithm to compute *approximate* best responses. Hence we find *approximate* regrets, and it turns out that this is sufficient for our algorithm to work.

Penalty Games. We then turn our attention to *penalty games*. In these games, the players play a strategic form game, and their utility is the payoff achieved in the game *minus* a penalty. The penalty function can be an arbitrary function that depends on the player's strategy. This is a general class of games that encompasses a number of games that have been studied before. The biased games studied by Caragiannis, Kurokawa, and Procaccia [6] are penalty games where the penalty is determined by the amount that a player deviates from a specified base strategy. The biased model was studied in the past by psychologists [22] and it is close to what they call *anchoring* [7,15]. In their seminal paper, Fiat and Papadimitriou [14] introduced a model for *risk prone* games, which resemble penalty games since the risk component can be encoded as a penalty. Mavronicolas and Monien [17] followed this line of research and provided results on the complexity of deciding if such games possess an equilibrium.

We again show that Lipschitz continuity helps us to find approximate equilibria. The only assumption that we make is that the penalty function is Lipschitz continuous in an L_p norm with $p \geq 2$. Again, this is a weak restriction, and it does not guarantee that exact equilibria exist. Even so, we give a quasi-polynomial time algorithm that either finds an ϵ-equilibrium, or verifies that the game has no exact equilibrium.

Our result can be seen as a generalisation of the QPTAS given by Lipton, Markakis, and Mehta [16] for bimatrix games. Their approach is to show the existence of an approximate equilibrium with a logarithmic support. They proved this via the probabilistic method: if we know an exact equilibrium of a bimatrix game, then we can take logarithmically many samples from the strategies, and playing the sampled strategies uniformly will be an approximate equilibrium with positive probability. We take a similar approach, but since our games are more complicated, our proof is necessarily more involved. In particular, for Lipton, Markakis, and Mehta, proving that the sampled strategies are an approximate equilibrium only requires showing that the expected payoff is close to the best response payoff. In penalty games, best response strategies are not necessarily pure, and so the events that we must consider are more complex.

Distance Biased Games. Finally, we consider distance biased games, which form a subclass of penalty games that have been studied recently by Caragiannis, Kurokawa, and Procaccia [6]. They showed that, under very mild assumptions on the bias function, biased games always have an exact equilibrium. Furthermore, for the case where the bias function is either the L_1 norm, or the L_2^2 norm, they give an exponential time algorithm for finding an exact equilibrium.

Our results for penalty games already give a QPTAS for biased games, but we are also interested in whether there are polynomial-time algorithms that can find non-trivial approximations. We give a positive answer to this question for games where the bias is the L_1 norm, the L_2^2 norm, or the L_∞ norm. We follow the well-known approach of Daskalakis, Mehta, Papadimitriou [11], who gave a simple algorithm for finding a 0.5-approximate equilibrium in a bimatrix game.

We show that this algorithm also works for biased games, although the generalisation is not entirely trivial. Again, this is because best responses cannot be trivially computed in biased games. For the L_1 and L_∞ norms, best responses can be computed via linear programming, and for the L_2^2 norm, best responses can be formulated as a quadratic program, and it turns out that this particular QP can be solved in polynomial time by the ellipsoid method. However, none of these algorithms are strongly polynomial. We show that, for each of the norms, best responses can be found by a simple strongly-polynomial combinatorial algorithm. We then analyse the quality of approximation provided by the technique of Daskalakis, Mehta, Papadimitriou [11]. We obtain a strongly polynomial algorithm for finding a 2/3 approximation in L_1 and L_∞ biased games, and a strongly polynomial algorithm for finding a 5/7 approximation in L_2^2 biased games. For the latter result, in the special case where the bias function is the inner product of the player's strategy we find a 13/21 approximation.

2   Preliminaries

We start by fixing some notation. For each positive integer n we use $[n]$ to denote the set $\{1, 2, \ldots, n\}$, we use Δ^n to denote the $(n-1)$-dimensional simplex, and $\|x\|_p^q$ to denote the (p,q)-norm of a vector $x \in \mathbb{R}^d$, i.e. $\|x\|_p^q = (\sum_{i \in [d]} |x_i|^p)^{q/p}$. When $q = 1$, then we will omit it for notation simplicity. Given a set $X = \{x_1, x_2, \ldots, x_n\} \subset \mathbb{R}^d$, we use $conv(X)$ to denote the convex hull of X. A vector $y \in conv(X)$ is said to be k-uniform with respect to X if there exists a size k multiset S of $[n]$ such that $y = \frac{1}{k} \sum_{i \in S} x_i$. When X is clear from the context we will simply say that a vector is k uniform without mentioning that uniformity is with respect to X.

Games and Strategies. A game with M players can be described by a set of available actions for each player and a utility function for each player that depends both on his chosen action and the actions the rest of the players chose. For each player $i \in [M]$ we use S_i to denote his set of available actions and we call it his *strategy space*. We will use $x_i \in S_i$ to denote a specific action chosen by player i and we will call it the *strategy* of player i, we use $\mathbf{x} = (x_1, \ldots, x_M)$ to denote a *strategy profile* of the game, and we will use $\mathbf{x}_{-i}$ to denote the strategy profile where the player i is excluded, i.e. $\mathbf{x}_{-i} = (x_1, \ldots, x_{i-1}, x_{i+1}, \ldots, x_M)$. We use $T_i(x_i, \mathbf{x}_{-i})$ to denote the utility of player i when he plays the strategy x_i and the rest of the players play according to the strategy profile $\mathbf{x}_{-i}$. A strategy $\hat{x}_i$ is a *best response* against the strategy profile $\mathbf{x}_{-i}$, if $T_i(\hat{x}_i, \mathbf{x}_{-i}) \geq T_i(x_i, \mathbf{x}_{-i})$ for all $x_i \in S_i$. The *regret* player i suffers under a strategy profile $\mathbf{x}$ is the difference between the utility of his best response and his utility under $\mathbf{x}$, i.e. $T_i(\hat{x}_i, \mathbf{x}_{-i}) - T_i(x_i, \mathbf{x}_{-i})$.

An $n \times n$ bimatrix game is a pair (R, C) of two $n \times n$ matrices: R gives payoffs for the *row* player and C gives the payoffs for the *column* player. We make the standard assumption that all payoffs lie in the range $[0, 1]$. If $\mathbf{x}$ and $\mathbf{y}$ are mixed strategies for the row and the column player, respectively, then the expected payoff for the row player under strategy profile $(\mathbf{x}, \mathbf{y})$ is given by $\mathbf{x}^T R \mathbf{y}$ and for the column player by $\mathbf{x}^T C \mathbf{y}$.

λ_p-Lipschitz Games. We will use the notion of the λ_p-Lipschitz continuity.

Definition 1 (λ_p-Lipschitz). A function $f : A \to \mathbb{R}$, with $A \subseteq \mathbb{R}^d$ is λ_p-Lipschitz continuous if for every x and y in A, it is true that $|f(x) - f(y)| \leq \lambda \cdot \|x - y\|_p$.

We call the game $\mathfrak{L} := (M, n, \lambda, p, \gamma, \mathcal{T})$ λ_p-*Lipschitz* if for each player $i \in [M]$ the strategy space S_i is the convex hull of n vectors $y_1, \ldots, y_n$ in $\mathbb{R}^d$, $\max_{x_i \in S_i} \|x_i\|_p \leq \gamma$, and the utility function $T_i(\mathbf{x}) \in \mathcal{T}$ is λ_p-Lipschitz continuous.

Two-Player Penalty Games. A two-player penalty game $\mathcal{P}$ is defined by a tuple $(R, C, \mathfrak{f}_r(\mathbf{x}), \mathfrak{f}_c(\mathbf{y}))$, where (R, C) is a bimatrix game and $\mathfrak{f}_r(\mathbf{x})$ and $\mathfrak{f}_c(\mathbf{y})$ are the penalty functions for the row and the column player respectively. The utilities for the players under a strategy profile $(\mathbf{x}, \mathbf{y})$, denoted by $T_r(\mathbf{x}, \mathbf{y})$ and $T_c(\mathbf{x}, \mathbf{y})$, are given by $T_r(\mathbf{x}, \mathbf{y}) = \mathbf{x}^T R \mathbf{y} - \mathfrak{f}_r(\mathbf{x})$ and $T_c(\mathbf{x}, \mathbf{y}) = \mathbf{x}^T C \mathbf{y} - \mathfrak{f}_c(\mathbf{y})$. We will use $\mathcal{P}_{\lambda_p}$ to denote the set of two-player penalty games with λ_p-Lipschitz penalty functions. A special class of penalty games is obtained when $\mathfrak{f}_r(\mathbf{x}) = \mathbf{x}^T \mathbf{x}$ and $\mathfrak{f}_c(\mathbf{y}) = \mathbf{y}^T \mathbf{y}$. We call these games as *inner product* penalty games.

Two-Player Biased Games. This is a subclass of penalty games, where extra constraints are added to the penalty functions $\mathfrak{f}_r(\mathbf{x})$ and $\mathfrak{f}_c(\mathbf{y})$ of the players. In this class of games there is a *base strategy* and for each player and the penalty they receive is increasing with the distance between the strategy they choose and their base strategy. Formally, the row player has a base strategy $\mathbf{p} \in \Delta^n$, the column player has a base strategy $\mathbf{q}$ and their strictly increasing penalty functions are defined as $\mathfrak{f}_r(\|\mathbf{x} - \mathbf{p}\|_t^s)$ and $\mathfrak{f}_c(\|\mathbf{y} - \mathbf{q}\|_m^l)$ respectively.

Two-Player Distance Biased Games. This is a special class of biased games where the penalty function is a fraction of the distance between the base strategy of the player and his chosen strategy. Formally, a two player distance biased game $\mathcal{B}$ is defined by a tuple $(R, C, \mathfrak{b}_r(\mathbf{x}, \mathbf{p}), \mathfrak{b}_c(\mathbf{y}, \mathbf{q}), d_r, d_c)$, where (R, C) is a bimatrix game, $\mathbf{p} \in \Delta^n$ is a base strategy for the row player, $\mathbf{q} \in \Delta^n$ is a base strategy for the column player, $\mathfrak{b}_r(\mathbf{x}, \mathbf{p}) = \|\mathbf{x} - \mathbf{p}\|_t^s$ and $\mathfrak{b}_c(\mathbf{y}, \mathbf{q}) = \|\mathbf{y} - \mathbf{q}\|_m^l$ are the penalty functions for the row and the column player respectively. The utilities for the players under a strategy profile $(\mathbf{x}, \mathbf{y})$, denoted by $T_r(\mathbf{x}, \mathbf{y})$ and $T_c(\mathbf{x}, \mathbf{y})$, are given by $T_r(\mathbf{x}, \mathbf{y}) = \mathbf{x}^T R \mathbf{y} - d_r \cdot \mathfrak{b}_r(\mathbf{x}, \mathbf{p})$ and $T_c(\mathbf{x}, \mathbf{y}) = \mathbf{x}^T C \mathbf{y} - d_c \cdot \mathfrak{b}_c(\mathbf{y}, \mathbf{q})$, where d_r and d_c are non negative constants.

Solution Concepts. A strategy profile is an equilibrium if no player can increase his utility by unilaterally changing his strategy. A relaxed version of this concept is the approximate equilibrium, or ϵ-equilibrium, in which no player can increase his utility more than ϵ by unilaterally changing his strategy. Formally, a strategy profile $\mathbf{x}$ is an ϵ-equilibrium in a game $\mathfrak{L}$ if for every player $i \in [M]$ it holds that $T_i(x_i, \mathbf{x}_{-i}) \geq T_i(x_i', \mathbf{x}_{-i}) - \epsilon$ for all $x_i' \in S_i$.

In [20] it was proven that, unless P = PPAD, there is no PTAS for computing an ϵ-NE in bimatrix games. The same result holds for the class of penalty games where the penalty functions $\mathfrak{f}(n, \mathbf{x})$ for the players depend on n, the size of the underlying bimatrix game, and $\lim_{n \to \infty} \mathfrak{f}(n, \mathbf{x}) = 0$ for every player, for every possible $\mathbf{x}$. Let $\mathcal{P}'$ denote this class of games.

Theorem 1. *Unless* P = PPAD, *there is no PTAS for computing an ϵ-equilibrium in penalty games in $\mathcal{P}'$.*

3   Approximate Equilibria in λ_p-Lipschitz Games

In this section, we give an algorithm for computing approximate equilibria in λ_p Lipschitz games. Note that, our definition of a λ_p-Lipschitz game does not guarantee that an equilibrium always exists. Our technique can be applied *irrespective* of whether an exact equilibrium exists. If an exact equilibrium does exist, then our technique will always find an ϵ-equilibrium. If an exact equilibrium does not exist, then our algorithm either finds an ϵ-equilibrium or reports that the game does not have an exact equilibrium.

We will utilize the following theorem that was recently proved by Barman [4].

Theorem 2 (Barman [4]). *Given a set of vectors $X = \{x_1, x_2, \ldots, x_n\} \subset \mathbb{R}^d$, let $conv(X)$ denote the convex hull of X. Furthermore, let $\gamma := \max_{x \in X} \|x\|_p$ for some $2 \leq p < \infty$. For every $\epsilon > 0$ and every $\mu \in conv(X)$, there exists an $\frac{4p\gamma^2}{\epsilon^2}$ uniform vector $\mu' \in conv(X)$ such that $\|\mu - \mu'\|_p \leq \epsilon$.*

Combining Theorem 2 with the Definition 1 we get the following lemma.

Lemma 1. *Let* $X = \{x_1, x_2, \ldots, x_n\} \subset \mathbb{R}^d$, *let* $f : conv(X) \to \mathbb{R}$ *be a* λ_p-*Lipschitz continuous function for some* $2 \leq p < \infty$, *let* $\epsilon > 0$ *and let* $k = \frac{4\lambda^2 p \gamma^2}{\epsilon^2}$, *where* $\gamma := \max_{x \in X} \|x\|_p$. *Furthermore, let* $f(\mathbf{x}^*)$ *be the optimum value of* f. *Then we can compute a* k-*uniform point* $\mathbf{x}' \in conv(X)$ *in time* $O(n^k)$, *such that* $|f(\mathbf{x}^*) - f(\mathbf{x}')| < \epsilon$.

We now prove our result about Lipschitz games. In what follows we will study a λ_p-Lipschitz game $\mathfrak{L} := (M, n, \lambda, p, \gamma, \mathcal{T})$. Assuming the existence of an exact Nash equilibrium, we establish the existence of a k-uniform approximate equilibrium in the game $\mathfrak{L}$, where k depends on M, λ, p and γ. Note that λ depends heavily on p and the utility functions for the players.

Since by the definition of λ_p-Lipschitz games the strategy space S_i for every player i is the convex hull of n vectors $y_1, \ldots, y_n$ in $\mathbb{R}^d$, any $x_i \in S_i$ can be written as a convex combination of y_js. Hence, $x_i = \sum_{j=1}^n \alpha_j y_j$, where $\alpha_j > 0$ for every $j \in [n]$ and $\sum_{j=1}^n \alpha_j = 1$. Then, $\alpha = (\alpha_1, \ldots, \alpha_n)$ is a probability distribution over the vectors $y_1, \ldots, y_n$, i.e. vector y_j is drawn with probability α_j. Thus, we can sample a strategy x_i by the probability distribution α.

So, let $\mathbf{x}^*$ be an equilibrium for $\mathfrak{L}$ and let $\mathbf{x}'$ be a sampled uniform strategy profile from $\mathbf{x}^*$. For each player i we define the following events

$$\phi_i = \left\{ |T_i(x_i', \mathbf{x}_{-i}') - T_i(x_i^*, \mathbf{x}_{-i}^*)| < \epsilon/2 \right\}$$

$$\pi_i = \left\{ T_i(x_i, \mathbf{x}_{-i}') < T_i(x_i', \mathbf{x}_{-i}') + \epsilon \right\} \quad \text{for all possible } x_i$$

$$\psi_i = \left\{ \|x_i' - x_i^*\|_p < \frac{\epsilon}{2M\lambda} \right\} \quad \text{for some } p > 1.$$

Notice that if all the events π_i occur at the same time, then the sampled profile $\mathbf{x}'$ is an ϵ-equilibrium. We will show that if for a player i the events ϕ_i and $\bigcap_j \psi_j$ hold, then the event π_i is also true.

Lemma 2. *For all* $i \in [M]$ *it holds that* $\bigcap_{j \in [M]} \psi_j \cap \phi_i \subseteq \pi_i$.

We are ready to prove the main result of the section.

Theorem 3. *In any* λ_p-*Lipschitz game* $\mathfrak{L}$ *that possess an equilibrium and any* $\epsilon > 0$, *there is a* k-*uniform strategy profile, with* $k = \frac{16M^2\lambda^2 p \gamma^2}{\epsilon^2}$ *that is an* ϵ-*equilibrium.*

Theorem 3 establishes the existence of a k-uniform approximate equilibrium, but this does not immediately give us our approximation algorithm. The obvious approach is to perform a brute force check of all k-uniform strategies, and then output the one that provides the best approximation. There is a problem with this, however, since computing the quality of approximation requires us to compute the regret for each player, which in turn requires us to compute a best response for each player. Computing an exact best response in a Lipschitz game is a hard problem in general, since we make no assumptions about the utility functions of the players. Fortunately, it is sufficient to instead compute an *approximate* best response for each player, and Lemma 1 can be used to do this.

Lemma 3. *Let* $\mathbf{x}$ *be a strategy profile for a* λ_p-*Lipschitz game* $\mathfrak{L}$, *and let* $\hat{x}_i$ *be a best response for player* i *against the profile* $\mathbf{x}_{-i}$. *There is a* $\frac{4\lambda^2 p\gamma^2}{\epsilon^2}$-*uniform strategy* x_i' *that is an* ϵ-*best response against* $\mathbf{x}_{-i}$.

Our goal is to *approximate* the approximation guarantee for a given strategy profile. More formally, given a strategy profile $\mathbf{x}$ that is an ϵ-equilibrium, and a constant $\delta > 0$, we want an algorithm that outputs a number within the range $[\epsilon - \delta, \epsilon + \delta]$. Lemma 3 allows us to do this. For a given strategy profile $\mathbf{x}$, we first compute δ-approximate best responses for each player, then we can use these to compute δ-approximate regrets for each player. The maximum over the δ-approximate regrets then gives us an approximation of ϵ with a tolerance of δ. This is formalised in the following algorithm.

Algorithm 1. Evaluation of approximation guarantee

Input: A strategy profile $\mathbf{x}$ for $\mathfrak{L}$, and a constant $\delta > 0$.
Output: An additive δ-approximation of the approximation guarantee $\alpha(\mathbf{x})$ for the strategy profile $\mathbf{x}$.

1. Set $l = \frac{4\lambda^2 p\gamma^2}{\delta^2}$.
2. For every player $i \in [M]$
    (a) For every l-uniform strategy x_i' of player i compute $T_i(x_i', \mathbf{x}_{-i})$.
    (b) Set $m^* = \max_{x_i'} T_i(x_i', \mathbf{x}_{-i})$.
    (c) Set $\mathcal{R}_i(\mathbf{x}) = m^* - T_i(x_i, \mathbf{x}_{-i})$.
3. Set $\alpha(\mathbf{x}) = \delta + \max_{i \in [M]} \mathcal{R}_i(\mathbf{x})$.
4. Return $\alpha(\mathbf{x})$.

Utilising the above algorithm, we can now produce an algorithm to find an approximate equilibrium in Lipschitz games. The algorithm checks all k-uniform strategy profiles, using the value of k given by Theorem 3, and for each one, computes an approximation of the quality approximation using the algorithm given above.

Algorithm 2. 3ϵ-equilibrium for λ_p-Lipschitz game $\mathfrak{L}$

Input: Game $\mathfrak{L}$ and $\epsilon > 0$.
Output: An 3ϵ-equilibrium for $\mathfrak{L}$.

1. Set $k > \frac{16\lambda^2 M p\gamma^2}{\epsilon^2}$.
2. For every k-uniform strategy profile $\mathbf{x}'$
    (a) Compute an ϵ-approximation of $\alpha(\mathbf{x}')$.
    (b) If the ϵ-approximation of $\alpha(\mathbf{x}')$ is less than 2ϵ, return $\mathbf{x}'$.

If the algorithm returns a strategy profile $\mathbf{x}$, then it must be a 3ϵ equilibrium. This is because we check that an ϵ-approximation of $\alpha(\mathbf{x})$ is less than 2ϵ, and therefore $\alpha(\mathbf{x}) \leq 3\epsilon$. Secondly, we argue that if the game has an exact Nash equilibrium, then this procedure will always output a 3ϵ-approximate equilibrium.

From Theorem 3 we know that if $k > \frac{16\lambda^2 M p \gamma^2}{\epsilon^2}$, then there is a k-uniform strategy profile $\mathbf{x}$ that is an ϵ-equilibrium for $\mathfrak{L}$. When we apply our approximate regret algorithm to $\mathbf{x}$, to find an ϵ-approximation of $\alpha(\mathbf{x})$, the algorithm will return a number that is less than 2ϵ, hence $\mathbf{x}$ will be returned by the algorithm.

To analyse the running time, observe that there are $\binom{n+k-1}{k} = O(n^k)$ possible k-uniform strategies for each player, thus $O(n^{Mk})$ k-uniform strategy profiles. Furthermore, our regret approximation algorithm runs in time $O(Mn^l)$, where $l = \frac{4\lambda^2 p \gamma^2}{\epsilon^2}$. Hence, we get the next theorem.

Theorem 4. *Given a λ_p-Lipschitz game $\mathfrak{L}$ that possess an equilibrium and any $\epsilon > 0$, a 3ϵ-equilibrium can be computed in time $O\left(Mn^{Mk+l}\right)$, where $k = O\left(\frac{\lambda^2 M p \gamma^2}{\epsilon^2}\right)$ and $l = O\left(\frac{\lambda^2 p \gamma^2}{\epsilon^2}\right)$.*

Although it might be hard to decide whether a game has an equilibrium, our algorithm can be applied in *any* λ_p-Lipschitz game. Notice that our algorithm never uses the fact that the game possess an equilibrium. If the game does not posses an exact equilibrium then our algorithm either finds an approximate equilibrium or determines that the game does not posses an exact equilibrium.

Theorem 5. *For any λ_p-Lipschitz game $\mathfrak{L}$ in time $O\left(Mn^{Mk+l}\right)$, we can either compute a 3ϵ-equilibrium, or decide that $\mathfrak{L}$ does not posses an exact equilibrium, where $k = O\left(\frac{\lambda^2 M p \gamma^2}{\epsilon^2}\right)$ and $l = O\left(\frac{\lambda^2 p \gamma^2}{\epsilon^2}\right)$.*

4   A Quasi-polynomial Algorithm for Penalty Games

In this section we present an algorithm that, for any $\epsilon > 0$, can compute an ϵ-equilibrium for any penalty game in $\mathcal{P}_{\lambda_p}$ that posses one in quasi-polynomial time. For the algorithm, we take the same approach as we did in the previous section for Lipschitz games: we show that if an exact equilibrium exists, then a k-uniform approximate equilibrium always exists too, and provide a brute-force search algorithm for finding it. Once again, since best response computation may be hard for this class of games, we must provide an approximation algorithm for finding the quality of an approximate equilibrium.

We first focus on penalty games that posses an exact equilibrium. So, let $(\mathbf{x}^*, \mathbf{y}^*)$ be an equilibrium of the game and let $(\mathbf{x}', \mathbf{y}')$ be a k-uniform strategy profile sampled from this equilibrium. We define the following four events:

$$\phi_r = \left\{ |T_r(\mathbf{x}', \mathbf{y}') - T_r(\mathbf{x}^*, \mathbf{y}^*)| < \epsilon/2 \right\}$$
$$\pi_r = \left\{ T_r(\mathbf{x}, \mathbf{y}') < T_r(\mathbf{x}', \mathbf{y}') + \epsilon \right\} \quad \text{for all } \mathbf{x}$$
$$\phi_c = \left\{ |T_c(\mathbf{x}', \mathbf{y}') - T_c(\mathbf{x}^*, \mathbf{y}^*)| < \epsilon/2 \right\}$$
$$\pi_c = \left\{ T_c(\mathbf{x}', \mathbf{y}) < T_c(\mathbf{x}', \mathbf{y}') + \epsilon \right\} \quad \text{for all } \mathbf{y}.$$

The goal is to derive a value for k such that all the four events above are true, or equivalently $Pr(\phi_r \cap \pi_r \cap \phi_c \cap \pi_r) > 0$.

Note that in order to prove that $(\mathbf{x}', \mathbf{y}')$ is an ϵ-equilibrium we *only* have to consider the events π_r and π_c. Nevertheless, as we show in Lemma 4, the events

ϕ_r and ϕ_c are crucial in our analysis. The proof of the main theorem boils down to the events ϕ_r and ϕ_c.

We will focus only on the row player, since the same analysis can be applied to the column player. Firstly we study the event π_r.

Lemma 4. *For all penalty games it holds that* $Pr(\pi_r^c) \leq n \cdot e^{-\frac{k\epsilon^2}{2}} + Pr(\phi_r^c)$.

With Lemma 4 in hand, we can see that in order to compute a value for k it is sufficient to study the event ϕ_r. We introduce the following auxiliary events that we will study separately: $\phi_{ru} = \{|\mathbf{x}'^T R \mathbf{y}' - \mathbf{x}^{*T} R \mathbf{y}^*| < \epsilon/4\}$ and $\phi_{rb} = \{|f_r(\mathbf{x}') - f_r(\mathbf{x}^*)| < \epsilon/4\}$. It is easy to see that if both ϕ_{rb} and ϕ_{ru} are true, then the event ϕ_r must be true too. So we have $\phi_{rb} \cap \phi_{ru} \subseteq \phi_r$. Using the analysis from [16] we can prove that $Pr(\phi_{ru}^c) \leq 2e^{-\frac{k\epsilon^2}{8}}$. Finally, we must prove an upper bound on the event ϕ_{rb}^c, which we provide in the following lemma.

Lemma 5. $Pr(\phi_{rb}^c) \leq \frac{8\lambda\sqrt{p}}{\epsilon\sqrt{k}}$.

Let us define the event $GOOD = \phi_r \cap \phi_c \cap \pi_r \cap \pi_c$. To prove our theorem it suffices to prove that $Pr(GOOD) > 0$. Notice that for the events ϕ_c and π_c the same analysis as for ϕ_r and π_r can be used. Then, using Lemmas 4, 5 and the analysis for ϕ_{ru} we get that $Pr(GOOD^c) < 1$ for the chosen value of k.

Theorem 6. *For any equilibrium* $(\mathbf{x}^*, \mathbf{y}^*)$ *of a penalty game from the class* $\mathcal{P}_{\lambda_p}$, *any* $\epsilon > 0$, *and any* $k \in \frac{\Omega(\lambda^2 \log n)}{\epsilon^2}$, *there exists a* k-*uniform strategy profile* $(\mathbf{x}', \mathbf{y}')$ *that:*

1. $(\mathbf{x}', \mathbf{y}')$ *is an* ϵ-*equilibrium for the game,*
2. $|T_r(\mathbf{x}', \mathbf{y}') - T_r(\mathbf{x}^*, \mathbf{y}^*)| < \epsilon/2$,
3. $|T_c(\mathbf{x}', \mathbf{y}') - T_c(\mathbf{x}^*, \mathbf{y}^*)| < \epsilon/2$.

Theorem 6 establishes the *existence* of a k-uniform strategy profile $(\mathbf{x}', \mathbf{y}')$ that is an ϵ-equilibrium, but as before, we must provide an efficient method for approximating the quality of approximation provided by a given strategy profile. To do so, we first give the following lemma, which shows that approximate best responses can be computed in quasi-polynomial time for penalty games.

Lemma 6. *Let* $(\mathbf{x}, \mathbf{y})$ *be a strategy profile for a penalty game* $\mathcal{P}_{\lambda_p}$, *and let* $\hat{\mathbf{x}}$ *be a best response against* $\mathbf{y}$. *There is an* l-*uniform strategy* $\mathbf{x}'$, *with* $l = \frac{17\lambda^2\sqrt{p}}{\epsilon^2}$, *that is an* ϵ-*best response against* $\mathbf{y}$, *i.e.* $T_r(\hat{\mathbf{x}}, \mathbf{y}) < T_r(\mathbf{x}', \mathbf{y}) + \epsilon$.

Given this lemma, we can reuse Algorithm 1, but with l set equal to $\frac{17\lambda^2\sqrt{p}}{\epsilon^2}$, to provide an algorithm that aproximates the quality of approximation of a given strategy profile. Then, we can reuse Algorithm 2 with $k = \frac{\Omega(\lambda^2 \log n)}{\epsilon^2}$ to provide a quasi-polynomial time algorithm that finds approximate equilibia in penalty games. Notice again that our algorithm can be applied in games in which it is computationally hard to verify whether an exact equilibrium exists. Our algorithm either will compute an approximate equilibrium or it will fail to find one, in which case the game does not posses an exact equilibrium.

Theorem 7. *In any penalty game* $\mathcal{P}_{\lambda_p}$ *and any* $\epsilon > 0$, *in quasi polynomial time we can either compute a* 3ϵ-*equilibrium, or decide that* $\mathcal{P}_{\lambda_p}$ *does not posses an exact equilibrium.*

5 Distance Biased Games

In this section, we focus on three particular classes of distance biased games, and we provide polynomial-time approximation algorithms when the penalty function is one of the L_1, L_2^2 and L_∞ norm. Our approach is to follow the technique of Daskalakis, Mehta, Papadimitriou [11] that finds a 0.5-NE in a bimatrix game. The algorithm that we will use for all three penalty functions is given below.

Algorithm 3. The Base Algorithm

1. Compute a best response $\mathbf{y}^*$ against $\mathbf{p}$.
2. Compute a best response $\mathbf{x}$ against $\mathbf{y}^*$.
3. Set $\mathbf{x}^* = \delta \cdot \mathbf{p} + (1 - \delta) \cdot \mathbf{x}$, for some $\delta \in [0,1]$.
4. Return the strategy profile $(\mathbf{x}^*, \mathbf{y}^*)$.

While this is a well-known technique for bimatrix games, it cannot immediately be applied to penalty games, because the algorithm requires us to compute two best responses. While computing a best-response is trivial in bimatrix games, this is not the case for penalty games. Best responses for L_1 and L_∞ penalties can be computed in polynomial-time via linear programming, and for L_2^2 penalties, the ellipsoid algorithm can be applied to a specialized quadratic program. However, these methods work as black boxes and do not provide strongly polynomial algorithms.

For each of the penalties we develop a simple combinatorial algorithm for computing best response strategies. We use the nature of these penalty functions and we provide strongly polynomial algorithms that compute best responses. More specifically, for the L_1 and L_∞ norms we compute the exact probability each pure strategy should be played in a best response by studying how the utility function increases. For the L_2^2 norm we use the KKT conditions of a quadratic program to produce a closed formula for the solution. Our algorithms, which are strongly polynomial, allow us to optimize the value of δ, and produce the following approximation guarantees.

Theorem 8. *In biased games with* L_1, L_2^2 *and* L_∞ *penalties a 2/3, 5/7 and 2/3-equilibrium respectively can be computed in polynomial time. For inner product games the approximation guarantee is 13/21.*

References

1. Azrieli, Y., Shmaya, E.: Lipschitz games. Math. Oper. Res. **38**(2), 350–357 (2013)
2. Babichenko, Y.: Best-reply dynamics in large binary-choice anonymous games. Games Econ. Behav. **81**, 130–144 (2013)

3. Babichenko, Y., Barman, S., Peretz, R.: Simple approximate equilibria in large games. In: Proceeding of EC, pp. 753–770 (2014)
4. Barman, S.: Approximating Nash equilibria and dense bipartite subgraphs via an approximate version of Caratheodory's theorem. In: Proceeding of STOC 2015, pp. 361–369 (2015)
5. Bosse, H., Byrka, J., Markakis, E.: New algorithms for approximate Nash equilibria in bimatrix games. Theor. Comput. Sci. **411**(1), 164–173 (2010)
6. Caragiannis, I., Kurokawa, D., Procaccia, A.D.: Biased games. In: Proceeding of AAAI, pp. 609–615 (2014)
7. Chapman, G.B., Johnson, E.J.: Anchoring, activation, and the construction of values. Organ. Behav. Hum. Decis. Process. **79**(2), 115–153 (1999)
8. Chen, X., Deng, X., Teng, S.-H.: Settling the complexity of computing two-player Nash equilibria. J. ACM **56**(3), 14:1–14:57 (2009)
9. Daskalakis, C., Goldberg, P.W., Papadimitriou, C.H.: The complexity of computing a Nash equilibrium. SIAM J. Comput. **39**(1), 195–259 (2009)
10. Daskalakis, C., Mehta, A., Papadimitriou, C.H.: Progress in approximate Nash equilibria. In: Proceeding of EC, pp. 355–358 (2007)
11. Daskalakis, C., Mehta, A., Papadimitriou, C.H.: A note on approximate Nash equilibria. Theor. Comput. Sci. **410**(17), 1581–1588 (2009)
12. Daskalakis, C., Papadimitriou, C.H.: Approximate Nash equilibria in anonymous games. J. Econ. Theory (2014, to appear)
13. Deb, J., Kalai, E.: Stability in large Bayesian games with heterogeneous players. J. Econ. Theor. **157**(C), 1041–1055 (2015)
14. Fiat, A., Papadimitriou, C.H.: When the players are not expectation maximizers. In: SAGT, pp. 1–14 (2010)
15. Kahneman, D.: Reference points, anchors, norms, and mixed feelings. Organ. Behav. Hum. Decis. Process. **51**(2), 296–312 (1992)
16. Lipton, R.J., Markakis, E., Mehta, A.: Playing large games using simple strategies. In: EC, pp. 36–41 (2003)
17. Mavronicolas, M., Monien, B.: The complexity of equilibria for risk-modeling valuations. CoRR, abs/1510.08980 (2015)
18. Nash, J.: Non-cooperative games. Ann. Math. **54**(2), 286–295 (1951)
19. Rosen, J.B.: Existence and uniqueness of equilibrium points for concave n-person games. Econometrica **33**(3), 520–534 (1965)
20. Rubinstein, A.: Settling the complexity of computing approximate two-player nash equilibria. CoRR, abs/1606.04550 (2016)
21. Tsaknakis, H., Spirakis, P.G.: An optimization approach for approximate Nash equilibria. Internet Math. **5**(4), 365–382 (2008)
22. Tversky, A., Kahneman, D.: Judgment under uncertainty: heuristics and biases. Science **185**(4157), 1124–1131 (1974)

The Parallel Complexity of Coloring Games

Guillaume Ducoffe[(⊠)]

Université Côte D'Azur, Inria, CNRS, I3S, Sophia Antipolis, France
`guillaume.ducoffe@inria.fr`

Abstract. We wish to motivate the problem of finding *decentralized lower-bounds* on the complexity of computing a Nash equilibrium in graph games. While the centralized computation of an equilibrium in polynomial time is generally perceived as a positive result, this does not reflect well the reality of some applications where the game serves to implement distributed resource allocation algorithms, or to model the social choices of users with limited memory and computing power. As a case study, we investigate on the parallel complexity of a game-theoretic variation of graph coloring. These "coloring games" were shown to capture key properties of the more general welfare games and Hedonic games. On the positive side, it can be computed a Nash equilibrium in polynomial-time for any such game with a local search algorithm. However, the algorithm is time-consuming and it requires polynomial space. The latter questions the use of coloring games in the modeling of information-propagation in social networks. We prove that the problem of computing a Nash equilibrium in a given coloring game is PTIME-hard, and so, it is unlikely that one can be computed with an efficient distributed algorithm. The latter brings more insights on the complexity of these games.

1 Introduction

In algorithmic game theory, it is often the case that a problem is considered "tractable" when it can be solved in polynomial time, and "difficult" only when it is NP-hard or it is PLS-hard to find a solution. On the other hand, with the growing size of real networks, it has become a boiling topic in (non game-theoretic) algorithmic to study on the finer-grained complexity of polynomial problems [16]. In our opinion, the same should apply to graph games when they serve as a basis for new distributed algorithms. We propose to do so in some cases when it can be easily computed a Nash equilibrium in polynomial time. The following case study will make use of well-established parallel and space complexity classes to better understand the hardness of a given graph game.

Precisely, we investigate on a "coloring game", first introduced in [14] in order to unify classical upper-bounds on the chromatic number. Since then it has been

This work is partially supported by ANR project Stint under reference ANR-13-BS02-0007 and ANR program "Investments for the Future" under reference ANR-11-LABX-0031-01.

© Springer-Verlag Berlin Heidelberg 2016
M. Gairing and R. Savani (Eds.): SAGT 2016, LNCS 9928, pp. 27–39, 2016.
DOI: 10.1007/978-3-662-53354-3_3

rediscovered many times, attracting attention on the way in the study of information propagation in wireless sensor networks [4] and in social networks [12]. We choose to consider this game since it is a good representative of the *separable welfare games*–proposed in [13] as a game-theoretic toolkit for distributed resource allocation algorithms–and the *additively separable symmetric Hedonic games* [3]. A coloring game is played on an undirected graph with each vertex being an agent (formal definitions will be given in the technical sections of the paper). Agents must choose a colour in order to construct a proper coloring of the graph. The individual goal of each agent is to maximize the number of agents with the same colour as hers. Furthermore, it can always be computed a Nash equilibrium in polynomial time with a simple local-search algorithm [6,12,14].

However, for n-vertex m-edge graphs, the above-mentioned algorithm has $\mathcal{O}(m + n\sqrt{n})$-time complexity and $\mathcal{O}(n + m)$-space complexity. Therefore, when the graph gets larger, potential applications of coloring games as a *computational mechanism design* (*e.g.*, in order to assign frequencies in sensor networks in a distributed fashion, or to model the behaviour of social network users with limited power and storage) can be questioned. In particular, the authors in [11] report on the limited abilities of human subject networks to solve a coloring problem. In this note, we will investigate on the belonging of our problem–the computation of a Nash equilibrium in coloring games–to some complexity classes that are related to parallel and space complexity. Our goal in doing so is to bring more insights on the complexity of the problem.

Related Work. Apart from lower-bounds in communication complexity [7], we are not aware of any analysis of decentralized complexity in game theory. Closest to our work are the studies on the sequential complexity of Hedonic games. Deciding whether a given Hedonic game admits a Nash equilibrium is NP-complete [1]. Every additively separable symmetric Hedonic games has a Nash equilibrium but it is PLS-complete to compute one [8]. Coloring games are a strict subclass where the local-search algorithm terminates on a Nash equilibrium within a polynomial number of steps. We will go one step further by considering their parallel complexity, something we think we are the first to study.

In [4], they introduced a distributed algorithm in order to compute the Nash equilibrium of a given coloring game. Their algorithm is a natural variation of the classical local-search algorithm for the problem, however, it does not speed up the computation of equilibria (at least theoretically). In addition, each agent needs to store locally the colouring of the graph at any given step, that implies *quadratic* space and communication complexity. Additional related work is [6,12], where it is studied the number of steps of more elaborate local-search algorithms when up to k players are allowed to *collude* at each step. Informally, collusion means that the players can simultaneously change their colours for the same new colour provided they all benefit from the process (note that the classical local-search algorithm corresponds to the case $k = 1$).

Contributions. We prove that the problem of computing a Nash equilibrium in a given coloring game is PTIME-hard (Theorem 2). This is hint that the problem is inherently *sequential, i.e.*, it is unlikely the computation of an equilibrium can be sped up significantly on a parallel machine with polynomially many processors. In particular, our negative result applies to the *distributed* setting since any distributed algorithm on graphs can be simulated on a parallel machine with one processor per edge and per vertex. By a well-known relationship between space and parallel complexity [15], Theorem 2 also extends to show that no *space efficient* algorithm for the problem (say, within logarithmic workspace) can exist. Altogether, this may be hint that coloring games are a too powerful computational mechanism design for "lightweight" distributed applications.

Our reduction is from the standard MONOTONE CIRCUIT VALUE problem. However, the gadgets needed are technically challenging, and we will need to leverage nontrivial properties of coloring games in order to prove its correctness. Definitions and useful background will be given in Sect. 2. We will detail our reduction in Sect. 3 before concluding this paper in Sect. 4.

2   Definitions and Notations

We use the graph terminology from [2]. Graphs in this study are finite, simple, and unweighted.

Coloring Games. Let $G = (V, E)$ be a graph. A *coloring* of G assigns a positive integer, taken in the range $\{1, \ldots, n\}$, to each of the n vertices in V. For every i, let L_i be the subset of vertices coloured i. We name L_i a *colour class* in what follows. Nonempty colour classes partition the vertex set V. The partition is a *proper coloring* when no two adjacent vertices are assigned the same colour, *i.e.*, for every $1 \leq i \leq n$ and for every $u, v \in L_i$, $\{u, v\} \notin E$.

Fig. 1. Proper coloring of a graph G. Each colour class is represented by an ellipse. Every agent receives unit payoff.

Every graph G defines a *coloring game* whose n agents are the vertices in V. The strategy of an agent is her colour. Furthermore, every $v \in L_i$ receives payoff: -1 if there is $u \in L_i$ s.t. $\{u, v\} \in E$ (in which case, the coloring is not proper), and $|L_i| - 1$ otherwise. We refer to Fig. 1 for an illustration. Finally, a *Nash equilibrium* of the coloring game is any coloring of G where no agent can increase her payoff by changing her strategy. In particular, the proper coloring in Fig. 1 is a Nash equilibrium. More generally, observe that a Nash equilibrium in this game is always a proper coloring of G. In what follows, we will focus on the computation of Nash equilibria in coloring games.

Theorem 1 ([6,12]). *For any coloring game that is specified by an n-vertex m-edge graph $G = (V, E)$, a Nash equilibrium can be computed in $\mathcal{O}(m + n\sqrt{n})$-time and $\mathcal{O}(n + m)$-space.*

Parallel Complexity. Computations are performed on a parallel random-access machine (PRAM, see [9]) with an unlimited amount of processors. However, as stated in the conclusion, our results also apply to more realistic parallel complexity classes. In what follows, we will use the fact that processors are numbered. We will handle with read/write conflicts between processors with the strategy CREW-PRAM (concurrent read, exclusive write). Let PTIME contain the decision problems that can be solved in *sequential* polynomial-time (that is, with a single processor). Problem A reduces to problem B if given an oracle to solve B, A can be solved in polylogarithmic-time with a polynomial number of processors. In particular, a problem B is PTIME-hard if every problem in PTIME reduces to B (this is formally defined as quasi-PTIME-hardness in [9]). Such reductions are finer-grained than the more standard logspace reductions.

3 Main Result

Theorem 2. *Computing a Nash equilibrium for coloring games is PTIME-hard.*

In order to prove Theorem 2, we will reduce from a variation of the well-known MONOTONE CIRCUIT VALUE problem, defined as follows.

---

*Problem 1 (*MONOTONE CIRCUIT VALUE*).*

Input: A boolean circuit $\mathcal{C}$ with m gates and n entries, a word $w \in \{0, 1\}^n$ such that:
- the gates are either AND-gates or OR-gates;
- every gate has exactly two entries (in-degree two);
- a *topological ordering* of the gates is given, with the mth gate being the output gate.

Question: Does $\mathcal{C}$ output 1 when it takes w as input?

MONOTONE CIRCUIT VALUE is proved to be PTIME-complete in [9].

3.1 The Reduction

Let $\langle \mathcal{C}, w \rangle$ be any instance of MONOTONE CIRCUIT VALUE. We will reduce it to a coloring game as follows. Let $\mathcal{G} := (g_1, g_2, \ldots, g_m)$ be the gates of the circuit, that are topologically ordered.

Construction of the Gate-Gadgets. For every $1 \leq j \leq m$, the jth gate will be simulated by a subgraph $G_j = (V_j, E_j)$ with $12(n+j) - 9$ vertices. We refer to Fig. 2 for an illustration. Let us give some intuition for the following construction of G_j. We aim at simulating the computation of the (binary) output of all the gates in $\mathcal{C}$ when it takes w as input. To do that, given a supergraph G of G_j (to be defined later), and a fixed Nash equilibrium for the coloring game that is defined on G, we aim at guessing the output of the jth gate from the subcoloring of G_j. More precisely, the subcoloring will encode a "local certificate" that indicates which values on the two entries of g_j cause the output.

Observe that to certify that an OR-gate outputs 1, it suffices to show that it receives 1 on any one of its two entries, whereas for an AND-gate it requires to show that it outputs 1 on its two entries. Since by de Morgan's laws, the negation of an AND-gate can be transformed into an OR-gate and vice-versa, therefore, we need to distinguish between three cases in order to certify the output of the gate. So, the vertices in V_j are partitioned in three subsets of equal size $4(n+j) - 3$, denoted by V_j^1, V_j^2, V_j^3. Furthermore, for every $1 \leq t \leq 3$, every vertex in V_j^t is adjacent to every vertex in $V_j \backslash V_j^t$.

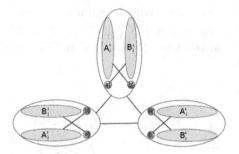

Fig. 2. Gadget subgraph G_j representing the jth gate. An edge between two subsets of vertices (delimited by an ellipse) denotes the existence of a complete bipartite subgraph.

Let us now describe the structure of the three (isomorphic) subgraphs $G_j[V_j^t] = (V_j^t, E_j^t)$ with $1 \leq t \leq 3$. Informally, we will need this internal structure in order to ensure that every of the three subsets V_j^t will behave as a "truthful" certificate to decide on the output of the gate; *i.e.*, only a few vertices of V_j will be used to certify the output of the jth gate, while all others will be divided into artificial aggregates that we name "private groups" whose role is to ensure "truthfulness" of the certificate (this will be made clearer in the following). There are two nonadjacent vertices $a_j^t, b_j^t \in V_j^t$ playing a special role. The other vertices in $V_j^t \backslash \{a_j^t, b_j^t\}$ are partitioned in two subsets A_j^t, B_j^t of respective size $2(n+j) - 3$ and $2(n+j) - 2$. The sets A_j^t, B_j^t are called the *private groups* of a_j^t, b_j^t. Furthermore, every vertex in A_j^t is adjacent to every vertex in $V_j^t \backslash (A_j^t \cup \{a_j^t\})$, similarly every vertex in B_j^t is adjacent to every vertex in $V_j^t \backslash (B_j^t \cup \{b_j^t\})$.

Since all edges are defined above independently the one from the other, the graph $G_j[V_j^1] = (V_j^1, E_j^1)$ (encoded by its adjacency lists) can be constructed with

$|V_j^1| + |E_j^1| = 4(n+j)^2 - 2(n+j) - 2$ processors simply by assigning the construction of each vertex and each edge to a different processor. Note that each processor can decide on the vertex, resp. the edge, it needs to compute from its number. Overall, it takes $\mathcal{O}(\log(n+j))$-time in order to construct $G_j[V_j^1]$ in parallel. The latter can be easily generalized in order to construct G_j in $\mathcal{O}(\log(n+j))$-time with $|V_j| + |E_j|$ processors. Therefore, the graphs $G_1, G_2, \ldots, G_m$ can be constructed in parallel in $\mathcal{O}(\log(n+m))$-time with $\sum_{j=1}^{m}(|V_j| + |E_j|)$ processors, that is polynomial in $n+m$.

Construction of the Graph. Let $X = \{x_1, x_1', \ldots, x_i, x_i', \ldots, x_n, x_n'\}$ contain $2n$ nonadjacent vertices, that are two vertices per letter in the binary word w. The graph $G = (V, E)$ for the reduction has vertex-set $V = X \cup \left(\bigcup_{j=1}^{m} V_j \right)$. In particular, it has $2n - 9m + 6m(m + 2n + 1)$ vertices. Furthermore, $G[V_j]$ is isomorphic to G_j for every $1 \leq j \leq m$. In order to complete our reduction, let us now describe how our gadgets are connected the one with the other.

For technical reasons, we will need to make adjacent every vertex in the private group A_j^t (resp. B_j^t), with $1 \leq j \leq m$ and $1 \leq t \leq 3$, to every vertex in $V \backslash V_j$. By doing so, note that every vertex in $V \backslash (A_j^t \cup \{a_j^t\})$ is adjacent to every vertex in A_j^t (resp., every vertex in $V \backslash (B_j^t \cup \{b_j^t\})$ is adjacent to every vertex in B_j^t). Furthermore, each edge is defined independently the one from the other. Hence, similarly as above, $\sum_{j=1}^{m} \sum_{t=1}^{3} (|A_j^t| + |B_j^t|)|V \backslash V_j|$ processors are sufficient in order to construct these edges in $\mathcal{O}(\log(n+m))$-time, that is polynomial in $n+m$.

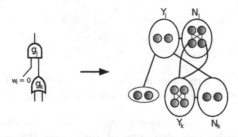

Fig. 3. Edges in G to simulate the two connections of an AND-gate in the circuit.

Finally, we recall that for every j, there are three cases to distinguish in order to decide on the output of the jth gate, with each case being represented with some subset V_j^t. The union of subsets representing a *positive* certificate (output 1) is named Y_j, while the union of those representing a *negative* certificate (output 0) is named N_j. In particular, if the jth gate is an OR-gate, let $Y_j := \{a_j^1, b_j^1, a_j^2, b_j^2\}$ and $N_j := \{a_j^3, b_j^3\}$ (it suffices to receive 1 on one input). Else, the jth gate is an AND-gate, so, let $Y_j := \{a_j^1, b_j^1\}$ and $N_j := \{a_j^2, b_j^2, a_j^3, b_j^3\}$.

Suppose the jth gate is an OR-gate (the case when it is an AND-gate follows by symmetry, up to interverting Y_j with N_j, see also Fig. 3). Let us consider the first entry of the gate. There are two cases. Suppose that it is the ith entry of the circuit, for some $1 \leq i \leq n$. If $w_i = 0$ then we make both x_i, x_i' adjacent to both a_j^1, b_j^1; else, $w_i = 1$, we make both x_i, x_i' adjacent to both a_j^3, b_j^3. Else, the entry is some other gate of the circuit, and so, since gates are topologically ordered, it is the kth gate for some $k < j$. We make every vertex in N_k adjacent to both a_j^1, b_j^1, and we make every vertex in Y_k adjacent to both a_j^3, b_j^3.

The second entry of the gate is similarly considered, up to replacing above the two vertices a_j^1, b_j^1 with a_j^2, b_j^2. We refer to Fig. 3 for an illustration. In particular, observe that there is only a constant number of edges that are added at this step for each gate. Furthermore, the construction of these new edges only requires to read the two in-neighbours of the gate in the circuit C. As a result, the last step can be done in parallel in $\mathcal{O}(\log(n+m))$-time with m processors.

3.2 Structure of a Nash Equilibrium

The graph $G = (V, E)$ of our reduction (constructed in Sect. 3.1) defines a coloring game. Let us fix any Nash equilibrium for this game (that exists by Theorem 1). We will show that it is sufficient to know the colour of every vertex in $Y_m \cup N_m$ in order to decide on the output of the circuit C (recall that the mth gate is the output gate). To prove it, we will need the following technical claims in order to gain more insights on the structure of the equilibrium.

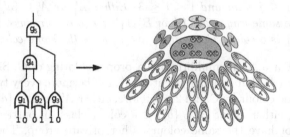

Fig. 4. A boolean circuit (left) with a Nash equilibrium of the coloring game from our reduction (right). Each colour class is represented with an ellipse. Intuitively, vertices in the central colour class simulate the computation of the output. Other colour classes contain a private group and they are "inactive".

More precisely, we will prove that there are exactly $6m + 1$ colour classes, that are one colour class per private group A_j^t or B_j^t and an additional colour for the vertices in X. The intuition is that there are $2(n + m)$ vertices in one special colour class (including X) that simulates the computation of the output of C, whereas all other vertices are "trapped" with the vertices in their respective private group. We refer to Fig. 4 for an illustration.

Claim 1. *For every j, any colour class does not contain more than two vertices in every $Y_j \cup N_j$. Furthermore, if it contains exactly two vertices in $Y_j \cup N_j$ then these are a_j^t, b_j^t for some $1 \leq t \leq 3$.*

Proof. A Nash equilibrium is a proper coloring of G. Therefore, since any two vertices in different subsets among V_j^1, V_j^2, V_j^3 are adjacent by construction, they cannot have the same colour. Since $Y_j \cup N_j = \{a_j^1, b_j^1, a_j^2, b_j^2, a_j^3, b_j^3\}$ and $a_j^t, b_j^t \in V_j^t$ for every $1 \leq t \leq 3$, the claim follows directly.   ◇

Claim 2. *Any two vertices that are in a same private group have the same colour. Similarly, x_i and x_i' have the same colour for every $1 \leq i \leq n$.*

Proof. Let S be either a private group ($S = A_j^t$ or $S = B_j^t$ for some $1 \leq j \leq m$ and $1 \leq t \leq 3$), or a pair representing the same letter of word w (*i.e.*, $S = \{x_i, x_i'\}$ for some $1 \leq i \leq n$). Let $v \in S$ maximize her payoff and let c be her colour. Note that v receives payoff $|L_c| - 1$ with L_c being the colour class composed of all the vertices with colour c. Furthermore, every $u \in S$ receives payoff lower than or equal to $|L_c| - 1$ by the choice of v. In such case, every $u \in S$ must be coloured c, or else, since the adjacency and the nonadjacency relations are the same for u and v (they are twins), furthermore u, v are nonadjacent, the agent u would increase her payoff to $|L_c|$ by choosing c as her new colour, thus contradicting the hypothesis that we are in a Nash equilibrium.   ◇

The argument we use in Claim 2 is that twin vertices must have the same colour. In what follows, we will use the same argument under different disguises.

Claim 3. *Let $1 \leq j \leq m$ and $1 \leq t \leq 3$. Either A_j^t or $A_j^t \cup \{a_j^t\}$ is a colour class, and in the same way either B_j^t or $B_j^t \cup \{b_j^t\}$ is a colour class. Furthermore, either $B_j^t \cup \{b_j^t\}$ is a colour class, or a_j^t and b_j^t have the same colour.*

Proof. Recall that a Nash equilibrium is a proper coloring of G. Since a_j^t is the only vertex in $V \setminus A_j^t$ that is nonadjacent to A_j^t, furthermore every two vertices in A_j^t have the same colour by Claim 2, therefore, either A_j^t or $A_j^t \cup \{a_j^t\}$ is a colour class. Similarly, either B_j^t or $B_j^t \cup \{b_j^t\}$ is a colour class. In particular, suppose that b_j^t does not have the same colour as her private group. Then, she must receive payoff at least $|B_j^t| = 2(n+j) - 2$ (else, she would increase her payoff by choosing the same colour as her private group, thus contradicting the hypothesis that we are in a Nash equilibrium). Furthermore, there can be only vertices in $V \setminus (A_j^t \cup B_j^t)$ with the same colour c as b_j^t. Suppose for the sake of contradiction that a_j^t does not have colour c. There are two cases to be considered.

Suppose that $A_j^t \cup \{a_j^t\}$ is a colour class. Then, a_j^t receives payoff $|A_j^t| = 2(n+j) - 3$. In such case, since a_j^t and b_j^t are twin vertices in $G \setminus (A_j^t \cup B_j^t)$, vertex a_j^t could increase her payoff to at least $2(n+j) - 1$ by choosing c as her new colour, thus contradicting the hypothesis that we are in a Nash equilibrium.

Else, a_j^t and b_j^t do not have the same colours as their respective private groups. In such case, A_j^t and B_j^t are colour classes, hence we can constrain ourselves to the subgraph $G \setminus (A_j^t \cup B_j^t)$. In particular, the constriction of the Nash equilibrium to the subgraph must be a Nash equilibrium of the coloring game defined on $G \setminus (A_j^t \cup B_j^t)$. Since a_j^t and b_j^t are twin vertices in $G \setminus (A_j^t \cup B_j^t)$, they must have the same colour by a similar argument as for Claim 2.

As a result, a_j^t must have colour c in both cases, that proves the claim.   ◇

We recall that we aim at simulating the computation of the output of all the gates in $\mathcal{C}$. To do that, we will prove the existence of a special colour class containing X and some pair in $Y_j \cup N_j$ for every j. Intuitively, the two vertices of $Y_j \cup N_j$ are used to certify the output of the jth gate. However, this certificate is "local" in the sense that it assumes the output of the $j-1$ smaller gates to be already certified. Therefore, we need to prove that there can be no "missing gate", i.e., every gate is represented in the special colour class.

Claim 4. *Let c be a colour such that $L_c \nsubseteq X$ and L_c does not intersect any private group (A_j^t or B_j^t for any $1 \le j \le m$ and $1 \le t \le 3$).*
  Then, $X \subseteq L_c$ and there exists an index j_0 such that the following holds true: $|L_c \cap (Y_j \cup N_j)| = 2$ for every $1 \le j \le j_0$, and $L_c \cap (Y_j \cup N_j) = \emptyset$ for every $j_0 + 1 \le j \le m$.

Proof. By the hypothesis $L_c \nsubseteq X$ and L_c does not intersect any private group, so, there is at least one vertex of $\bigcup_{j=1}^m (Y_j \cup N_j)$ with colour c. Let j_0 be the largest index j such that there is a vertex in $Y_j \cup N_j$ with colour c. Since by Claim 1, there can be no more than two vertices of $Y_j \cup N_j$ that are in L_c for every j, therefore, by maximality of j_0 we get $|L_c| \le |X| + 2j_0 = 2(n+j_0)$. In particular, observe that if $|L_c| = 2(n+j_0)$ then $X \subseteq L_c$ and for every $1 \le j \le j_0$ there are exactly two vertices in $Y_j \cup N_j$ with colour c. So, let us prove that $|L_c| = 2(n+j_0)$, that will prove the claim. By the choice of j_0, there is some $1 \le t \le 3$ such that $a_{j_0}^t \in L_c$ or $b_{j_0}^t \in L_c$. In particular, $|L_c| \ge \min\{|A_{j_0}^t|, |B_{j_0}^t|\} + 1 = 2(n+j_0) - 2$ or else, every vertex $v_{j_0}^t \in L_c \cap \{a_{j_0}^t, b_{j_0}^t\}$ would increase her payoff by choosing the colour of the vertices in her private group (that is a colour class by Claim 3), thus contradicting the hypothesis that we are in a Nash equilibrium.

We prove as an intermediate subclaim that for any $1 \le j \le j_0 - 1$ such that $L_c \cap (Y_j \cup N_j) \ne \emptyset$, there is some $1 \le t' \le 3$ such that $a_j^{t'}, b_j^{t'} \in L_c$. Indeed, in this situation, there is some t' such that $a_j^{t'} \in L_c$ or $b_j^{t'} \in L_c$. If $b_j^{t'} \in L_c$ then we are done as by Claim 3, $a_j^{t'} \in L_c$. Otherwise, $b_j^{t'} \notin L_c$ and we prove this case cannot happen. First observe that $a_j^{t'} \in L_c$ in this case. Furthermore, since $a_j^{t'}$ and $b_j^{t'}$ do not have the same colour we have by Claim 3 that $B_j^{t'} \cup \{b_j^{t'}\}$ is a colour class. In this situation, $b_j^{t'}$ receives payoff $2(n+j) - 2 \le 2(n+j_0-1) - 2 < |L_c|$. Since in addition $a_j^{t'}$ and $b_j^{t'}$ are twins in $G \setminus (A_j^{t'} \cup B_j^{t'})$, vertex $b_j^{t'}$ could increase her payoff by choosing colour c, thus contradicting that we are in a Nash equilibrium. This proves $a_j^{t'}, b_j^{t'} \in L_c$, and so, the subclaim.

By the subclaim, there is an even number $2k$ of vertices in $\bigcup_{j=1}^{j_0-1} (Y_j \cup N_j)$ with colour c, for some $k \le j_0 - 1$. Similarly, since by Claim 2 the vertices x_i, x_i' have the same colour for every $1 \le i \le n$, $|X \cap L_c| = 2n'$ for some $n' \le n$. Now there are two cases to be considered.

Suppose that $b_{j_0}^t \in L_c$. Then, by Claim 3 $a_{j_0}^t \in L_c$. Furthermore $|L_c| \ge 2(n+j_0) - 1$ or else, vertex $b_{j_0}^t$ would increase her payoff by choosing the colour of the vertices in $B_{j_0}^t$ (that is a colour class by Claim 3), thus contradicting the hypothesis that we are in a Nash equilibrium. As a result, $|L_c| = 2(n'+k+1) \ge 2(n+j_0) - 1$, that implies $n' + k \ge n + j_0 - 1$, and so, $|L_c| \ge 2(n+j_0)$, as desired.

Else, $b_{j_0}^t \notin L_c$ and we prove this case cannot happen. First observe that $a_{j_0}^t \in L_c$. Furthermore, $|L_c| = 2(n' + k) + 1 \geq 2(n + j_0) - 2$, that implies $n' + k \geq n + j_0 - 1$, and so, $|L_c| \geq 2(n + j_0) - 1$. However, since $a_{j_0}^t$ and $b_{j_0}^t$ do not have the same colour, $B_{j_0}^t \cup \{b_{j_0}^t\}$ is a colour class by Claim 3. In particular, $b_{j_0}^t$ receives payoff $2(n + j_0) - 2 < |L_c|$. Since $a_{j_0}^t, b_{j_0}^t$ are twins in $G \backslash (A_{j_0}^t \cup B_{j_0}^t)$, vertex $b_{j_0}^t$ could increase her payoff by choosing colour c, thus contradicting that we are in a Nash equilibrium.

Altogether, $|L_c| \geq 2(n + j_0)$, that proves the claim.   ◇

We point out that by combining Claim 1 with Claim 4, one obtains that for every $1 \leq j \leq m$, there are either zero or two vertices in $Y_j \cup N_j$ in each colour class not containing a private group, and in case there are two vertices then these are a_j^t, b_j^t for some $1 \leq t \leq 3$.

Claim 5. *Any two vertices in X have the same colour. Furthermore, for every $1 \leq j \leq m$, every vertex in $Y_j \cup N_j$ either has the same colour as vertices in X or as vertices in her private group.*

Proof. Let L_c be any colour class with at least one vertex in $\bigcup_{j=1}^m (Y_j \cup N_j)$. Let j_0 be the largest index j such that there is a vertex in $Y_j \cup N_j$ with colour c. In order to prove the claim, there are two cases to be considered. Suppose that $L_c \neq A_{j_0}^t \cup \{a_{j_0}^t\}$ and $L_c \neq B_{j_0}^t \cup \{b_{j_0}^t\}$ for any $1 \leq t \leq 3$. We will prove that $X \subseteq L_c$, that will imply that L_c is unique in such a case, and so, will prove the claim. By the choice of colour c, $L_c \not\subseteq X$. Further, observe that there can be no private group with a vertex in L_c. As a result, this case follows directly from Claim 4.

Else, either $L_c = A_{j_0}^t \cup \{a_{j_0}^t\}$ or $L_c = B_{j_0}^t \cup \{b_{j_0}^t\}$ for some $1 \leq t \leq 3$, and we may assume that it is the case for any colour class L_c that contains at least one vertex in $\bigcup_{j=1}^m (Y_j \cup N_j)$ (or else, we are back to the previous case). So, let us constrain ourselves to the subgraph $G[X]$. In particular, the constriction of the Nash equilibrium to the subgraph must be a Nash equilibrium of the coloring game defined on $G[X]$. Since the vertices in X are pairwise nonadjacent, they must form a unique colour class in such case, that proves the claim.   ◇

We will need a "truthfulness" property to prove correctness of our reduction. Namely, the value of the output of any gate in the circuit must be correctly guessed from the agents with the same colour as vertices in X.

Claim 6. *Let $1 \leq j_0 \leq m$ such that for every $1 \leq j \leq j_0$, there is at least one vertex in $Y_j \cup N_j$ with the same colour c_0 as all vertices in X. Then for every $1 \leq j \leq j_0$, $L_{c_0} \cap Y_j \neq \emptyset$ if and only if the output of the jth gate is 1.*

Proof. In order to prove the claim by contradiction, let $1 \leq j_1 \leq j_0$ be the smallest index j such that either $Y_j \cap L_{c_0} = \emptyset$ and the output of the jth gate is 1 (false negative) or $Y_j \cap L_{c_0} \neq \emptyset$ and the output of the jth gate is 0 (false positive). We will show that in such case, there is an edge with two endpoints of colour c_0, hence the coloring is not proper, thus contradicting the hypothesis that we are in a Nash equilibrium. Note that since by de Morgan's laws, the negation of an AND-gate can be transformed into an OR-gate and vice-versa, both cases are symmetrical, and so, we can assume w.l.o.g. that the j_1^{th} gate is an OR-gate. There are two subcases to be considered.

Suppose that the output of the j_1^{th} gate is 0 (false positive). In such case, $Y_{j_1} \cap L_{c_0} \neq \emptyset$. Let us consider the first entry of the gate. If it is the ith entry of the circuit for some $1 \leq i \leq n$ then $w_i = 0$ (because the output of the j_1^{th} gate is 0) and so, by construction, $x_i, x_i' \in L_{c_0}$ are adjacent to $a_{j_1}^1, b_{j_1}^1$. Else, it is the kth gate of the circuit for some $k < j_1$. By minimality of j_1, since the output of the kth gate must be 0 (because the output of the j_1^{th} gate is 0), $Y_k \cap L_{c_0} = \emptyset$, and so, $N_k \cap L_{c_0} \neq \emptyset$. By construction, every vertex in N_k is adjacent to $a_{j_1}^1, b_{j_1}^1$. As a result, $a_{j_1}^1, b_{j_1}^1$ have a neighbour in L_{c_0} in this subcase. We can prove similarly (by considering the second entry of the gate) that $a_{j_1}^2, b_{j_1}^2$ have a neighbour in L_{c_0} in this subcase. The latter implies the existence of an edge with both endpoints in L_{c_0} since $Y_{j_1} = \{a_{j_1}^1, b_{j_1}^1, a_{j_1}^2, b_{j_1}^2\}$.

Else, the output of the j_1^{th} gate is 1 (false negative). In such case, $Y_{j_1} \cap L_{c_0} = \emptyset$, hence $N_{j_1} \cap L_{c_0} \neq \emptyset$. Since the output of the gate is 1, there must be an entry of the gate such that: either it is the ith entry of the circuit for some $1 \leq i \leq n$, and $w_i = 1$ (in which case, the two vertices $x_i, x_i' \in Lc_0$ are adjacent to both $a_{j_1}^3, b_{j_1}^3$ by construction); or it is the kth gate of the circuit for some $k < j_1$ and this gate outputs 1. In the latter case, by minimality of j_1, $Y_k \cap L_{c_0} \neq \emptyset$. By construction, every vertex in Y_k is adjacent to $a_{j_1}^3, b_{j_1}^3$. As a result, $a_{j_1}^3, b_{j_1}^3$ have a neighbour in L_{c_0} in this subcase. The latter implies the existence of an edge with both endpoints in L_{c_0} since $N_{j_1} = \{a_{j_1}^3, b_{j_1}^3\}$. ◇

3.3 Proof of Theorem 2

Proof of Theorem 2. Let $\langle C, w \rangle$ be any instance of MONOTONE CIRCUIT VALUE. Let $G = (V, E)$ be the graph obtained with our reduction from Sect. 3.1, which can be constructed in polylogarithmic-time with a polynomial number of processors. The graph G defines a coloring game. We fix any Nash equilibrium for this game, that exists by Theorem 1. By Claim 5, any two vertices in X have the same colour c_0. We will prove that there is at least one vertex in Y_m with colour c_0 if and only if the circuit C outputs 1 when it takes w as input. Since MONOTONE CIRCUIT VALUE is PTIME-complete [9], the latter will prove that computing a Nash equilibrium for coloring games is PTIME-hard.

By Claim 6, we only need to prove that for every $1 \leq j \leq m$, there is at least one vertex in $Y_j \cup N_j$ with colour c_0. To prove it by contradiction, let j_0 be the smallest index j such that no vertex in $Y_j \cup N_j$ has colour c_0. By Claim 5, every vertex in $Y_{j_0} \cup N_{j_0}$ has the same colour as her private group. In particular, the three of $a_{j_0}^1, a_{j_0}^2, a_{j_0}^3$ receive payoff $2(n + j_0) - 3$. We will prove that one of these three agents could increase her payoff by choosing c_0 as her new colour, thus contradicting that we are in a Nash equilibrium. Indeed, by the minimality of j_0, it follows by Claim 4 that for any $1 \leq j \leq j_0 - 1$, there are exactly two vertices of $Y_j \cup N_j$ with colour c_0, while for every $j_0 \leq j \leq m$ there is no vertex in $Y_j \cup N_j$ with colour c_0. As a result, $|L_{c_0}| = 2(n + j_0) - 2$. In particular, any agent among

$a_{j_0}^1, a_{j_0}^2, a_{j_0}^3$ could increase her payoff by choosing c_0 as her new colour—provided she is nonadjacent to every vertex in L_{c_0}. We will show it is the case for at least one of the three vertices, that will conclude the proof of the theorem. Assume w.l.o.g. that the j_0^{th} gate is an OR-gate (indeed, since by de Morgan's laws, the negation of an AND-gate can be transformed into an OR-gate and vice-versa, both cases are symmetrical). There are two cases.

Suppose that the output of the j_0^{th} gate is 1. In such case, there must be an entry of the gate such that: it is the ith entry of the circuit, for some $1 \leq i \leq n$, and $w_i = 1$; or it is the kth gate of the circuit for some $k < j_0$ and the output of that gate is 1. In the latter case, we have by Claim 6 that the two vertices of $Y_k \cup N_k$ with colour c_0 are in the set Y_k. Assume w.l.o.g. that the above-mentioned entry is the first entry of the gate. By construction, the two vertices $a_{j_0}^1, b_{j_0}^1$ are nonadjacent to every vertex in L_{c_0}. Else, the output of the j_0^{th} gate is 0. Therefore, for every entry of the gate: either it is the ith entry of the circuit, for some $1 \leq i \leq n$, and $w_i = 0$; or it is the kth gate of the circuit for some $k < j_0$ and the output of that gate is 0. In the latter case, we have by Claim 6 that the two vertices of $Y_k \cup N_k$ with colour c_0 are in the set N_k. By construction, the two vertices $a_{j_0}^3, b_{j_0}^3$ are nonadjacent to every vertex in L_{c_0}. In both cases, it contradicts that we are in a Nash equilibrium.   □

4   Conclusion and Open Perspectives

We suggest through this case study a more in-depth analysis of the complexity of computational mechanism designs. We would find it interesting to pursue similar investigations for other games. Experiments in the spirit of [11] could be helpful for our purposes. Further, we note that PRAM is seen by some as a too unrealistic model for parallel computation. Thus, one may argue that proving our reduction in this model casts a doubt on its reach. However, we can leverage on the stronger statement that MONOTONE CIRCUIT VALUE is *strictly* PTIME-hard [5]. It implies roughly that the sequential time and the parallel time to solve this problem cannot differ by more than a moderate polynomial-factor (unless the solving of *all* problems in PTIME can be sped up on a parallel machine by at least a polynomial-factor). Our reduction directly shows the same holds true for the problem of computing a Nash equilibrium in a given coloring game, that generalizes our hardness result to more recent parallel complexity classes (*e.g.*, [10]).

References

1. Ballester, C.: NP-completeness in hedonic games. Games Econ. Behav. **49**(1), 1–30 (2004)
2. Bondy, J.A., Murty, U.S.R.: Graph Theory. Graduate Texts in Mathematics. Springer, London (2008)
3. Burani, N., Zwicker, W.S.: Coalition formation games with separable preferences. Math. Soc. Sci. **45**(1), 27–52 (2003)

4. Chatzigiannakis, I., Koninis, C., Panagopoulou, P.N., Spirakis, P.G.: Distributed game-theoretic vertex coloring. In: OPODIS 2010, pp. 103–118 (2010)
5. Condon, A.: A theory of strict P-completeness. Comput. Complex. **4**(3), 220–241 (1994)
6. Ducoffe, G., Mazauric, D., Chaintreau, A.: The complexity of hedonic coalitions under bounded cooperation (submitted)
7. Feigenbaum, J., Papadimitriou, C.H., Shenker, S.: Sharing the cost of multicast transmissions. J. Comput. Syst. Sci. **63**(1), 21–41 (2001)
8. Gairing, M., Savani, R.: Computing stable outcomes in hedonic games. In: SAGT 2010, pp. 174–185 (2010)
9. Greenlaw, R., Hoover, H.J., Ruzzo, W.L.: Limits to Parallel Computation: P-Completeness Theory. Oxford University Press, Oxford (1995)
10. Karloff, H., Suri, S., Vassilvitskii, S.: A model of computation for MapReduce. In: SODA 2010, pp. 938–948 (2010)
11. Kearns, M., Suri, S., Montfort, N.: An experimental study of the coloring problem on human subject networks. Science **313**(5788), 824–827 (2006)
12. Kleinberg, J., Ligett, K.: Information-sharing in social networks. Games Econ. Behav. **82**, 702–716 (2013)
13. Marden, J.R., Wierman, A.: Distributed welfare games. Oper. Res. **61**(1), 155–168 (2013)
14. Panagopoulou, P.N., Spirakis, P.G.: A game theoretic approach for efficient graph coloring. In: ISAAC 2008, pp. 183–195 (2008)
15. Papadimitriou, C.H.: Computational Complexity. Wiley, Reading (2003)
16. Vassilevska Williams, V.: Fine-Grained algorithms and complexity (invited talk). In: STACS 2016 (2016)

Complexity and Optimality of the Best Response Algorithm in Random Potential Games

Stéphane Durand[1] and Bruno Gaujal[2(⊠)]

[1] Univ. Grenoble Alpes, 38000 Grenoble, France
stephane.durand@inria.fr
[2] Inria, Grenoble, France
bruno.gaujal@inria.fr

Abstract. In this paper we compute the worst-case and average execution time of the Best Response Algorithm (BRA) to compute a pure Nash equilibrium in finite potential games. Our approach is based on a Markov chain model of BRA and a coupling technique that transform the average execution time of this discrete algorithm into the solution of an ordinary differential equation. In a potential game with N players and A strategies per player, we show that the worst case complexity of BRA (number of moves) is exactly NA^{N-1}, while its average complexity over random potential games is equal to $e^\gamma N + O(N)$, where γ is the Euler constant. We also show that the effective number of states visited by BRA is equal to $\log N + c + O(1/N)$ (with $c \leqslant e^\gamma$), on average. Finally, we show that BRA computes a pure Nash Equilibrium faster (in the strong stochastic order sense) than any local search algorithm over random potential games.

1 Introduction

The question of computing Nash Equilibria (NE) in games is a central question in algorithmic game theory and has been investigated of many papers. The most classical result is in [1], showing that the problem of computing NE in finite games is PPAD complete.

Potential games have been introduced in [2] and have proven very useful, especially in the context of routing problems in networks, first mentioned in [3] and exhaustively studied ever since, in the transportation as well as computer science literature, see for example [4–6]. They have also been heavily investigated in the context of distributed optimization (see for example [7]). In [8,9] the authors show that the computation of NE for general potential games is PLS complete (Polynomial Local Search complete). As for PPAD, this complexity class is believed to be different from P.

The best response dynamics is one of the most basic tool in game theory. The original proof of the existence of a Nash Equilibrium by Nash [10] can be seen as the proof of existence of a fixed point of the best response correspondence (best response is called *countering* in [10]). It has been well-known for a long time that

© Springer-Verlag Berlin Heidelberg 2016
M. Gairing and R. Savani (Eds.): SAGT 2016, LNCS 9928, pp. 40–51, 2016.
DOI: 10.1007/978-3-662-53354-3_4

the Best Response Algorithm converges in finite time to a pure NE in potential games [11]. So BRA is a natural candidate for computing Nash equilibria.

In this paper, we analyze the performance of BRA over potential games with N players, each with A possible strategies. It is well known that the convergence time of BRA over potential games can be exponential in the number of players (see for example [8]). Here, we confirm this by showing that the worst case complexity of BRA (number of plays) is exactly NA^{N-1}. Special cases, such as graphical potential games have been analyzed in [12] by showing an equivalence between the potential of such games and Markov fields. In other special cases such as scheduling congestion games with identical tasks, it has been show that BRA takes at most N steps before finding a NE [13]. Extensions with positive and negative externalities also have a linear complexity [14].

However the *average* complexity of BRA over all potential games has attracted surprisingly little attention. Random (non potential) games with two players have been studied in [15]: With two IID utility matrices of size $A \times A$, the computation of a NE is $O(A^3 \log \log A)$ with high probability using a rather sophisticated algorithm.

Our main contribution is to show that for potential games with N players, $\mathbb{E}[M_{BRA}]$, the average number of strategy profiles visited by BRA before convergence, is $\mathbb{E}[M_{BRA}] = \log(N) + C + O(1/N)$ (where $C \leqslant e^\gamma$, γ being the Euler constant). We also show that the average number of comparisons performed by the algorithm is equal to $e^\gamma(A-1)(N-1) + o(AN)$. This could be intuitively explained by the fact that random potential games have a lot of pure NE [16]. In our framework, potentials are IID random variables so that, on average, one action profile out of $(A-1)N+1$ profiles is a NE while in the worst case, a potential game may have a single pure NE. This is only a partial explanation, however. This does not explain the fact that the complexity does not depend on the number of actions, nor the value of the constant factor, $e^\gamma \approx 1.78$.

We further show that the Best Response Algorithm computes a pure Nash Equilibrium faster than any algorithm based on player's local information, not only in average but also in the strong stochastic order sense.

Missing proofs and additional details (numerical simulations, analysis of alternative algorithms) are given in a long version of this paper, available in HAL Archive [17].

1.1 Coupling and Markovian Analysis

The main idea of our approach is to see the evolution of BRA in a random environment as a dynamical system, whose behavior can be computed using differential equations. This will allow us to compute the exact asymptotics of the average complexity in N and A, not only $O(.)$ bounds. Second moments of T_{BRA} and of M_{BRA} can also be computed by the same approach (see [17]).

The first step (Sect. 4.2) is to construct an approximation of the behavior of BRA over a potential game. This approximation is called IFA in the paper, for Intersection-Free Approximation because it discards strategies already explored by BRA. We show that the execution time of BRA is smaller than the execution time of its IFA approximation for the strong stochastic order. This is done by constructing a non-trivial *coupling* between both executions. This powerful technique is exploited to our great benefit here.

The second and most important step (Sect. 4.4) is to consider one run of the IFA approximation of BRA as a trajectory of a Markov chain over the continuous space of potentials. Doing so, the average complexity is transformed into the average hitting time of an absorbing state of the Markov chain. The theory of Markov chains implies that this average hitting time satisfies a Poisson differential equation. Thus, the average complexity of BRA is given by the solution of a system of ordinary differential equations. This system happens to have a solution in closed form whose asymptotics in N and A can be computed by taking integrals over initial states.

As for the proof of optimality of BRA among all local search algorithms (Sect. 5), our approach is based once again on a coupling argument. While using coupling techniques is more classical in this context (comparison of algorithms), this particular case retains some originality because the coupling used here is not built off-line but is being constructed on the fly while the algorithm runs.

2   Best Response Algorithm and Potential Games

We consider a game with a finite number N of players, each with A strategies.

Definition 1 (*N*-player game). *A game is a tuple* $\mathfrak{G} \stackrel{\text{def}}{=} \mathfrak{G}(\mathcal{N}, \mathcal{A}, u)$ *with*

- *a finite set of* **players** $\mathcal{N} = \{1, \dots, N\}$;
- *a finite set* $\mathcal{A}_k$ *of pure* **strategies** *for each player* $k \in \mathcal{N}$.

  The set of **strategy profiles** *or* **states** *of the game is* $\mathcal{A} \stackrel{\text{def}}{=} \mathcal{A}_1 \times \mathcal{A}_2 \times \cdots \mathcal{A}_N$.
- *The players'* **payoff** *functions* $u_k : \mathcal{A} \to \mathbb{R}$.

  We define the *best response correspondence* $\mathbf{br}_k(x)$ as the set of all strategies that maximizes the payoff for player k under profile $x = (x_1, \dots, x_N)$: $\mathbf{br}_k(x) \stackrel{\text{def}}{=} \left\{ \underset{\alpha \in \mathcal{A}_k}{\mathrm{argmax}} \ u_k(\alpha; x_{-k}) \right\}$.

  A *Nash equilibrium* (NE) is a fixed point of this correspondence, i.e. a profile x^* such that $x_k^* \in \mathbf{br}_k(x^*)$ for every player k.

Definition 2 (Potential games and its generalizations). *A game is an (exact)* potential game *[11] if it admits a function (called the potential)* $\Phi : \mathcal{A} \to \mathbb{R}$ *such that for any player k and any* unilateral *deviation of k from strategy profile x to x':* $u_k(\alpha, x_{-k}) - u_k(\alpha', x_{-k}) = \Phi(\alpha, x_{-k}) - \Phi(\alpha', x_{-k})$.

A game is a generalized ordinal potential game *[11] (or G-potential game for short) if there exists a potential function* $\Phi : \mathcal{A} \to \mathbb{R}$ *such that, for any player k and any state x,* $u_k(\alpha, x_{-k}) > u_k(\alpha', x_{-k}) \Rightarrow \Phi(\alpha, x_{-k}) > \Phi(\alpha', x_{-k})$.

A game is a best-response potential game *[18] (or BR-potential game for short) if there is* $\Phi : \mathcal{A} \to \mathbb{R}$ *such that for any player k and strategy profile x,*

$$\mathbf{br}_k(x) = \left\{ \underset{\alpha \in \mathcal{A}_k}{\mathrm{argmax}} \ \Phi(\alpha, x_{-k}) \right\}.$$

As shown in [18], exact potential games are BR-potential games, but there exist G-potential games that are not BR-potential games. In the following, we will consider the most general case (*i.e.* all games that are either BR-potential or G-potential games) and call them potential games for simplicity.

We consider a general version of the *Best Response Algorithm* (BRA) with uniform choice over all possible best responses when ties occur and where the next player is selected according to a *revision function* $R(.)$, that may depend of the whole past of the algorithm. We assume that this function is *weakly fair*: each player appears infinitely often in the sequence of plays induced by R, almost surely. This revision function can be deterministic (for example, round-robin: $R(t) = t \bmod N$) or random (for example, Bernoulli where the next player is chosen according to an probability distribution ρ (the *revision law*): $\forall k \in \mathcal{N}, \mathbb{P}(R(t) = k) = \rho_k$). In that case, weak fairness implies that the probability of choosing any player k is strictly positive ($\forall k \in \mathcal{N}, \rho_k > 0$).

Algorithm 1. Best Response Algorithm (BRA)

Input: Game utilities $(u_k(\cdot))$; Initial state $(x := x(0))$;
Weakly fair revision function R;
List of satisfied customers, initially empty: $L := \emptyset$;

repeat
   Pick next player $k := R(t); t := t + 1$;
   if $x_k \notin \mathbf{br}_k(x)$ **then**
      Update strategy for player k to any $x_k \in \mathbf{br}_k(x)$;
      $L := \emptyset$;
   $L := L \cup \{k\}$;
until *size(L) = N*;

It is well known (see [11]) that for any potential game $\mathfrak{G}$, Algorithm 1 converges in finite time, almost surely, to a Nash Equilibrium of $\mathfrak{G}$.

3   Worst Case Complexity

In this section, we analyze the time complexity of BRA. More precisely, we consider three measures (related to each other). The first one is T_{BRA}, the number of iterations (or the number of times that the function **br** is called) before BRA reaches a Nash equilibrium. A related measure is the total number of comparisons used by BRA (denoted C_{BRA}). One should expect that $C_{BRA} \approx (A - 1)T_{BRA}$. Finally, another interesting quantity is the number of different states visited by BRA (denoted M_{BRA}). This is called the number of *moves* done by BRA before convergence to a Nash equilibrium (NE). Of course, $M_{BRA} \leqslant T_{BRA}$.

These quantities depend on the game over which BRA is run, on the initial state x^0 and on the infinite sequence of revision players R. It should also be clear that they are functions of the game only through the potential Φ, so we denote by $T_{BRA}(\Phi, x^0, R)$ the number of steps before convergence of Algorithm BRA for a game with potential Φ, starting in state x^0, under the condition that the sequence of players is given by R.

In the worst case, for some weakly fair revision functions R, $T_{BRA}(\Phi, x^0, R)$ can be unbounded because the revision sequence induced by R can be arbitrarily bad: one player might appear too few times to guarantee convergence in any bounded time. When R is the round-robin function, the time for convergence is finite but can still be very large, as shown in the following theorem.

Theorem 1. *In the worst case,* $T_{BRA}(\Phi, x^0, round\text{-}robin) = NA^{N-1}$.

It is well known that the worst case complexity of BRA is exponential in the number of players (see for example [8]). The version of this result given here (for round robin revision and generalized potential games) is only given for the record (the proof is given in [17]).

4 Average Complexity of BRA

4.1 Randomization

In the following we will randomize over the potential games over which BRA is used. Since the behavior of BRA only depends on the potential function, we randomize directly over the potential Φ.

We consider a randomization over all games, uniformly over all possible orders for the potentials. On one hand, this is the classical average complexity approach when no additional information is known about the games (the same approach is used in [15] for 2 player games for example). This yields IID potential for all profiles, as explained below. On the other hand, some may argue that uniformly random games are not generic in some sense and a good performance of BRA on average does not necessarily translate in good performances for "real word" games. In any case, this is a first step that must be taken in absence of additional information about specific games that one may want to study.

There are several equivalent ways to do this randomization. The first one is based on the fact that the complexity of the algorithm does not depend on the actual values of the potential of the states but only on the comparisons between them. When two potentials are equal, a strict order between them is chosen uniformly. Therefore, the natural randomization is to consider the linear extensions (total orders) of all possible partial orders over the set $\mathcal{A}$ and pick one uniformly. The number of total orders on $\mathcal{A}$ is the number of permutations on $\mathcal{A}$, namely $(A^N)!$.

The second (equivalent) randomization is the following: The potentials of all states x are chosen independent, identically distributed according to an arbitrary distribution F admitting a density w.r.t. the Lebesgue measure.

Both randomizations are equivalent. Indeed, take any k states $x_1, \ldots, x_k$ in $\mathcal{A}$. In both cases, $\mathbb{P}(\Phi(x_1) > \Phi(x_2) > \cdots > \Phi(x_k)) = 1/k!$. Now, since F is increasing, F^{-1} is well-defined and we get $\mathbb{P}(\Phi(x) > \Phi(x')) = \mathbb{P}(F^{-1}(\Phi(x)) > F^{-1}(\Phi(x')))$. Note that $F^{-1}(\Phi(x))$ is uniformly distributed on $[0,1]$. Therefore, with no loss of generality, one can assume that the potential of all the states are i.i.d., uniformly distributed on $[0,1]$. This randomization is used in the following.

4.2   Intersection-Free Approximation

The direct analysis of the behavior of BRA over a random potential is difficult because, over time, more and more states have been visited by the algorithm. Thus, its behavior is non-homogeneous in time. To avoid this difficulty, we consider a new model, called the *Intersection-Free Approximation* (IFA) in the following. Under the Intersection-Free Approximation, every time a new player (say k) has to compute its best response in a state (say x), it compares $\Phi(x)$ with the potential of its $A - 1$ other possible strategies, as for the real BRA. Here however, we assume that those $A - 1$ states have not yet been visited during the previous steps of the algorithm. Note that under the real behavior of BRA, it could happen that some of these states have already been compared at a previous step of the algorithm, by another player (this will be called an *intersection* in the following). Under the Intersection-Free Approximation, the states visited by the algorithm are always "new" states, never compared before with any other states.

More formally, the algorithm BRA under IFA can be written as follows.

Algorithm 2. BRA algorithm under IFA

Input: Initial state $(x(0))$; Revision function R;
Set of satisfied players, initially empty $L := \emptyset$.

repeat
 | Pick next player $k := R(t)$; $t := t + 1$;
 | **if** $k \notin L$ **then**
 | | **Generate** IID potentials $\Phi(\alpha, x_{-k})$, $\alpha \in \mathcal{A}_k \setminus \{x_k\}$ unif. on $[0,1]$;
 | | Compute best response: $\alpha_k := \underset{\beta \in \mathcal{A}_k}{\operatorname{argmax}} \Phi(\beta; x_{-k})$;
 | | **if** $\alpha_k = x_k$ **then**
 | | | $L := L \cup \{k\}$
 | | **else**
 | | | $L := \{k\}$; $x_k := \alpha_k$;
until $size(L) = N$;

Let us recall that C_{BRA} (resp. T_{BRA}, M_{BRA}) is the number of comparisons (resp. number of steps, number of moves) taken by BRA before convergence and let us define C_{IFA} (resp. T_{IFA}, M_{IFA}) to be the number of comparisons (steps, moves) of BRA under the intersection-free approximation. By definition, the worst case complexity of IFA under a round-robin revision sequence is infinite. However, its average complexity is the same as for BRA, as shown by the following lemma.

Lemma 1 (BRA and IFA are asymptotically equivalent).   *Under the foregoing notations and using a round-robin revision function, the following comparisons hold, where $\leqslant_{st}$ is the strong stochastic order:*

1. $C_{BRA} \leqslant_{st} C_{IFA}$ *(equivalently, [19] $\forall t \in \mathbb{R}$, $\mathbb{P}(C_{BRA} > t) \leqslant \mathbb{P}(C_{IFA} > t)$).*
2. *If I is the total number of intersections in BRA, then $T_{BRA} \leqslant_{st} T_{IFA} + \frac{I}{A-1}$.*
3. $\mathbb{E}[T_{BRA}] = \mathbb{E}[T_{IFA}] + o(1)$ *and* $\mathbb{E}[C_{BRA}] = \mathbb{E}[C_{IFA}] + o(1)$,
4. $\mathbb{E}[M_{BRA}] = \mathbb{E}[M_{IFA}] + o(1)$.

The proof of the lemma is available in [17]. It is based on the construction of a *coupling* between the executions of BRA with and without IFA. The assumption that the revision function is round-robin for BRA and for IFA does not play a big role in the proof, and it could be removed. However, the following section, asserting the optimality of round-robin implies that extending the proof to more general revision functions has a limited interest.

4.3   Round-Robin and Other Revision Sequences

As for the worst case analysis, the revision sequence influences the average time complexity of the algorithm. We show that on average round-robin is asymptotically the best one.

Lemma 2 (Asymptotic optimality of round-robin).   *For any revision function R, $\mathbb{E}\left[T_{BRA}(\Phi, x^0, round\text{-}robin)\right] \leqslant \mathbb{E}\left[T_{BRA}(\Phi, x^0, R)\right] + \epsilon(N)$, where the expectation is taken over all potentials Φ and all initial states x^0 and $\epsilon(N)$ goes to zero when N goes to infinity.*

The proof is again available in [17]. It uses the comparison with IFA. In the rest, we focus on round-robin revision functions and omit it in the notations, unless specified otherwise.

4.4   Complexity Analysis

We will be analyzing the intersection-free approximation of the behavior of BRA, under a round-robin revision sequence, with no further reference to this.

Let us consider the intersection-free approximation and let y be the potential of the current state x: ($y \overset{\text{def}}{=} \Phi(x)$). Let k be the number of players that have already played best response without changing the profile. This number of "satisfied" players can replace the explicit set L used in Algorithm 2 when the revision sequence is round-robin. The evolution at the next step of BRA under IFA is as follows. The kth player computes its best response. The player has $A - 1$ new strategies whose potential must be compared with the current potential (y). As mentioned before, we can assume that the potentials of those $a \overset{\text{def}}{=} A - 1$ strategies are IID, uniformly distributed in $[0, 1]$.

With probability y^a none of the new strategies beat the current choice. The state remains at y, one more player is satisfied and it is the turn of the $k + 1$-st player to try its best response.

With probability $1 - y^a$, one of the new strategies is the best response. The current state moves to a new state where the number of satisfied players is set back to 1 and the potential increases to a value larger than $u > y$ with probability $1 - u^a$.

Let Y_t be the potential at step t ($Y_t \in [0,1]$) and K_t be the current number of consecutive players whose best response did not change the current potential ($K_t \in \{1, 2, \ldots, N\}$) (number of satisfied players). The previous discussion says that the couple (Y_t, K_t) is a discrete-time, continuous-space Markov chain whose kernel is:

$$\mathbb{P}\left((Y_{t+1}, K_{t+1}) = (y, k+1) \,\middle|\, (Y_t, K_t) = (y, k) \right) = y^a,$$

and, for any $u > y$,

$$\mathbb{P}\left((Y_{t+1}, K_{t+1}) \in ([u, 1], 1) \,\middle|\, (Y_t, K_t) = (y, k) \right) = 1 - u^a.$$

All the other transitions have a null probability.

Let $m(y, k)$ be the number of moves of IFA before convergence when the current state of the Markov chain is equal to (y, k).

With probability y^a, the next player does not change its choice so that $m(y, k) = m(y, k+1)$.

With probability density au^{a-1} the next player finds a new best response with potential u so that one move is taken and $m(y, k) = 1 + m(u, 1)$.

Let $M(y, k) = \mathbb{E}[m(y, k)]$. The previous one step analysis of $m(y, k)$ makes $M(y, k)$ satisfy a forward Poisson equation:

$$M(y, k) = y^a M(y, k+1) + \int_y^1 au^{a-1}(M(u, 1) + 1)du.$$

By definition, the boundary conditions are: $\forall y, M(y, N) = 0$ (the current state is NE when all players agree on this) and $\forall k, M(1, k) = 0$ (the potentials are all bounded by 1, so a state with potential 1 is guaranteed to be a NE).

By setting $B(y) \stackrel{\text{def}}{=} \int_y^1 au^{a-1}(M(u, 1) + 1)du$, we get the following system of integral equations

$$\begin{cases} M(y, 1) & = y^a M(y, 2) + B(y), \\ M(y, 2) & = y^a M(y, 3) + B(y), \\ \vdots & = \vdots \\ M(y, N-2) & = y^a M(y, N-1) + B(y), \\ M(y, N-1) & = B(y). \end{cases} \tag{1}$$

Successive substitution of $M(y, 2), \ldots, M(y, N-1)$ in the first equality yields $M(y, 1) = B(y)H(y)$ where $H(y) \stackrel{\text{def}}{=} 1 + y^a + \cdots + y^{a(N-2)}$. Differentiating w.r.t. y, one gets an ordinary differential equation in $M(y, 1)$:

$$\frac{dM(y, 1)}{dy} + (ay^{a-1}H - \frac{1}{H}\frac{dH}{dy})M(y, 1) = -ay^{a-1}H.$$

The equation is of the form $\dot{f} + gf = h$. Using the boundary condition $M(1, 1) = 0$, its generic solution is

$$M(y, 1) = e^{-Q(y)} \int_y^1 au^{a-1} H(u) e^{Q(u)} du. \tag{2}$$

where

$$Q(y) \overset{def}{=} \int_0^y \left(au^{a-1} H(u) - \frac{1}{H(u)} \frac{dH(u)}{du} \right) du = -\log(H(y)) + \int_0^y au^{a-1} H(u) du. \tag{3}$$

The average number of profile changes in the execution of the algorithm starting from an arbitrary profile is $\mathbb{E}[M_{IFA}] = \int_0^1 M(y, 1) dy$. Since $M(y, 1)$ is decreasing in y, $\mathbb{E}[M_{IFA}]$ is upper-bounded by $M(0, 1)$. Using $Q(0) = 0$, $H(0) = 1$ and replacing Q and H by their values,

$$M(0, 1) = \int_0^1 \exp\left(\sum_{i=0}^{N-2} \frac{u^{a(i+1)}}{i+1} \right) au^{a-1} du = \int_0^1 \exp\left(\sum_{i=0}^{N-2} \frac{v^{i+1}}{i+1} \right) dv \quad \text{(with } v = u^a)$$

$$= \int_0^{1-\frac{1}{N}} \exp\left(\sum_{i=1}^{N-1} \frac{v^i}{i} \right) dv + \int_{1-\frac{1}{N}}^1 \exp\left(\sum_{i=1}^{N-1} \frac{v^i}{i} \right) dv$$

$$\leqslant \int_0^{1-\frac{1}{N}} \exp\left(\sum_{i=1}^{\infty} \frac{v^i}{i} \right) dv + \frac{1}{N} \exp\left(\sum_{i=1}^{N-1} \frac{1}{i} \right) \tag{4}$$

$$= \int_0^{1-\frac{1}{N}} \frac{dv}{1-v} + e^\gamma + O(1/N) = \log(N) + e^\gamma + O(1/N). \tag{5}$$

Furthermore, this bound is tight, up to an additive constant (see [17]).

Let us now consider the average number of comparisons made by BRA under the intersection-free assumption. Let $C(y, k)$ be the average number of comparisons starting in a state with potential y and k players have played without changing their strategy. The Poisson equation for $C(y, k)$ is :

$$C(y, k) = y^a (C(y, k+1) + a) + \int_y^1 au^{a-1}(C(u, 1) + a) du,$$

with the boundary conditions $C(1, 1) = a(N-1)$ and $C(y, N) = 0$.

The solution of this differential system can be obtained in closed form, using a similar approach as for $M(y, 1)$.

$$C(y, 1) = a \left(\sum_{i=0}^{N-2} y^{ai} \right) \exp\left(-\sum_{i=1}^{N-1} \frac{y^{ai} - 1}{i} \right).$$

The average number of comparisons is $\mathbb{E}[C_{IFA}] = \int_0^1 C(y, 1) dy$.

For all $y < 1$,

$$C(y,1) = a \left(\sum_{i=0}^{\infty} y^{ai} \right) \exp \left(\sum_{i=1}^{N-1} 1/i \right) \exp \left(-\sum_{i=1}^{\infty} y^{ia}/i \right) + o(aN) \qquad (6)$$

$$= a \frac{1}{1-y^a}(N-1)e^{\gamma}(1-y^a) + o(aN) + O(1) \qquad (7)$$

$$= a(N-1)e^{\gamma} + o(aN), \qquad (8)$$

where γ is the Euler constant ($\gamma \approx 0.5772...$). Therefore, the same equality holds for the integral, equal to $\mathbb{E}[C_{IFA}]$.

The results of this section, together with Lemma 1, lead to the following theorem, the main result of the section.

Theorem 2 (Average complexity of BRA). *Under the round-robin revision sequence, the average complexity of BRA over a potential game satisfies:*
(i) Average number of moves: $\mathbb{E}[M_{BRA}] = \log(N) + c + O(1/N).$, *where* $c \leqslant e^{\gamma}$
(ii) Average number of comparisons : $\mathbb{E}[C_{BRA}] = e^{\gamma}AN + o(AN).$
(iii) Average number of steps: $\mathbb{E}[T_{BRA}] = e^{\gamma}N + o(N).$

The average complexity $\mathbb{E}[T_{BRA}]$ can be split into two parts: The number of plays before reaching a NE and the number of plays needed to check if a state is indeed a NE. This last part takes exactly $N-1$ steps in Algorithm 1: The players have to play one by one to fill up set L. This means that a NE equilibrium is reached on average as soon as $e^{\gamma} - 1 \approx 78\%$ of the players have played once. The second moments of the number of steps and the number of moves under IFA can be computed similarly (see [17]). In both cases, the standard deviations are of the same order as the means.

5 Optimality of BRA

In this section, we prove that BRA finds a Nash equilibrium faster than any local search algorithm (defined in Sect. 5), in the strong stochastic order sense.

By definition a *Local Search Algorithm* can only access the payoff matrix, one player at a time. This access is often called a *query* in the literature. Once the payoff of a strategy profile has been obtained, it is stored in memory and can re-used later by the algorithm without an additional query.

In addition to queries, a local search algorithm can use any arithmetic operation, draw random variables and choose a strategy for all players.

Any local search algorithm can be written in the following form, based on the history of the execution, $\mathcal{H}_t$, that corresponds to the amount of information gathered by the algorithm up to step t.

Algorithm 3. A general local search algorithm

Initial storage reduced to the initial profile: $\mathcal{H}_0 := \{(x(0)\}$.
repeat

  Select next player: $k := R(\mathcal{H}_t)$;
  Query payoff vector of k under current state: $u_k(\cdot, x_{-k}(t))$;
  Store the new visited states and their payoffs in memory:
  $\mathcal{H}_{t+1} := \mathcal{H}_t \cup \{((\alpha, x_{-k}(t)), u_k(\alpha, x_{-k}(t)))_{\alpha \in A_k}\}$;
  jump to next state $x(t+1) := J(\mathcal{H}_{t+1})$;
  Set $stop := 1$ if the current state is a NE;
  $t := t + 1$;
until $stop$;

The functions J and R used in the inner loop are arbitrary functions that choose the next state as well as the next player to play, according to the whole history of the process. These functions can be deterministic or random. Testing if $x(t+1)$ is NE is not detailed. Notice, however, that it can only be done when all the payoff vectors for all the players in state $x(t+1)$ have been stored in memory.

The complexity of a local search algorithm A is defined as the total number of its payoff vector queries (denoted T_A).

Theorem 3 (Optimality of BRA). *Let A be any local search algorithm that computes a Nash Equilibria in potential games. Under the foregoing random-ization, and choosing the starting point x^0 uniformly among all states, $\forall t \geqslant 0$, $\mathbb{P}(T_{BRA} \geqslant t | R = R_A) \leqslant \mathbb{P}(T_A \geqslant t)$, where R_A is the revision sequence con-structed in A.*

The proof is reported in [17]. Combining this theorem with Lemma 2 estab-lishes the optimality of BRA with round-robin.

6   Conclusion and Perspectives

The best response algorithm is one on the most basic object in game theory. In this paper, we prove it has a linear complexity on average over uniformly random-ized potential games. Furthermore, BRA is optimal in the class of local search algorithms when one has no information about the structure of the potential game.

Does all this make BRA the perfect algorithm to compute NE in general? We believe that the answer is no because BRA suffers from several drawbacks. First, it does not tolerate simultaneous plays. Second, it requires to know the entire payoff vector of a player before choosing its strategy. Other drawbacks include high sensitivity on the order of play and on noisy perturbations on the payoffs. Designing algorithms that do not suffer from these drawbacks is the object of our future investigations.

Acknowledgement. This work was partially supported by LabEx Persyval-Lab.

References

1. Daskalakis, P.G.C., Papadimitriou, C.: The complexity of computing a nash equilibrium. SIAM J. Comput. **39**(3), 195–259 (2009)
2. Rosenthal, R.W.: A class of games possessing pure-strategy nash equilibria. Int. J. Game Theory **2**(1), 65–67 (1973)
3. Beckman, M., McGuire, C.B., Winsten, C.B.: Studies in the Economics of Transportation. Yale University Press, New Haven (1956)
4. Orda, A., Rom, R., Shimkin, N.: Competitive routing in multi user communication networks. IEEE/ACM Trans. Netw. **1**(5), 510–521 (1993)
5. Gallager, R.G.: A minimum delay routing algorithm using distributed computation. IEEE Trans. Commun. **25**(1), 73–85 (1977)
6. Wardrop, J.: Some theoretical aspects of road traffic research, part ii. In: Proceedings of the Institute of Civil Engineers, vol. 1, pp. 325–378 (1954)
7. Roughgarden, T.: Selfish Routing and the Price of Anarchy. MIT Press, Cambridge (2005)
8. Fabrikant, A., Papadimitriou, C., Talwar, K.: The complexity of pure nash equilibria. In: Proceedings of the Thirty-Sixth Annual ACM Symposium on Theory of Computing, ser. (STOC 2004), pp. 604–612. ACM (2004)
9. Ackermann, H., Röglin, H., Vöcking, B.: On the impact of combinatorial structure on congestion games. In: FOCS 2006, pp. 613–622 (2006)
10. Nash, J.: Equilibrium points in n-person game. Proc. Nat. Acad. Sci. **38**, 48–49 (1950)
11. Monderer, D., Shapley, L.: Potential games. Games Econ. Behav. **14**(1), 124–143 (1996). Elsevier
12. Babichenko, Y., Tamuz, O.: Graphical potential games. arXiv.org, Tech. Rep.: 1405.1481v2
13. Even-Dar, E., Kesselman, A., Mansour, Y.: Convergence time to nash equilibria. In: Baeten, J.C.M., Lenstra, J.K., Parrow, J., Woeginger, G.J. (eds.) ICALP 2003. LNCS, vol. 2719. Springer, Heidelberg (2003)
14. Feldman, M., Tamir, T.: Convergence of best-response dynamics in games with conflicting congestion effects. In: Goldberg, P.W. (ed.) WINE 2012. LNCS, vol. 7695, pp. 496–503. Springer, Heidelberg (2012)
15. Imre, B., Vempala, S., Vetta, A.: Nash equilibria in random games. Random Structures Algorithms **31**(4), 391–405 (2007)
16. Rinott, Y., Scarsini, M.: On the number of pure strategy nash equilibria in random games. Games Econ. Behav. **33**(2), 274–293 (2000)
17. Durand, S., Gaujal, B.: Complexity, optimality of the best response algorithm in random potential games. Inria, Research Report RR-8925 (2016). https://hal.inria.fr/hal-01330805
18. Voorneveld, M.: Best-response potential games. Econ. Lett. **66**(3), 289–295 (2000)
19. Müller, A., Stoyan, D.: Comparison Methods for Stochastic Models and Risks. Series in Probability and Statistics. Wiley, Chichester (2002)

Deciding Maxmin Reachability
in Half-Blind Stochastic Games

Edon Kelmendi[1(✉)] and Hugo Gimbert[2]

[1] LaBRI, Bordeaux, France
edon.kelmendi@labri.fr
[2] LaBRI & CNRS, Bordeaux, France
hugo.gimbert@labri.fr

Abstract. Two-player, turn-based, stochastic games with reachability conditions are considered, where the maximizer has no information (he is blind) and is restricted to deterministic strategies whereas the minimizer is perfectly informed. We ask the question of whether the game has maxmin value of 1 in other words we ask whether for all $\epsilon > 0$ there exists a deterministic strategy for the (blind) maximizer such that against all the strategies of the minimizer, it is possible to reach the set of final states with probability larger than $1 - \epsilon$. This problem is undecidable in general, but we define a class of games, called leaktight half-blind games where the problem becomes decidable. We also show that mixed strategies in general are stronger for both players and that optimal strategies for the minimizer might require infinite-memory.

1 Introduction

Two-player stochastic games are a natural framework for modeling and verification in the presence of uncertainty, where the problem of control is reduced to the problem of optimal strategy synthesis [9]. There is a variety of two-player stochastic games that have been studied, depending on the information available to the players (perfect information or partial information), the winning objective (safety, reachability, etc.), the winning condition (surely, almost-surely, or limit-surely winning; probability higher than some quantity), whether the players choose actions concurrently or whether they take turns. Stochastic games with partial observation are particularly well suited for modeling many scenarios occurring in practice; normally we do not know the exact state of the system we are trying to model, e.g. we are aided by noisy sensors or by a software interface that provides only a partial picture. Unfortunately, compared to perfect information games, algorithmic problems on partial information games are substantially harder and often undecidable [3,13,16]. Assuming one player to be perfectly informed while the other player is partially informed (semiperfect-information games [4,5]) brings some relief to the computational hardness as opposed to general partial information games.

This work was partially supported by the French ANR project "Stoch-MC" and "LaBEX CPU" of Université de Bordeaux.

M. Gairing and R. Savani (Eds.): SAGT 2016, LNCS 9928, pp. 52–63, 2016.
DOI: 10.1007/978-3-662-53354-3_5

In the present paper we consider half-blind stochastic games: one player has no information (he is blind) and plays deterministically while the other player is perfectly informed. We study half-blind games for the reachability objective and maxmin winning condition: we want to decide if for every $\epsilon > 0$ there exists a deterministic strategy for the maximizer such that against all strategies of the minimizer, the final states are reached with probability at least $1 - \epsilon$.

The maxmin condition for half-blind games is a generalization of the value 1 problem for probabilistic finite automata [17]. Most decision problems on probabilistic finite automata are undecidable, notably language emptiness [1,13,16], and the value 1 problem [13]. Consequently, stochastic games with partial information and quantitative winning conditions (the probability of fulfilling the winning objective is larger than some quantity) are undecidable. Nevertheless recently there has been some effort on characterizing decidable classes of probabilistic automata [2,6,10,12,13], with the leaktight class [12] subsuming the others [11].

The interest of this model is twofold. First, it can be considered as a probabilistic finite automaton where the transition probabilities are not fixed but controlled by an adversary with some constraints. In this sense, it is a more robust notion of a probabilistic automaton. Second interest is the study how much more difficult a problem becomes when another player is added, in this case the problem of limit-sure reachability.

Our Results. In the present paper we show that a subclass of half-blind games called leaktight games have a decidable maxmin reachability problem. The game is abstracted through a finite algebraic structure called the belief monoid. This is an extension to the Markov monoid used in [12]. Indeed the elements of the belief monoid are sets of elements of the Markov monoid, and they contain information on the outcome of the game when one strategy choice is fixed. The algorithm builds the belief monoid and searches for particular elements which are witnesses that the set of final states is maxmin reachable. The proof of the correctness of the algorithm uses k-decomposition tree, a data structure used in [8] that is related to Simon's factorization forests. The k-decomposition trees are used to prove lower and upper bounds on certain outcomes of the game and show that it behaves as predicted by the belief monoid.

Comparison with Previous Work. The proof methods extends those developed in [12] in three aspects. First, we define a new monoid structure on top of the Markov monoid structure introduced in [12]. Second, we rely on the extension of Simon's factorization forest theorem [18] to k-factorization trees instead of 2-factorization trees in [12] in order to derive upper and lower bound on the actual probabilities abstracted by the belief monoid. Third, we rely on the leaktight hypothesis to prove both completeness and soundness, while in the case of probabilistic automata the soundness of the abstraction by the Markov monoid was for free.

Outline of the Paper. We start by fixing some notions and notation in Sect. 2 as well as providing a couple of examples. In Sect. 3 we introduce the belief monoid algorithm and the Markov and belief monoids themselves. In Sect. 4 the class of leaktight games is defined using the notion of a leak. The correctness of the algorithm is sketched in Sect. 5, and finally we discuss the power of different types of strategies in Sect. 6 and conclude. The details and proofs can be found in [15].

2   Half-Blind Games and the Maxmin Reachability Problem

Given a set X, we denote by $\Delta(X)$ the set of distributions on X, i.e. functions $f : X \to [0,1]$ such that $\sum_{x \in X} f(x) = 1$.

A *half-blind* game is a two-player, zero-sum, stochastic, turn-based game, played on a finite bipartite graph, where the maximizer has no information, whereas the minimizer has perfect information. Formally a game G is given by the tuple $G = (\mathbf{S_1}, \mathbf{S_2}, \mathbf{A_1}, \mathbf{A_2}, p, F)$. The finite set $\mathbf{S_i}$ is the states controlled by Player i, the finite set $\mathbf{A_i}$ is the actions available to Player i $(i = 1, 2)$. Player 1 is the maximizer and Player 2 is the minimizer. The function p mapping $(\mathbf{S_1}, \mathbf{A_1})$ to $\Delta(\mathbf{S_2})$ and $(\mathbf{S_2}, \mathbf{A_2})$ to $\Delta(\mathbf{S_1})$ gives the dynamics of the game. The sets $\mathbf{S_1}, \mathbf{S_2}$ and $\mathbf{A_1}, \mathbf{A_2}$ are disjoint, i.e. $\mathbf{S_1} \cap \mathbf{S_2} = \emptyset$ and $\mathbf{A_1} \cap \mathbf{A_2} = \emptyset$. The set $F \subseteq \mathbf{S_1}$ is the set of final states.

A play of such a game takes place in turns. Initially the game is in some state $s_1 \in \mathbf{S_1}$, then the maximizer (a.k.a. player 1) chooses some action $a_1 \in \mathbf{A_1}$ which moves the game to some state $t_1 \in \mathbf{S_2}$ selected randomly according to the lottery $p(s_1, a_1)$. It is up to the minimizer (a.k.a. player 2) now to choose some action $b_1 \in \mathbf{A_2}$ which moves the game to some state $s_2 \in S_1$. Then again maximizer chooses some action $a_2 \in \mathbf{A_1}$ and so on, until the maximizer decides to stop, at which point, if the game is in a state that belongs to the set of final states F, the maximizer wins, otherwise it is the minimizer who wins. The maximizer is totally blind and does not know what happens, he does not know in which state the game is nor the actions played by the minimizer. Moreover the maximizer plays in a deterministic way, he is not allowed to use a random generator to select his actions. As a consequence, the decisions of maximizer only depend on the time elapsed and can be represented as words on $\mathbf{A_1}$. On the other hand, the minimizer has full information and is allowed to plays actions selected randomly.

Formally, the set of strategies for the maximizer is denoted by Σ_1 they consist of finite words, i.e. $\Sigma_1 = \mathbf{A_1}^*$. In order to emphasize that the strategies of the maximizer are words, elements of Σ_1 are usually denoted by w.

The minimizer's strategies are functions from $\mathcal{H} = (\mathbf{S_1 A_1 S_2 A_2})^* \mathbf{S_1}$ to $\Delta(\mathbf{A_2})$. Let Σ_2 be the set of such strategies. Its elements are typically denoted by τ.

Fixing strategies $w \in \Sigma_1$ of length n, $\tau \in \Sigma_2$ and an initial state $s \in \mathbf{S_1}$ gives a probability measure on the set $\mathcal{H}_n = (\mathbf{S_1 A_1 S_2 A_2})^n \mathbf{S_1}$ which is denoted by $\mathbb{P}_s^{w,\tau}$: for a history $h = s_1 a_1 t_1 b_1 \cdots s_n a_n t_n b_n s_{n+1} \in \mathcal{H}_n$,

$$\mathbb{P}_s^{w,\tau}(h) = \prod_{i=1}^{n} p(s_i, a_i)(t_i) \cdot \tau(h_i)(b_i) \cdot p(t_i, b_i)(s_{i+1})$$

if $s = s_1$ and $w = a_1 \cdots a_n$, and 0 otherwise, where $h_i = s_1 a_1 t_1 b_1 \cdots s_i a_i t_i$, $1 \leq i \leq n$.

For $t \in \mathbf{S_1}$, we will denote by $\mathbb{P}_s^{w,\tau}(t)$ the chance of ending up in state t after starting from state s and playing the respective strategies, i.e. $\mathbb{P}_s^{w,\tau}(t) = \sum_{ht \in \mathcal{H}} \mathbb{P}_s^{w,\tau}(ht)$. Whereas for a set of states $R \subseteq \mathbf{S_1}$ let $\mathbb{P}_s^{w,\tau}(R) = \sum_{t \in R} \mathbb{P}_s^{w,\tau}(t)$.

2.1 The Maxmin Reachability Problem

Now we can introduce the maxmin reachability and for half-blind games, using the notation and notions just defined. Given a game with initial state $s \in \mathbf{S_1}$ and final states $F \subseteq \mathbf{S_1}$, the maxmin value $\underline{val}(s)$ is defined by

$$\underline{val}(s) = \sup_{w \in \Sigma_1} \inf_{\tau \in \Sigma_2} \mathbb{P}_s^{w,\tau}(F).$$

In case $\underline{val}(s) = 1$, we say that F is maxmin reachable from s.

Problem 1 (Maxmin reachability). Given a game, is the set of final states F maxmin reachable from the initial state s?

There is no hope to decide this problem in general. The reason is that in the special case where the minimizer has no choice in any of the states that she controls, then Problem 1 is equivalent to the value one problem for *probabilistic finite automata* which is already known to be undecidable [13]. However, in the present paper, we establish that Problem 1 is decidable for a subclass of half-blind games called leaktight games.

2.2 Deterministic Strategies for the Minimizer

In general, strategies of the minimizer are functions from $\mathcal{H} = (\mathbf{S_1 A_1 S_2 A_2})^* \mathbf{S_1}$ to $\Delta(\mathbf{A_2})$. However, because in the present paper we focus on the maxmin reachability problem, we can assume that strategies of the minimizer have a much simpler form: the choice of action by the minimizer is deterministic and only depends on the current state and on how much time has elapsed since the beginning of the play. Formally, we assume that minimizer strategies are functions $\mathbb{N} \rightarrow (\mathbf{S_2} \rightarrow \mathbf{A_2})$. Denote Σ_2^p the set of all such strategies. This restriction of the set of minimizer strategies does not change the answer to the maxmin reachability problem (Theorem 1, [15]).

2.3 Two Examples

The graph on which a half-blind game is played is visualized as in Figs. 1 and 2. The circle states are controlled by the maximizer, and the square states are controlled by the minimizer, so for the example in Fig. 1, $\mathbf{S_1} = \{i, f\}$ and $\mathbf{S_2} = \{1, 2\}$. We represent only edges (s, t) such that $p(s, a)(t) > 0$ for some action a and we label the edge (s, t) by a if $p(s, a)(t) = 1$ and by $(a, p(s, a, t))$ otherwise.

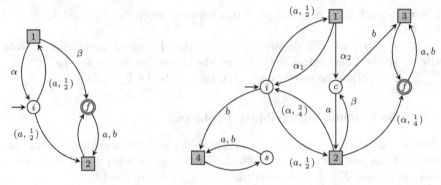

Fig. 1. A half-blind game with $\underline{val}(i) = 1$.

Fig. 2. A half-blind game with $\underline{val}(i) < 1$.

For the game in Fig. 1 it is easy to see that $\underline{val}(i) = 1$, since if the maximizer plays the strategy a^n, no matter what strategy the minimizer chooses the probability to be on the final state is at least $1 - \frac{1}{2^n}$. On the other hand in the game depicted on Fig. 2, $\{f\}$ is not maxmin reachable from i. If the maximizer plays a strategy of only a's then the minimizer always plays the action β and α_1 for example and the probability to be in the final state will be 0. Therefore the maximizer has to play a b at some point. But then the strategy of the minimizer will be to play β except against the action just before b, against that action the minimizer plays α letting at most $1/4$ of the chance to go to the final state, but making sure that the rest of the probability distribution is stuck in the sink state s. Consequently $\underline{val}(s) = 1/4$. We reuse the examples above to illustrate the belief monoid algorithm in the next section.

3   The Belief Monoid Algorithm

We abstract the game using two (finite) monoid structures that are constructed, one on top of the other. Given that the game belongs to the class of leaktight games, the monoids will contain enough information to decide maxmin reachability.

3.1   The Markov Monoid

The Markov monoid is a finite algebraic object that is in fact richer than a monoid; it is a *stabilization* monoid (see [7]). The Markov monoid was used in [12] to decide the value 1 problem for leaktight probabilistic automata on finite words.

Elements of the Markov monoid are $\mathbf{S_1} \times \mathbf{S_1}$ binary matrices. They are typically denoted by capital letters such as U, V, W. The entry that corresponds to the states $s, t \in \mathbf{S_1}$ is denoted by $U(s, t)$. We will make use of the notation $s \xrightarrow{U} t$ in place of $U(s, t) = 1$, when it is helpful.

We define two operations on these matrices: the product and the iteration.

Definition 1. *Given two* $S_1 \times S_1$ *binary matrices* U, V, *their* product *(denoted* UV*) is defined for all* $s, t \in S_1$ *as*

$$UV(s,t) = 1 \iff \exists s' \in S_1, \; s \xrightarrow{U} s' \wedge s' \xrightarrow{V} t.$$

Given a $S_1 \times S_1$ *binary matrix* U *that is idempotent, i.e.* $U^2 = U$, *its* iteration *(denoted* $U^\#$*) is defined for all* $s, t \in S_1$ *as*

$$U^\#(s,t) = 1 \iff s \xrightarrow{U} t \text{ and } t \text{ is } U\text{-recurrent}.$$

We say that some state $t \in S_1$ *is* U-recurrent, *if for all* $t' \in S_1$, $t \xrightarrow{U} t' \implies t' \xrightarrow{U} t$. *Otherwise we say that* t *is* U-transient.

For a set X of binary matrices, we denote $\langle X \rangle$ the smallest set of binary matrices containing X and closed under product and iteration. Let $B^{a,\tau}$, $a \in A_1$, $\tau \in \Sigma_2^p$ be a matrix defined by $s \xrightarrow{B^{a,\tau}} t \iff \mathbb{P}_s^{a,\tau}(t) > 0$, $s, t \in S_1$. Now the definition of the Markov monoid can be given.

Definition 2 (Markov monoid). *The Markov monoid denoted* $\mathcal{M}$ *is*

$$\mathcal{M} = \langle \{B^{a,\tau} \mid a \in A_1, \tau \in \Sigma_2^p\} \cup \{1\} \rangle,$$

where **1** *is the unit matrix.*

3.2 The Belief Monoid

Roughly speaking, while the elements of the Markov monoid try to abstract the outcome of the game when both strategies are fixed, the belief monoid tries to abstract the *possible outcomes* of the game when only the strategy of the maximizer is fixed. Hence the elements of the belief monoid are subsets of $\mathcal{M}$, and they are typically denoted by boldfaced lowercase letters such as **u**, **v**, **w**.

Given two elements of the belief monoid **u** and **v**, their product is the product of their elements, while the iteration of some idempotent **u** is the sub-Markov monoid that is generated by **u** minus the elements in **u** that are not iterated.

Definition 3. *Given* $\mathbf{u}, \mathbf{v} \subseteq \mathcal{M}$, *their* product *(denoted* **uv**) *is defined as*

$$\mathbf{uv} = \{UV \mid U \in \mathbf{u}, V \in \mathbf{v}\}.$$

Given $\mathbf{u} \subseteq \mathcal{M}$ *that is idempotent, i.e.* $\mathbf{u}^2 = \mathbf{u}$, *its iteration (denoted* $\mathbf{u}^\#$*) is defined as*

$$\mathbf{u}^\# = \langle \{UE^\# V \mid U, E, V \in \mathbf{u}, EE = E\} \rangle \;.$$

Given $a \in A_1$, let $\mathbf{a} = \{B^{a,\tau} \mid \tau \in \Sigma_2^p\}$, we give a definition of the belief monoid.

Definition 4 (Belief Monoid). *The belief monoid, denoted* $\mathcal{B}$, *is the smallest subset of* $2^{\mathcal{M}}$ *that is closed under product and iteration and contains* $\{\mathbf{a} \mid a \in A_1\} \cup \{\{1\}\}$, *where* **1** *is the unit matrix.*

For the proofs we use *extended* versions of the monoids denoted $\widetilde{\mathcal{M}}, \widetilde{\mathcal{B}}$, where the edges that are deleted by the iteration operation are saved for book-keeping, the definitions can be found in the long version of the paper [15].

We are interested in a particular kind of elements in the belief monoid, called *reachability witnesses*.

Definition 5 (Reachability Witness). *An element $\mathbf{u} \in \mathcal{B}$ is called a reachability witness if for all $U \in \mathbf{u}$, $s \xrightarrow{U} t \implies t \in F$, where s is the initial state of the game and F is the set of final states.*

We give an informal description of the way that the belief monoid abstracts the outcomes of the game. Roughly speaking the strategy choice of the maximizer corresponds to choosing an element $\mathbf{u} \in \mathcal{B}$ while the strategy choice of the minimizer corresponds to picking some $U \in \mathbf{u}$. Consequently under those strategy choices, U will tell us the outcome of the game, that is to say if for some $s, t \in \mathbf{S_1}$, if we have $s \xrightarrow{U} t$ then there is some positive probability (larger than a uniform bound) of going from the state s to the state t. In case of $s \xarrownot{U} t$ we will be ensured that the probability of reaching the state t from s can be made arbitrarily small. Therefore if a reachability witness is found then we will know that for any strategy that the minimizer picks the probability of going to some non-final state from the initial state can be made to be arbitrarily small.

3.3 The Belief Monoid Algorithm

The belief monoid associated with a given game is computed by the belief monoid algorithm, see Algorithm 1. We will see later that under some conditions, the belief monoid algorithm decides the maxmin reachability problem.

Algorithm 1. The belief monoid algorithm.

Data: A leaktight half-blind game.
Result: Answer to the Maxmin reachability problem.
$\mathcal{B} \leftarrow \{\mathbf{a} \mid a \in \mathbf{A_1}\}$.
Close $\mathcal{B}$ by product and iteration
Return true iff there is a reachability witness in $\mathcal{B}$

We illustrate the computation of the belief monoid with an example. Consider the game represented on Fig. 2. The minimizer has four pure stationary strategies $\tau_{\alpha_1\alpha}$, mapping 1 to α_1 and 2 to α, and similarly the strategies $\tau_{\alpha_1\beta}, \tau_{\alpha_2\alpha}, \tau_{\alpha_2\beta}$. Now we compute $B^{a,\tau}$ where τ is one of the strategies above. Assume that we have the following order on the states: $i < c < s < f$, then $B^{a,\tau_{\alpha_1\alpha}} = \begin{bmatrix} 1 & 0 & 0 & 1 \\ 1 & 0 & 0 & 1 \\ 0 & 0 & 1 & 0 \\ 0 & 0 & 0 & 1 \end{bmatrix}$,

$B^{a,\tau_{\alpha_1\beta}} = \begin{bmatrix} 1 & 1 & 0 & 0 \\ 0 & 1 & 0 & 0 \\ 0 & 0 & 1 & 0 \\ 0 & 0 & 0 & 1 \end{bmatrix}$, $B^{a,\tau_{\alpha_2\alpha}} = \begin{bmatrix} 1 & 1 & 0 & 1 \\ 1 & 0 & 0 & 1 \\ 0 & 0 & 1 & 0 \\ 0 & 0 & 0 & 1 \end{bmatrix}$, and $B^{a,\tau_{\alpha_2\beta}} = \begin{bmatrix} 0 & 1 & 0 & 0 \\ 0 & 1 & 0 & 1 \\ 0 & 0 & 1 & 0 \\ 0 & 0 & 0 & 1 \end{bmatrix}$. The set that contains these matrices is the set $\mathbf{a}$. We can verify that $\mathbf{a}$ is not idempotent but the set $\mathbf{a}^2$ on the other hand is closed under taking products, i.e. $\mathbf{a}^4 = \mathbf{a}^2$. Therefore we can take it's iteration and compute the element $(\mathbf{a}^2)^{\#}$. The reader can verify that $(\mathbf{a}^2)^{\#}$ contains $(B^{a,\tau_{\alpha_1\alpha}})^{\#} = \begin{bmatrix} 0&0&0&1 \\ 0&0&0&1 \\ 0&0&1&0 \\ 0&0&0&1 \end{bmatrix}$, $(B^{a,\tau_{\alpha_1\beta}})^{\#} = \begin{bmatrix} 0&1&0&0 \\ 0&1&0&0 \\ 0&0&1&0 \\ 0&0&0&1 \end{bmatrix}$,

and $B^{a,\tau_{\alpha_2\beta}}$. But it also contains $(B^{a,\tau_{\alpha_1\beta}})^{\#}B^{a,\tau_{\alpha_1\alpha}} = B^{a,\tau_{\alpha_1\alpha}}$. Therefore $(\mathbf{a}^2)^{\#}\mathbf{b}$ is not a reachability witness because if we pick $A = B^{a,\tau_{\alpha_1\alpha}}$ in $(\mathbf{a}^2)^{\#}$ and some $B \in \mathbf{b}$, we will have $i \xrightarrow{AB} s$, and s is a sink state.

This roughly tells us that maximizer cannot win with the strategies $((a^{2n}b))_n$, because against $a^{2n}b$ the minimizer plays the strategy $\tau_{\alpha_1\beta}$ for the first $2n - 1$ turns and then plays the strategy $\tau_{\alpha_1\alpha}$ against the last a, making sure that after the b is played the we end up in the sink state s with at least $3/4$ probability. Continuing the computation we can verify that the belief monoid of the game in Fig. 2 does not contain a reachability witness.

4 Leaks

Leaks were first introduced in [12] to define a decidable class of instances for the value 1 problem for probabilistic automata on finite words. The decidable class of *leaktight automata* is general enough to encompass all known decidable classes for the value 1 problem [11] and is optimal in some sense [10]. We extend the notion of leak from probabilistic automata to half-blind games and prove that when a game does not contain any leak then the belief monoid algorithm decides the maxmin reachability problem.

Intuitively a leak happens when there is some communication between two recurrence classes with transitions that have a small probability of occurring. Whether this small probability builds up to render one of the recurrence classes transient is a computationally hard question to answer — and in fact impossible in general. Examples of leaks can be found in [11] and the link between leaks and convergence rates are discussed further in [10].

Definition 6 (Leaks). *An element of the extended Markov monoid* $(U,\widetilde{U}) \in \widetilde{\mathcal{M}}$ *is a leak if it is idempotent and there exist* $r, r' \in \mathbf{S_1}$, *such that: (1)* r, r' *are* U-*recurrent, (2)* $r \xrightarrow{U} \not\,\, r'$ *and (3)* $r \xrightarrow{\widetilde{U}} r'$.

An element of the extended belief monoid $\mathbf{u} \in \widetilde{\mathcal{B}}$ *is a leak if it contains* $(U,\widetilde{U}) \in \mathbf{u}$ *such that* $(U,\widetilde{U})$ *is a leak.*

A game is leaktight if its extended belief monoid does not contain any leaks.

Note also that the question of whether a game is leaktight is decidable, since this information can be found in the belief monoid itself.

5 Correctness of the Belief Monoid Algorithm

This section is dedicated to proving that when the game is leaktight the belief monoid algorithm is both sound (a reachability witness is found implies $\underline{val}(s) = 1$) and complete (no reachability witness is found implies $\underline{val}(s) < 1$).

Theorem 1. *The belief monoid algorithm solves the maxmin reachability problem for half-blind leaktight games.*

Theorem 1 is a direct consequence of Theorems 2 and 3 which are given in the next two sections.

5.1 Soundness

In this section we give the main ideas to prove soundness of the belied monoid algorithm.

Theorem 2 (Soundness). *Assume that the game is leaktight and that its extended belief monoid contains a reachability witness. Then the set of final states is maxmin reachable from the initial state.*

Theorem 2 is justifying the *yes* instances of the belief monoid algorithm, i.e. if the algorithm replies yes, then indeed $\underline{val}(s) = 1$. It is interesting to note that the equivalent soundness theorem for probabilistic automata in [12] does not make use of the leaktight hypothesis. Theorem 2 follows as a corollary of:

Lemma 1. *Given a game whose extended belief monoid is leaktight, with every element $\mathbf{u} \in \mathcal{B}$ of its belief monoid we can associate a sequence $(u_n)_n$, $u_n \in \Sigma_1$ such that for all $(\tau_n)_n$, $\tau_n \in \Sigma_2^p$ there exists $U \in \mathbf{u}$ and a subsequence $((u_n', \tau_n'))_n \subset ((u_n, \tau_n))_n$ for which*

$$U(s, t) = 0 \implies \lim_n \mathbb{P}_s^{u_n', \tau_n'}(t) = 0,$$

for all $s, t \in \mathbf{S_1}$.

We can prove Theorem 2 as follows. We are given a game that is leaktight and has a reachability witness $\mathbf{u} \in \mathcal{B}$, to whom we can associate a sequence of words $(u_n)_n$ according to Lemma 1. If on the contrary there exists $\epsilon > 0$ such that $\underline{val}(s) \leq 1 - \epsilon$ then there exists a sequence of strategies $(\tau_n)_n$ such that for all $n \in \mathbb{N}$, $\mathbb{P}_s^{u_n, \tau_n}(F) \leq 1 - \epsilon'$, for some $\epsilon' > 0$. This contradicts Lemma 1 because for the reachability witness we have by definition that for all $U \in \mathbf{u}$, $U(s, t) = 1$ implies $t \in F$.

We give a short sketch of the main ideas utilized into proving Lemma 1.

To $\mathbf{a} \in \mathcal{B}$, $a \in \mathbf{A_1}$ we associate the constant sequence of words $(a)_n$. To the product of two elements in $\mathcal{B}$ we associate the concatenation of their respective sequences, and to $\mathbf{u}^{\#} \in \mathcal{B}$ the sequence $(u_n^n)_n$ is associated, given that $(u_n)_n$ is coupled with $\mathbf{u}$. Then we consider words whose letters are pairs (a, τ), where $a \in \mathbf{A_1}$ and τ is a strategy that maps $\mathbf{S_2}$ to $\mathbf{A_2}$, i.e. a pure and stationary strategy, and give a morphism from these words to the extended Markov monoid $\widetilde{\mathcal{M}}$. This allows us to construct k-decomposition trees of such words with respect to $\widetilde{\mathcal{M}}$ (see [15]) . Then the k-decomposition trees are used to prove lower and upper bounds on the outcomes of the game under the strategy choices given by the word of pairs. The main idea is that we can construct for longer and longer words, k-decomposition trees for larger and larger k, thereby making sure that the iteration nodes have a large enough number of children which enables us to show that the probability of being in transient states is bounded above by a quantity that vanishes in the limit.

5.2 Completeness

Before introducing the main theorem of this section let us give a definition.

Definition 7 (μ-faithful abstraction). *Let $u \in \Sigma_1$ be a word, and $\mu > 0$ a strictly positive real number. We say that $\mathbf{u} \in \widetilde{\mathcal{B}}$ is a μ-faithful abstraction of the word u if for all $(U, \widetilde{U}) \in \mathbf{u}$ there exists $\tau \in \Sigma_2^p$ such that for all $s, t \in \mathbf{S_1}$,*

$$\widetilde{U}(s,t) = 1 \iff \mathbb{P}_s^{u,\tau}(t) > 0 \tag{1}$$
$$U(s,t) = 1 \implies \mathbb{P}_s^{u,\tau}(t) \geq \mu. \tag{2}$$

This section is devoted to giving the main ideas behind the proof of the following theorem.

Theorem 3. *Assume that the game is leaktight. Then there exists $\mu > 0$ such that for all words $u \in \Sigma_1$ there is some element $\mathbf{u} \in \widetilde{\mathcal{B}}$ that is a μ-faithful abstraction of u.*

The notion of μ-faithful abstraction is compatible with product in the following sense.

Lemma 2. *Let $\mathbf{u}, \mathbf{v} \in \widetilde{\mathcal{B}}$ be μ-faithful abstractions of $u \in \Sigma_1$ and $v \in \Sigma_1$ respectively. Then $\mathbf{uv}$ is a μ^2-faithful abstraction of $uv \in \Sigma_1$.*

A naïve use of Lemma 2 shows that any word w has a μ_w-faithful abstraction in $\widetilde{\mathcal{B}}$, where μ_w converges to 0 as the length of w increases. However we need μ_w to depend only on $\widetilde{\mathcal{B}}$, independently of $|w|$. For that we make use of k-decomposition trees. More precisely we build N-decomposition trees for words in Σ_1 where $N = 2^{3 \cdot |\widetilde{\mathcal{M}}|}$. We can construct N-decomposition trees for any word $u \in \Sigma_1$ whose height is at most $3 \cdot |\widetilde{\mathcal{B}}|^2$ and since N is fixed we will be able to propagate the constant μ, it only remains to take care that the constant does not shrink as a function of the number of children in iteration nodes, hence the following lemma.

Lemma 3. *Let $u \in \Sigma_1$ be a word factorized as $u = u_1 \cdots u_n$ where $n > 2^{3 \cdot |\widetilde{\mathcal{M}}|} = N$, and $\mathbf{u} \in \widetilde{\mathcal{B}}$ an idempotent element such that $\mathbf{u}$ is a μ-faithful abstraction of u_i, $1 \leq i \leq n$, for some $\mu > 0$. If $\mathbf{u}$ is not a leak then $\mathbf{u}^{\#}$ is a μ'-faithful abstraction of u, where $\mu' = \mu^{N+1}$.*

Then Theorem 3 is an easy consequence from the lemmata above, which can be shown as follows. We construct a N-decomposition tree for the word $u \in \Sigma_1$, and propagate the lower bound from the leaf nodes, for which we have the bound $\nu > 0$ (where ν is the smallest transition probability appearing in the game) up to the root node. If we know that a bound $\mu > 0$ holds for the children, for the parents we have the following lower bounds as a function of the kind of the node: (1) *product node*: μ^2; (2) *idempotent node* μ^N; (3) *iteration node* μ^{N+1}. Since the length of the tree is at most $h = 3 \cdot |\widetilde{\mathcal{B}}|^2$ we have the lower bound $\mu = \nu^{h(N+1)}$ that holds for all $u \in \Sigma_1$.

6   Complexity of Optimal Strategies

The maxmin reachability problem solved by the belief monoid algorithm concerns games where the maximizer is restricted to pure strategies, and decides whether $\underline{val}(s) = \sup_{w \in \Sigma_1} \inf_{\tau \in \Sigma_2} \mathbb{P}_s^{w,\tau}(F) = 1$, where $\Sigma_1 = \mathbf{A_1}^*$. If we extend further the set Σ_1 of strategies of the maximizer and allow him to have mixed strategies too, then half-blind games have a value [14]. Let $\Sigma_1^m = \Delta(\mathbf{A_1}^*)$ be the set of mixed words then

$$val(s) = \sup_{w \in \Sigma_1^m} \inf_{\tau \in \Sigma_2} \mathbb{P}_s^{w,\tau}(F) = \inf_{\tau \in \Sigma_2} \sup_{w \in \Sigma_1^m} \mathbb{P}_s^{w,\tau}(F).$$

Define Σ_2^f to be the set of finite-memory strategies for the minimizer. These are strategies that are stochastic finite-state probabilistic transducers reading histories and outputting elements of $\Delta(\mathbf{A_2})$, mixed actions.

Let $val^f(s) = \inf_{\tau \in \Sigma_2^f} \sup_{w \in \Sigma_1} \mathbb{P}_s^{w,\tau}(F)$. In general $\underline{val}(s) \leq val(s) \leq val^f(s)$.

A natural question is whether the inequalities above are strict in general, i.e. whether mixed strategies are strictly more powerful for the maximizer and whether infinite-memory strategies are strictly more powerful for the minimizer. The answer to both questions is positive; the relevant examples and further details can be found in [15].

Conclusion

We have defined a class of stochastic games with partial observation where the maxmin-reachability problem is decidable. This holds under the assumption that maximizer is restricted to deterministic strategies. The extension of this result to the value 1 problem where maximizer is allowed to use mixed strategies seems rather challenging.

References

1. Bertoni, A.: The solution of problems relative to probabilistic automata in the frame of the formal languages theory. In: Siefkes, D. (ed.) Gl-4.Jahrestagung: LNCS, vol. 26, pp. 107–112. Springer, Heidelberg (1975)
2. Chadha, R., Sistla, A.P., Viswanathan, M.: Power of randomization in automata on infinite strings. In: Bravetti, M., Zavattaro, G. (eds.) CONCUR 2009. LNCS, vol. 5710, pp. 229–243. Springer, Heidelberg (2009)

3. Chatterjee, K., Doyen, L.: Partial-observation stochastic games: how to win when belief fails. ACM Trans. Comput. Logic (TOCL) **15**(2), 16 (2014)
4. Chatterjee, K., Doyen, L., Nain, S., Vardi, M.Y.: The complexity of partial-observation stochastic parity games with finite-memory strategies. In: Muscholl, A. (ed.) FOSSACS 2014 (ETAPS). LNCS, vol. 8412, pp. 242–257. Springer, Heidelberg (2014)
5. Chatterjee, K., Henzinger, T.A.: Semiperfect-information games. In: Sarukkai, S., Sen, S. (eds.) FSTTCS 2005. LNCS, vol. 3821, pp. 1–18. Springer, Heidelberg (2005)
6. Chatterjee, K., Tracol, M.: Decidable problems for probabilistic automata on infinite words. In: 2012 27th Annual IEEE Symposium on Logic in Computer Science (LICS), vol. 2, pp. 185–194. IEEE (2012)
7. Colcombet, T.: The theory of stabilisation monoids and regular cost functions. In: Albers, S., Marchetti-Spaccamela, A., Matias, Y., Nikoletseas, S., Thomas, W. (eds.) ICALP 2009, Part II. LNCS, vol. 5556, pp. 139–150. Springer, Heidelberg (2009)
8. Colcombet, T.: Regular cost functions, part I: logic and algebra overwords. Logical Methods Comput. Sci. **9**(3), 47 (2013)
9. de Alfaro, L., Henzinger, T.A., Kupferman, O.: Concurrent reachability games. Theor. Comput. Sci. **386**(3), 188–217 (2007)
10. Fijalkow, N.: Profinite techniques for probabilistic automata and the optimality of the markov monoid algorithm. CoRR, abs/1501.02997 (2015)
11. Fijalkow, N., Gimbert, H., Kelmendi, E., Oualhadj, Y: Deciding the value 1 problem for probabilistic leaktight automata. Logical Methods Comput. Sci. **11**(2) (2015)
12. Fijalkow, N., Gimbert, H., Oualhadj, Y.: Deciding the value 1 problem for probabilistic leaktight automata. In: LICS, pp. 295–304. IEEE Computer Society (2012)
13. Gimbert, H., Oualhadj, Y.: Probabilistic automata on finite words: decidable and undecidable problems. In: Abramsky, S., Gavoille, C., Kirchner, C., Meyer auf der Heide, F., Spirakis, P.G. (eds.) ICALP 2010. LNCS, vol. 6199, pp. 527–538. Springer, Heidelberg (2010)
14. Gimbert, H., Renault, J., Sorin, S., Venel, X., Zielonka, W.: On the values of repeated games with signals. CoRR, abs/1406.4248 (2014)
15. Kelmendi, E., Gimbert, H.: Deciding maxmin reachability in half-blind stochastic games. CoRR, abs/1605.07753 (2016). http://arxiv.org/abs/1605.07753
16. Paz, A.: Some aspects of probabilistic automata. Inf. Control **9**(1), 26–60 (1966)
17. Rabin, M.O.: Probabilistic automata. Inf. Control **6**(3), 230–245 (1963)
18. Simon, I.: On semigroups of matrices over the tropical semiring. ITA **28**(3–4), 277–294 (1994)

The Big Match in Small Space
(Extended Abstract)

Kristoffer Arnsfelt Hansen[1], Rasmus Ibsen-Jensen[2], and Michal Koucký[3]([⊠])

[1] Aarhus University, Aarhus, Denmark
arnsfelt@cs.au.dk
[2] IST Austria, Klosterneuburg, Austria
ribsen@ist.ac.at
[3] Charles University, Prague, Czech Republic
koucky@iuuk.mff.cuni.cz

Abstract. We study repeated games with absorbing states, a type of two-player, zero-sum concurrent mean-payoff games with the prototypical example being the Big Match of Gillete (1957). These games may not allow *optimal* strategies but they always have ε-optimal strategies. In this paper we design ε-optimal strategies for Player 1 in these games that use only $O(\log \log T)$ space. Furthermore, we construct strategies for Player 1 that use space $s(T)$, for an arbitrary small unbounded non-decreasing function s, and which guarantee an ε-optimal value for Player 1 in the limit superior sense. The previously known strategies use space $\Omega(\log T)$ and it was known that no strategy can use constant space if it is ε-optimal even in the limit superior sense. We also give a complementary lower bound. Furthermore, we also show that no Markov strategy, even extended with finite memory, can ensure value greater than 0 in the Big Match, answering a question posed by Neyman [11].

1 Introduction

In game theory there has been considerable interest in studying the complexity of strategies in infinitely repeated games. A natural way how to measure the complexity of a strategy is by the number of states of a finite automaton implementing the strategy. A common theme is to consider what happens when some or all players are restricted to play using a strategy given by an automaton of a certain bounded complexity.

Asymptotic View. Previous works have mostly been limited to dichotomy results: either there is a good strategy implementable by finite automaton or there is no such strategy. Our goal here is to refine this picture. We do this by taking the asymptotic view: measuring the complexity as a function of the number of rounds played in the game. Now when the strategy no longer depends just on a finite amount of information about the history of the game it could even

---

Extended Abstract of Hansen et al. [6].

M. Gairing and R. Savani (Eds.): SAGT 2016, LNCS 9928, pp. 64–76, 2016.
DOI: 10.1007/978-3-662-53354-3_6

be a computationally difficult problem to decide the next move of the strategy. But we focus on investigating how much information a good strategy must store about the play so far to decide on the next move; in other words, we study how much space the strategy needs.

Game Classes. The class of games we study is that of repeated zero-sum games with absorbing states. These form a special case of undiscounted stochastic games. Stochastic games were introduced by Shapley [12], and they constitute a very general model of games proceeding in rounds. We consider the basic version of two-player zero-sum stochastic games with a constant number of states and a constant number of actions. In a given round t the two players simultaneously choose among a number of different *actions* depending on the current *state*. Based on the choice of the pair (i, j) of actions as well as the current state k, Player 1 receives a *reward* $r_t = a_{ij}^k$ from Player 2, and the game proceeds to the next state ℓ according to probabilities $p_{ij}^{k\ell}$.

Limit-Average Rewards. In Shapley's model, in every round the game stops with non-zero probability, and the payoff assigned to Player 1 by a play is simply the sum of rewards r_i. The stopping might be viewed as *discounting* later rewards by a discounting factor $0 < \beta < 1$. Gillette [4] considered the more general model of undiscounted stochastic games where all plays are infinite. He is interested in the average reward $\frac{1}{T} \sum_{t=1}^{T} r_t$ to Player 1 as T tends to infinity. As the limit may not exist one needs to consider lim inf, lim sup, or some Banach limit [13] of the sums. In many cases the particular choice of the limit does not matter much, but it turns out that for our results it has interesting consequences. For this reason we consider both $\liminf_{T\to\infty} \frac{1}{T} \sum_{t=1}^{T} r_t$ and $\limsup_{T\to\infty} \frac{1}{T} \sum_{t=1}^{T} r_t$.

Note that both these notions have natural interpretations. For instance, the lim inf notion suits the setup where the infinite repeated game actually models a game played repeatedly for an unspecified (but large) number of rounds, where one thus desires a guarantee on the average reward after a certain number of rounds. The lim sup notion on the other hand models the ability to always recover from arbitrary losing streaks in the repeated game.

The Big Match. A prototypical example of an undiscounted stochastic game is the well-known Big Match of Gillette [4] (see Fig. 1 for an illustration of the Big Match, together with the adjacent description). This game fits also into an important special subclass of undiscounted stochastic games: the *repeated games with absorbing states*, defined by Kohlberg [9]. In a repeated game with absorbing states there is only one state that can be left; all the other states are *absorbing*, i.e., the probability of leaving them is zero regardless of the actions of the players. Even in these games, as for general undiscounted stochastic games, there might not be an optimal strategy for the players [4]. On the other hand there always exist ε-optimal strategies [9], which are strategies guaranteeing the value of the game up to an additive term ε. The Big Match provides such an example: the value of the game is $1/2$, but Player 1 does not have an optimal strategy, and must settle for an ε-optimal strategy [2]. However, it is known that such ε-optimal strategies in the Big Match must have a certain level of complexity.

More precisely, for any $\varepsilon < \frac{1}{2}$, an ε-optimal strategy can neither be implemented by a finite automaton nor take the form of a Markov strategy (a strategy whose only dependence on the history is the number of rounds played) [14].

In this paper we consider the Big Match in particular and then generalize our results to general repeated games with absorbing states.

The Model Under Consideration. We are interested in the *space complexity* of ε-optimal strategies in repeated games with absorbing states. A general strategy of a player in a game might depend on the whole history of the play up to the current time step. Moreover the decision about the next move might depend arbitrarily on the history. This provides the strategies with lots of power. There are two natural ways how to restrict the strategies: one can put computational restrictions on how the next move is decided based on the history of the play, or one can put a limit on how much information can the strategy remember about the history. One can also combine both types of restrictions, which leads to an interactive Turing machine based model, modelling a dynamic algorithm.

In this paper we mainly focus on restricting the amount of information the strategy can remember. This restriction is usually studied in the form of how *large size a finite automaton* (transducer) for the strategy has to be, and we follow this convention. By the size of a finite automaton we mean the number of states. The automatons we consider can make use of probabilistic transitions, and we will not consider the description of these probabilities as part of the size of the automaton. We do address these separately, however.

History of the Model. The idea of measuring complexity of strategies in repeated games in terms of automata was proposed by Aumann [1]. The survey by Kalai [8] further discuss the idea in several settings of repeated games. However in this line of research the finite automata is assumed to be fixed for the duration of the game. This represents a considerable restriction as for many games there is no good strategy that could be described in this setting. Hence we consider strategies in which the automata can grow with time. To be more precise we consider infinite automata and measure how many different states we could have visited during the first T steps of the play. The logarithm of this number corresponds to the amount of space one would need to keep track of the current state of the automaton. We are interested in how this space grows with the number of rounds of the play.

Comparison of our Model with a Turing Machine Based Model. To impose also computational restrictions on the model, one can consider the usual Turing machine with one-way input and output tapes that work in lock-step and that record the play: whenever the machine writes its next action on the output tape it advances the input head to see the corresponding move of the other player. The space usage of the model is then the work space used by the machine, growing with the number of actions processed. The Turing machine can be randomized to allow for randomized strategies. The main differences between this model and the automaton based model we focus on in this paper is that in the case of infinite automata the strategy can be *non-uniform* and use *arbitrary* probabilities on its transitions whereas the Turing machine is *uniform* in the sense that it has a finite program that is fixed for the duration play and in particular, all transition probabilities are explicitly generated by the machine.

Bounds for Strategies with Deterministic Update. Trivially, any strategy needs space at most $O(T)$, since such memory would suffice to remember the whole history of the play. It is not hard to see (cf. [7, Chapter 3.2.1]) that if a strategy is not restricted to a finite number of states, then the number of reachable states by round T must be at least T, thus the space usage is at least $\log T$, by definition of space. However this provides only worst-case answer to our question, since for randomized strategies it might happen that only negligible fraction of the states can be reached with reasonable probability. Indeed, it might be that with probability close to 1 the strategy reaches only a very limited number of states. This is the setup we are interested in. As we will see in a moment, the strategies we consider use substantially less space than $O(\log T)$ with high probability (and $O(\log T)$ space in the worst case).

Our Results. We provide two types of results. We show that there are ε-optimal strategies for repeated games with absorbing states, and we also show that there are limits on how small space such strategies could possibly use. Our strategies are first constructed for the Big Match. Then, following Kohlberg [9] these strategies are extended to general repeated games with absorbing states.

Upper Bounds on Space Usage. Our first results concern the Big Match. We show that for all $\varepsilon > 0$, there exists an ε-optimal strategy that uses $O(\log \log T)$ space with probability $1 - \delta$ for any $\delta > 0$. We note that the previous constructed strategies of Blackwell and Ferguson [2] and Kohlberg [9] uses space $\Theta(\log T)$.

Theorem 1. *For all $\varepsilon > 0$, there is an ε-optimal strategy σ_1 for Player 1 in the Big Match such that for any $\delta > 0$ with probability at least $1 - \delta$, the strategy σ_1 uses $O(\log \log T)$ space in round T.*

Remark 2. We would like to stress the order of quantification above and their impact on the big-O notation used above for conciseness. The strategy we build depends on the choice of ε, but only for the actions made – the memory updates are independent thereof, and thus likewise is the space usage. The dependence of δ is also very benign. More precisely, there exists a constant $C > 0$ independent of ε and δ, and an integer T_0 depending on δ, but independent of ε, in such a way that with probability at least $1 - \delta$, the strategy σ_1 uses at most space $C \log \log T$, for all $T \geq T_0$. The same remark holds elsewhere in our statements.

Our Results Translated to the Turing Based Model. After a slight modification our ε-optimal strategy can be implemented by a Turing machine so that (1) it processes T actions in time $O(T)$; and (2) each time it processes an action, all randomness used comes from at most 1 unbiased coin flip; and (3) for all $\delta > 0$, it uses $O(\log \log T + \log \log \varepsilon^{-1})$ space with probability $1 - \delta$, before round T. See Theorem 9.

Arbitrary Small, but Growing Space for lim sup. For the case of lim sup evaluation of the average rewards we can design strategies that uses even less space, in fact arbitrarily small, but growing, space.

Theorem 3. *For any non-decreasing unbounded function s, there exists an ε-supremum-optimal strategy σ_1 for Player 1 in the Big Match such that for each $\delta > 0$, with probability at least $1 - \delta$, strategy σ_1 uses $O(s(T))$ space in round T.*

We may for instance think of s as the inverse of the Ackermann function. For $s = o(\log \log T)$ such a strategy cannot be implemented by a Turing machine as it would require generating probabilities that cannot be achieved on any Turing machine using only an unbiased coin as implied by our Theorem 10.

Our strategy that is ε-optimal is actually an instantiation of the ε-supremum-optimal strategy by setting $s(T)$ to $O(\log \log T)$. We are unable to achieve ε-optimality in less space, and this seems to be inherent to our techniques.

Generalization to Repeated Games with Absorbing States. We can extend the above results to the case of general repeated games with absorbing states.

Theorem 4. *For all $\varepsilon > 0$ and any repeated game with absorbing states G, there is an ε-optimal strategy σ_k for Player k in G such that, for each $\delta > 0$, with probability at least $1 - \delta$, the strategy σ_k uses $O(\log \log T + \log 1/\varepsilon \cdot \mathrm{poly}(|G|))$ space in round T.*

Theorem 5. *For all $\varepsilon > 0$, any repeated game with absorbing states G, and any non-decreasing unbounded function s, there exists an ε-supremum optimal strategy σ_k for Player k in G such that for each $\delta > 0$, with probability at least $1 - \delta$, the strategy σ_1 uses $O(s(T) + \log 1/\varepsilon \cdot \mathrm{poly}(|G|))$ space in round T.*

These strategies are obtained by reducing to a special simple case of repeated games with absorbing states, generalized Big Match games, to which our Big Match strategies can be generalized. This reduction can furthermore be done effectively by a polynomial time algorithm.

Lower Bound on Space Usage. We provide two lower bounds on space addressing different aspects of our strategies. One property of our strategies is that the smaller the space used is, the smaller the probabilities of actions employed are. The reciprocal of the smallest non-zero probability is the *patience* of a strategy. This is a parameter of interest for strategies. We show that the patience of our strategies is close to optimal. In particular, we show that the first $f(T)$ memory states must use probabilities close to $1/T^{f(T)}$, where $s(T) = \log f(T)$ is the space usage. We can almost match this bound by our strategies.

Finite-Memory Deterministic-Update Markov Strategies are no Good. Beside the lower bound on patience we investigate the possibility of using a good strategy for Player 1 which would use only a constant number of states but where the actions could also depend on the round number. This

is what we call a *finite-memory Markov strategy*. We show that such a strategy
which also updates its memory state deterministically cannot exist. This answers
a question posed by Abraham Neyman [11].

Theorem 6. *For all $\varepsilon < \frac{1}{2}$, there exists no finite-memory deterministic-update
ε-optimal Markov strategy for Player 1 in the Big Match.*

Our Techniques. The previously given strategies for Player 1 in the Big Match
[2,9] use space $\Theta(\log T)$ as they maintain the count of the number of different
actions taken by the other player. There are two principal ways to decrease
the number of states for such randomized strategies: either to use approximate
counters [3,10], or to sub-sample the stream of actions of the other player and use
a good strategy on the sparse sample. In this paper we use the latter approach.

Overview Over our Strategy for the Big Match. Our strategies for Player 1
proceed by observing the actions of Player 2 and collecting statistics on the
payoff. Based on these statistics Player 1 adjusts his actions. The statistics is
collected at random sample points and Player 1 plays according to a "safe"
strategy on the points not sampled and plays according to a good (but space-
inefficient) strategy on the sample points. If the space of Player 1 is at least
$\log \log T$ then Player 1 is able to collect sufficient statistics to accurately estimate
properties of the actions of Player 2. Namely, substantial dips in the average
reward given to Player 1 can be detected with high probability and Player 1 can
react accordingly. Thus that during infinite play, the average reward will not
be able drop for extended periods of time, and this will guarantee that $\liminf$
evaluation of the average rewards is close to the value of the game.

The Bottleneck in the lim inf Case. However, if our space is considerably less
than $\log \log T$ we do not know how to accurately estimate these properties of the
actions of Player 2. Thus, long stretches of actions of Player 2 giving low average
rewards might go undetected as long as they are accompanied by stretches of
high average rewards. Thus one could design a strategy for Player 2 that has
low $\liminf$ value of the average rewards, but has large $\limsup$ value. Against
such a strategy, our space-efficient strategy for Player 1 is unlikely to stop. So
during infinite play, while our strategy guarantees that the $\limsup$ evaluation of
the average rewards is close to the value of the game, it performs poorly under
$\liminf$ evaluation. It is not clear whether this is an intrinsic property of all very
small space strategies for Player 1 or whether one could design a very small space
strategy achieving that the $\liminf$ evaluation of the average rewards is close to
the value of the game. We leave this as an interesting open question.

Generalizing to Repeated Games with Absorbing States. Our extension to general repeated games with absorbing states follow closely the work of
Kohlberg [9]. He showed that all such games have a value and constructed ε-
optimal strategies for them, building on the work of Blackwell and Ferguson [2].
His construction is in two steps: The question of value and of ε-optimal strategies
are solved for a special case of repeated games with absorbing states, generalized

Big Match games, that are sufficiently similar to the Big Match game that one of the strategies given by Blackwell and Ferguson [2] can be extended to this more general class of games. Having done this, Kohlberg shows how to reduce general repeated games with absorbing states to generalized Big Match games.

In a similar way we can extend our small-space strategies for the Big Match to the larger class of generalized Big Match games. These can then directly be used for Kohlberg's reduction. This reduction is however only given as an existence statement. We show how the reduction can be made explicit and computed by a polynomial time algorithm. This is done using linear programming formulations and fundamental root bounds of univariate polynomials. This also provides explicit bounds on the bitsize of the reduced generalized Big Match games. We also give a simple polynomial time algorithm for approximating the value of any repeated game with absorbing states based on bisection and linear programming.

2 Definitions

A *probability distribution* over a finite set S, is a map $d : S \to [0,1]$, such that $\sum_{s \in S} d(s) = 1$. Let $\Delta(S)$ denote the set of all probability distributions over S.

Repeated Games with Absorbing States. The games we consider are special cases of two player, zero-sum concurrent mean-payoff games in which all states except at most one are *absorbing*, i.e. never left if entered (note also that an absorbing state can be assumed to have just a single action for each player). We restrict our definitions to this special case, introduced by Kohlberg [9] as repeated games with absorbing states. Such a game G is given by sets of actions A_1 and A_2 for each player together with maps $\pi : A_1 \times A_2 \to \mathbb{R}$ (the stage payoffs) and $\omega : A_1 \times A_2 \to [0,1]$ (the absorption probabilities).

The game G is played in rounds. In every round $T = 1,2,3,\ldots$, each player $k \in \{1,2\}$ independently picks an action $a_k^T \in A_k$. Player 1 then receives the stage payoff $\pi(a_1^T, a_2^T)$ from Player 2. Then, with probability $\omega(a_1^T, a_2^T)$ the game stops and all payoffs of future rounds are fixed to $\pi(a_1^T, a_2^T)$ (think of this as the game proceeding to an absorbing state where the stage payoff for future rounds is $\pi(a_1^T, a_2^T)$). Otherwise, the game just proceeds to the next round.

The sequence $(a_1^1, a_2^1), (a_1^2, a_2^2), (a_1^3, a_2^3), \ldots$ of actions taken by the two players is called a *play*. A finite play occurs when the game stops after the last pair of actions. Otherwise the play is infinite. To a given play P we associate an infinite sequence of rewards $(r_T)_{T \geq 1}$ received by Player 1. If $P = (a_1^1, a_2^1), (a_1^2, a_2^2), \ldots, (a_1^\ell, a_2^\ell)$ is a finite play of length ℓ we let $r_T = \pi(a_1^T, a_2^T)$ for $1 \leq T \leq \ell$, and $r_T = \pi(a_1^\ell, a_2^\ell)$ for $T > \ell$. In this case we say that the game stops with *outcome* r_ℓ. Otherwise, if $P = (a_1^1, a_2^1), (a_1^2, a_2^2), \ldots$ is infinite we simply let $r_T = \pi(a_1^T, a_2^T)$ for all $T \geq 1$.

To evaluate the sequence of the rewards we consider both the lim inf and lim sup value of the average reward $\frac{1}{T} \sum_{t=1}^T r_t$. We thus define the limit-infimum payoff to Player 1 of the play as $u_{\inf}(P) = \liminf_{n \to \infty} \frac{1}{n} \sum_{T=1}^n r_T$, and similarly we define the limit-supremum payoff to Player 1 of the play as $u_{\sup}(P) = \limsup_{n \to \infty} \frac{1}{n} \sum_{T=1}^n r_T$.

Strategies. A *strategy* for Player k is a function $\sigma_k : (A_1 \times A_2)^* \to \Delta(A_k)$ describing the probability distribution of the next chosen action after each finite play. We say that Player k *follows* a strategy σ_k if for every finite play P of length $T - 1$, at round T Player k picks the next action according to $\sigma_k(P)$. We say that a strategy σ_k is *pure* if for every finite play P the distribution $\sigma_k(P)$ assigns probability 1 to one of the actions of A_k. Also, we say that a strategy σ_k is a *Markov strategy* if for every T and every play P of length $T - 1$, the distribution $\sigma_k(P)$ does not depend on the particular actions during the first $T - 1$ rounds but is just a function of T. Thus Markov strategy σ_k can be viewed as a map $\mathbb{Z}_+ \to \Delta(A_k)$ or simply a sequence of distributions over A_k.

A *strategy Profile* σ is a pair of strategies (σ_1, σ_2), one for each player. A strategy profile σ defines a probability measure on plays in the natural way. We define the expected limit-infimum payoff to Player 1 of the strategy profile $\sigma = (\sigma_1, \sigma_2)$ as $u_{\inf}(\sigma) = u_{\inf}(\sigma_1, \sigma_2) = \mathrm{E}_{P \sim (\sigma_1, \sigma_2)}[u_{\inf}(P)]$ and similarly the expected limit-supremum payoff to Player 1 of the strategy profile σ as $u_{\sup}(\sigma) = u_{\sup}(\sigma_1, \sigma_2) = \mathrm{E}_{P \sim (\sigma_1, \sigma_2)}[u_{\sup}(P)]$.

Values and Near-Optimal Strategies. We define the *lower values* of G by $\underline{v}_{\inf} = \sup_{\sigma_1} \inf_{\sigma_2} u_{\inf}(\sigma_1, \sigma_2)$ and $\underline{v}_{\sup} = \sup_{\sigma_1} \inf_{\sigma_2} u_{\sup}(\sigma_1, \sigma_2)$, and the *upper values* of G by $\overline{v}_{\inf} = \inf_{\sigma_2} \sup_{\sigma_1} u_{\inf}(\sigma_1, \sigma_2)$ and $\overline{v}_{\sup} = \inf_{\sigma_2} \sup_{\sigma_1} u_{\sup}(\sigma_1, \sigma_2)$. Clearly $\underline{v}_{\inf} \leq \underline{v}_{\sup} \leq \overline{v}_{\sup}$ and $\underline{v}_{\inf} \leq \overline{v}_{\inf} \leq \overline{v}_{\sup}$. Kohlberg showed that all these values coincide and we call this common number $v(G)$ the value v of G.

Theorem 7 (Kohlberg, Theorem 2.1). $\underline{v}_{\inf} = \overline{v}_{\sup}$.

This implies that for the purpose of defining the value of G the choice of the limit of the average rewards does not matter. But a given strategy σ_1 for Player 1 could be close to guaranteeing the value with respect to lim sup evaluation of the average rewards, while being far from doing so with respect to the lim inf evaluation. We shall hence distinguish between these different guarantees.

Let $\varepsilon > 0$ and let σ_1 be a strategy for Player 1. We say that σ_1 is *ε-supremum-optimal*, if $v(G) - \varepsilon \leq \inf_{\sigma_2} u_{\sup}(\sigma_1, \sigma_2)$ and that σ_1 is *ε-optimal*, if $v(G) - \varepsilon \leq \inf_{\sigma_2} u_{\inf}(\sigma_1, \sigma_2)$.

Observation 1. *Clearly it is sufficient to take the infimum over just pure strategies σ_2 for Player 2, and hence when showing that a particular strategy σ_1 is ε-supremum-optimal or ε-optimal we may restrict our attention to pure strategies σ_2 for Player 2.*

One can naturally make similar definitions for Player 2, where the roles of lim inf and lim sup would then be interchanged, but we shall restrict ourselves here to the perspective of Player 1.

If the strategy σ_1 is 0-supremum-optimal (0-optimal) we simply say that σ_1 is supremum-optimal (optimal). The Big Match gives an example where Player 1 does not have a supremum-optimal strategy [2].

Memory and Memory-Based Strategies. A *memory configuration* or *state* is simply a natural number. We will often think of memory configurations as representing discrete objects such as tuples of integers. In such a case we will always have a specific encoding of these objects in mind.

Let $\mathcal{M} \subseteq \mathbb{N}$ be a set of memory states. A *memory-based strategy* σ_1 for Player 1 consists of a *starting state* $m_s \in \mathcal{M}$ and two maps, the *action map* $\sigma_1^a : \mathcal{M} \to \Delta(A_1)$ and the *update map* $\sigma_1^u : A_1 \times A_2 \times \mathcal{M} \to \Delta(\mathcal{M})$. We say that Player 1 *follows* the memory-based strategy σ_1 if in every round T when the game did not stop yet, he picks his next move a_1^T at random according to $\sigma_1^a(m_T)$, where the sequence $m_1, m_2, \ldots$ is given by letting $m_1 = m_s$ and for $T = 1, 2, 3, \ldots$ choosing m_{T+1} at random according to $\sigma_1^u(a_1^T, a_2^T, m_T)$, where a_2^T is the action chosen by Player 2 at round T. The strategies we construct in this paper have the property that their action maps do not depend on the action a_1^T of Player 1. In these cases we simplify notation and write just $\sigma_1^u(a_2^T, m^T)$.

Since each finite play can be encoded by a binary string, and thus a natural number, we can view any strategy σ_k for Player k as a memory-based strategy. One can find similarly defined types of strategies in the literature, but typically, the function corresponding to the update function is deterministic.

Memory Sequences and Space Usage of Strategies. Let σ_1 be a memory-based strategy for Player 1 on memory states $\mathcal{M}$ and σ_2 be a strategy for Player 2. Assume that Player 1 follows σ_1 and Player 2 follows σ_2. The strategy profile (σ_1, σ_2) defines a probability measure on (finite and infinite) sequences over $\mathcal{M}$ in the natural way. For a (finite) sequence $M \in \mathcal{M}^*$, let $\omega_1(M)$ be the probability that Player 1 follows this sequence of memory states during the first $|M|$ rounds of the game, while the game does not stop before round $|M|$.

Fix a non-decreasing function $f : \mathbb{N} \to \mathbb{N}$ and a probability p. The strategy σ_1 *uses* $\log f(T)$ *space with probability at least* p against σ_2, if for all T, the probability $\Pr_{(\sigma_1, \sigma_2)}[\forall i \leq T : M_i \leq f(T)] \geq p$ (i.e., with probability at least p, the current memory has stayed below that of $f(T)$ before round T, for all T). If σ_1 uses $\log f(T)$ space with probability at least p against every strategy σ_2', then we say that σ_1 *uses* $\log f(T)$ *space with probability at least* p.

The Big Match. The Big Match, introduced by Gillette [4] is a simply defined repeated game with absorbing states, where each player has only two actions. In each round Player 1 has the choice to stop the game (action **R**), or continue with the next round (action **L**). Player 2 has the choice to declare the round safe (action **L**) or unsafe (action **R**). If play continues in a round declared safe, or if play stops in a round declared unsafe, Player 2 must give Player 1 a reward 1. In the other two cases no reward is given.

Formally, action sets are $A_1 = A_2 = \{\mathbf{L}, \mathbf{R}\}$. The rewards are $\pi(a_1, a_2) = 1$ if $a_1 = a_2$ and $\pi(a_1, a_2) = 0$ if $a_1 \neq a_2$. The stopping probabilities are $\omega(\mathbf{R}, a_2) = 1$ and $\omega(\mathbf{L}, a_2) = 0$.

We can illustrate this game succinctly in a matrix form as shown in Fig. 1, where rows are indexed by the actions of Player 1, columns are indexed by the actions of Player 2, entries give the rewards, and a star on the reward means that the game stops with probability 1.

	L	**R**
L	1	0
R	0*	1*

Fig. 1. The Big Match in matrix form.

3 Small Space ε-supremum-optimal Strategies in the Big Match

For any non-decreasing and *unbounded* function $f : \mathbb{Z}_+ \rightarrow \mathbb{Z}_+$, we will now give an ε-supremum optimal strategy σ_1^* for Player 1 in the Big Match that for all $\delta > 0$ with probability $1 - \delta$ uses $O(\log f(T))$ space. Let $\overline{f}$ be a strictly increasing unbounded function from $\mathbb{Z}_+$ to $\mathbb{R}_+$, such that $\overline{f}(x) \leq f(x)$ for all $x \in \mathbb{Z}_+$, and let F be the inverse of $\overline{f}$. For simplicity, and without loss of generality, we assume that $F(1) = 1$ and $F(T + 1) \geq 2 \cdot F(T)$.

Intuitive Description of the Strategy and Proof. The main idea for building the strategy is to partition the rounds of the game into epochs, such that epoch i has expected length $F(i)$. The i'th epoch is further split into i sub-epochs. In each sub-epoch j of the i-th epoch we sample i^2 rounds uniformly at random. In every round not sampled we simply stay in the same memory state and play **L** with probability 1. We view the i^2 samples as a stream of actions chosen by Player 2. We then follow a particular ε^2-optimal base strategy $\sigma_1^{i,\varepsilon}$ for the Big Match on the samples of sub-epoch j. This strategy $\sigma_1^{i,\varepsilon}$ is a suitably modified version of a strategy by Blackwell and Ferguson [2] and Kohlberg [9].

More precisely, if $\sigma_1^{i,\varepsilon}$ stops in its k-th round when run on the samples of sub-epoch j, the strategy σ_1^* stops on the k-th sample in sub-epoch j. This will ensure that if σ_1^* stops with probability at least ε, the outcome is at least $\frac{1}{2} - \varepsilon$.

Also, for any $0 < \delta < \frac{1}{2}$ and for sufficiently large i, depending on δ, if the samples have density of **L** at most $\frac{1}{2} - \delta$ then $\sigma_1^{i,\varepsilon}$ stops on the samples with a positive probability depending only on ε, namely ε^8. For $f(T) = \Theta(\log T)$, the division into sub-epochs ensures that if $\liminf_{T \to \infty} \text{dens}(\sigma^T) < \frac{1}{2}$ then infinitely many sub-epochs have density of **L** smaller than $1/2$, and thus the play stops with probability 1 in one of such epochs. This is not necessarily true for $f(T)$ smaller than $\log T$.

The Base Strategy. The important inner part of our strategy is a ε^2-optimal strategy $\sigma_1^{i,\varepsilon}$ parametrized by a non-negative integer i. These strategies are similar to ε-optimal strategies given by Blackwell and Ferguson [2] and Kohlberg [9].

The strategy $\sigma_1^{i,\varepsilon}$ uses deterministic updates of memory, and uses integers as memory states (we think of the memory as an integer counter). The memory update and action function is given by

$$\sigma_1^{i,u}(a, \ell) = \begin{cases} \ell + 1 & \text{if } a = \mathbf{L} \\ \ell - 1 & \text{if } a = \mathbf{R} \end{cases} \qquad \sigma_1^{i,a}(\ell)(\mathbf{R}) = \begin{cases} \varepsilon^8 (1 - \varepsilon)^{2(i+\ell)} & \text{if } i + \ell > 0 \\ \varepsilon^8 & \text{if } i + \ell \leq 0 \end{cases}$$

The Complete Strategy. We are now ready to define σ_1^*. The memory states of this strategy are 5-tuples $(i, j, k, \ell, b) \in \mathbb{Z}_+ \times \mathbb{Z}_+ \times \mathbb{N} \times \mathbb{Z} \times \{0, 1\}$. Here i denotes the current epoch and j denotes the current sub-epoch of epoch i. The number of samples already made in the current sub-epoch is k. The memory state of the inner strategy is stored as ℓ. Finally b is 1 if and only if the strategy will sample to the inner strategy in the next step.

The memory update function $\sigma_1^{*,u}$ is as follows. Let (i, j, k, ℓ, b) be the current memory state and let a be the action of Player 2 in the current step. We then describe the distribution of the next memory state (i', j', k', ℓ', b').

- The current step is not sampled if $b = 0$. In that case we keep $i' = i$, $j' = j$, $k' = k$, and $\ell' = \ell$.
- The current epoch is ending if $j = i$, $k = i^2 - 1$, and $b = 1$. In that case $i' = i + 1$, $j' = 1$, $k' = 0$, and $\ell' = 0$.
- A sub-epoch is ending within the current epoch if $j < i$, $k = i^2 - 1$, and $b = 1$. In that case $i' = i$, $j' = j + 1$, $k' = 0$, and $\ell' = 0$.
- We sample within a sub-epoch if $k < i^2 - 1$ and $b = 1$. In that case $i' = i$, $j' = j$, $k' = k + 1$, and $\ell' = \sigma_1^{i,u}(a, \ell)$.

Finally, in every case, we make a probabilistic choice whether to sample in the next step by letting $b' = 1$ with probability $\frac{(i')^3}{F(i')}$.

The action function $\sigma_1^{*,a}$ is given by

$$\sigma_1^{*,a}((i, j, k, \ell, b))(a) = \begin{cases} \sigma_1^{i,\varepsilon}(\ell)(a) & \text{if } b = 1 \\ 1 & \text{if } b = 0 \text{ and } a = \mathbf{L} \\ 0 & \text{otherwise} \end{cases}.$$

In other words, if the current step is sampled, Player 1 follows the current base strategy, and otherwise always plays $\mathbf{L}$. The starting memory state is $m_s = (1, 1, 1, 0, 0)$. The states that can be reached in sub-epoch j of epoch i are of the form (i, j, k, ℓ, b) where $0 \le k < i^2$ and $-i^2 < \ell < i^2$. Thus at most $4i^4$ states can be reached. The states are mapped to the natural numbers as follows: The memory $(1, 1, 1, 0, 0)$ is mapped to 0 and for each epoch i, all states in epoch i are mapped to the numbers (in an arbitrary order) following the numbers mapped to by epoch $i - 1$. We state our main theorem.

Theorem 8. *The strategy σ_1^* is ε-supremum-optimal, and for all $\delta > 0$, with probability at least $1 - \delta$ it uses space $O(\log f(T))$.*

By instantiating the strategy σ_1^* for $f(T) = \lceil \log T \rceil$ we obtain an ε-optimal strategy σ_1^* for Player 1 in the Big Match that for all $\delta > 0$ with probability $1 - \delta$ uses $O(\log \log T)$ space. We can now conclude with the other main result.

Theorem 9. *For any natural number k, there is a strategy which is 2^{-k}-optimal, has patience 2 and can be implemented on a Turing machine, using at most 1 random bit and amortized constant time per round and with probability at least $1 - \delta$ does it use tape space $O(\log \log(T) + \log k)$ up to round T.*

4   Lower Bound on Patience

When considering a strategy of a player one may want to look at how small or large the probabilities occurring in that strategy are. The parameter of interest is the *patience* of the strategy which is the reciprocal of the smallest non-zero probability occurring in the strategy. Patience is closely related to the expected length of finite plays as small probability events will not occur if the play is too short so they will have little influence on the overall outcome [5,9]. Care has to be taken how to define patience for strategies with infinitely many possible events. Note that for our space efficient strategies, the patience of the states in which we are with high probability during the first T steps is approximately T, for rounds close to the end of an epoch. This patience can be further improved to roughly $T^{1/f(T)}$. In this section we show that this is essentially necessary.

We use the following definitions to deal with the fact that our strategies use infinitely many transitions so their overall patience is infinite. For a memory based strategy σ_1 of Player 1 in a repeated zero-sum game with absorbing states, the *patience* of a set of memory states M is defined as:

$$\text{pat}(M) = \max\left\{\frac{1}{\sigma_1^a(m)(a_1)}, \frac{1}{\sigma_1^u(a_1, a_2, m)(m')}, \ m, m' \in M, a_1 \in A_1, a_2 \in A_2\right\}.$$

Theorem 10. *Let $\delta, \varepsilon > 0$ be reals and $f : \mathbb{N} \to \mathbb{N}$ be an unbounded non-decreasing function such that $f(T) \leq \frac{1}{4}\log_{1/\varepsilon} T$ for all large enough T. If a strategy σ_1 of Player 1 in the Big Match uses space $\log f(T)$ before time T with probability at least $1 - \delta$, and the patience of the set of lexicographically first $f(T)$ memory states is at most $T^{1/(2f(T))}$ for all T large enough, then there is a strategy σ_2 of Player 2 such that $u_{\sup}(\sigma_1, \sigma_2) \leq \delta + 2\varepsilon$.*

Acknowledgements. The research leading to these results has received funding from the European Research Council under the European Union's Seventh Framework Programme (FP/2007-2013) / ERC Grant Agreement n. 616787. The first author acknowledges support from the Danish National Research Foundation and The National Science Foundation of China (under the grant 61361136003) for the Sino-Danish Center for the Theory of Interactive Computation and from the Center for Research in Foundations of Electronic Markets (CFEM), supported by the Danish Strategic Research Council. The second author was partly supported by Austrian Science Fund (FWF) NFN Grant No S11407-N23 (RiSE/SHiNE), Vienna Science and Technology Fund (WWTF) through project ICT15-003, and ERC Start grant (279307: Graph Games). The third author was supported in part by grant from Neuron Fund for Support of Science, and by the Center of Excellence CE-ITI under the grant P202/12/G061 of GA ČR.

References

1. Aumann, R.J.: Survey of repeated games. In: Bohm, V. (ed.) Essays in Game Theory and Mathematical Economics in Honor of Oskar Morgenstern, Gesellschaft, Recht, Wirtschaft, vol. 4, pp. 11–42. Bibliographisches Institut, Mannheim (1981)
2. Blackwell, D., Ferguson, T.S.: The big match. Ann. Math. Stat. **39**(1), 159–163 (1968)
3. Flajolet, P.: Approximate counting: a detailed analysis. BIT **25**(1), 113–134 (1985)
4. Gillette, D.: Stochastic games with zero stop probabilities. Contrib. Theor. Games III, Ann. Math. Stud. **39**, 179–187 (1957)

5. Hansen, K., Koucký, M., Miltersen, P.: Winning concurrent reachability games requires doubly exponential patience. In: Proceedings of LICS, pp. 332–341 (2009)
6. Hansen, K.A., Ibsen-Jensen, R., Koucký, M.: The big match in small space (full version) (2016). CoRR abs/1604.07634
7. Ibsen-Jensen, R.: Strategy complexity of two-player, zero-sum games. Ph.D. thesis, Aarhus University (2013)
8. Kalai, E.: Bounded rationality and strategic complexity in repeated games. In: Game Theory and Applications, pp. 131–157. Academic Press (1990)
9. Kohlberg, E.: Repeated games with absorbing states. Ann. Statist. 2(4), 724–738 (1974)
10. Morris, R.: Counting large numbers of events in small registers. Commun. ACM 21(10), 840–842 (1978)
11. Neyman, A.: Personal communication (2015)
12. Shapley, L.: Stochastic games. Proc. Natl. Acad. Sci. U. S. A. 39, 1095–1100 (1953)
13. Sorin, S.: Repeated games with complete information. In: Handbook of Game Theory with Economic Applications, vol. 1, chap. 4, pp. 71–107. Elsevier, 1edn. (1992)
14. Sorin, S.: A First Course on Zero Sum Repeated Games. Springer, Heidelberg (2002)

History-Independent Distributed Multi-agent Learning

Amos Fiat[1], Yishay Mansour[1(✉)], and Mariano Schain[2]

[1] Tel Aviv University, Tel Aviv, Israel
mansour.yishay@gmail.com
[2] Google, Mountain View, USA

"Rumor is not always wrong"
De vita et moribus Iulii Agricolae
— *Publius Cornelius TACITUS (56 - 117)*

Abstract. How should we evaluate a rumor? We address this question in a setting where multiple agents seek an estimate of the probability, b, of some future binary event. A common uniform prior on b is assumed. A rumor about b meanders through the network, evolving over time. The rumor evolves, not because of ill will or noise, but because agents incorporate private signals about b before passing on the (modified) rumor. The loss to an agent is the (realized) square error of her opinion.

Our setting introduces strategic behavior based on evidence regarding an exogenous event to current models of rumor/influence propagation in social networks.

We study a simple Exponential Moving Average (EMA) for combining experience evidence and trusted advice (rumor), quantifying its resulting performance and comparing it to the optimal achievable using Bayes posterior having access to the agents private signals.

We study the quality of p_T, the prediction of the last agent along a chain of T rumor-mongering agents. The prediction p_T can be viewed as an aggregate estimator of b that depends on the private signals of T agents. We show that

- When agents know their position in the rumor-mongering sequence, the expected mean square error of the aggregate estimator is $\Theta(\frac{1}{T})$. Moreover, with probability $1-\delta$, the aggregate estimator's deviation from b is $\Theta\left(\sqrt{\frac{\ln(1/\delta)}{T}}\right)$.

- If the position information is not available, and agents act strategically, the aggregate estimator has a mean square error of $O(\frac{1}{\sqrt{T}})$. Furthermore, with probability $1 - \delta$, the aggregate estimator's deviation from b is $\tilde{O}\left(\sqrt{\frac{\ln(1/\delta)}{\sqrt{T}}}\right)$.

1 Introduction

According to McKinsey, word-of-mouth is the primary factor behind up to 50 % of purchasing decisions in developing markets, with 'experimental' (that is, based on personal experience) advice being most common and powerful.

© Springer-Verlag Berlin Heidelberg 2016
M. Gairing and R. Savani (Eds.): SAGT 2016, LNCS 9928, pp. 77–89, 2016.
DOI: 10.1007/978-3-662-53354-3_7

The question we address herein is how does one balance word of mouth "rumors" with other, private, sources of information. At issue is the providence of the rumor, what is the impact of not knowing the history of a rumor?

Consider predictions on the probability, b, that some newly purchased product will malfunction out of the box. Some acquaintance, Bob, tells Alice his own opinion on the *probability b*. Moreover, Alice may experience the product herself, forming a private signal that is one with probability b and zero otherwise. Alice now forms her own opinion on b by merging Bob's opinion with her private signal. Subsequently, she may reveal her opinion to other acquaintance[s], and so on. The loss (of face) for giving an erroneous opinion is the realized mean squared error.

The main issue in this paper is that it is important for Alice to know just how much weight to give to Bob's opinion, versus how much to trust her own individual signal. Clearly, if Alice believes that Bob's opinion is well founded then Alice should give little weight to her own individual signal. Contra-wise, if Alice believes that Bob's opinion has little statistical evidence behind it (e.g., is only based upon Bob's private signal) then Alice should give her own private signal much greater weight. Critically, Alice is uncertain regarding how many agents have influenced Bob's opinion.

Our solution concept is that of strategic *individually optimal* agents, in a symmetric Nash Equilibrium, where agents seek to minimize the mean squared error of their prediction. We give a variety of results in this context. In particular, we compare how well the predictions produced by individually optimal agents compare to the predictions that would have been produced if more information were available to the agents. That is, how much better could Alice predict if the actual signals of those agents along the rumor-mongering sequence, up to and including Bob, were revealed? How much better would the prediction be if Alice knew the length of the rumor-mongering chain of opinions that culminated with Bob's opinion? In fact, it turns out that knowing the actual signals or knowing the length of the preceding chain gives the same precision (and is optimal). However, not knowing the length of the preceding chain gives weaker estimates, where the error made by Alice grows from $\frac{1}{T}$ to $\frac{1}{\sqrt{T}}$ where T is the true length of the whole chain. We also consider confidence/precision measures, see Fig. 1 which explains how not knowing the providence of the rumor impacts the quality of the learning process.

1.1   Our Model

We study the following scenario. Some future event is to occur with (unknown) probability, and the individual agents share a common prior regarding this probability. The agents get independent binary signals correlated with the future event (specifically, they have the same occurrence probability as the event). In addition to their private signal, each agent observes the prediction of the preceding agent. Agents arrive in a random order and predict exactly once (unaware of their position in the sequence). The agents use a simple strategy that combines their private signal and the prediction of the previous agent. (The first agent observes a "dummy" prediction.) Finally, the loss of the agents, determined upon the realization of the event, is a quadratic loss. The agents either try to minimize the loss selfishly or cooperatively. One of the main goals of the work is to understand the difference in the outcomes between the selfish and the cooperative behavior of the agents.

To be able to analyze the dynamics in our setting we need to designate the class of simple strategies that the agents utilize. We focus on linear combination of the private signal and the observed prediction of the previous agent, namely, exponential moving averages. This has the effect that more recent signals (and, in particular, the agent's private signal) are given more weight than older signals whose effect decays with time. Also, no access to the past updates' history (or its length) is needed to implement the exponentially moving average update, making it an appropriate strategy class for this task.

By the inherent symmetry of our setting, all the agents use the same prediction update rule strategy. The prediction update rule may depend on the total number of agents T, and has a decay parameter γ which governs the exponential weighting. We consider two cases. First, the *cooperative socially optimal* update, aiming at minimal expected sum of losses. This benchmark is important to see the loss due to the uncertainty regarding the agents position in the sequence, on the one hand, and the loss due to strategic behavior, on the other hand. Second, the *individually optimal* update, where each selfish agent predicts in order to minimize its own loss, ignoring the effect it will have on future agents and their loss. When doing the comparison, we consider the last prediction made as an estimator for the unknown bias, and our goal is to study the quadratic loss between the last prediction and the true unknown bias.

1.2 Our Results

Our main results are summarized in Fig. 1. We considered three measures of quality for a predictor. The worst case mean square error for any choice of $b \in [0, 1]$ (Column 2), the expected mean square error where b is uniformly distributed in $[0, 1]$ before the signals are generated (Column 3), and the guaranteed accuracy for a given level of confidence (Column 4). The rows of Fig. 1 are for Bayes updates (the equivalent to agents knowing the complete history, namely, the count and values of previous updates performed) and values of γ for exponential moving averages in two settings: Selfish agents (where $\gamma = \frac{1}{\sqrt{2T}}$ is a symmetric equilibrium), and for cooperative agents $\gamma = \frac{\ln T}{2T}$ minimizes the mean square error. We also give the general form of the estimators performance metrics for any value of γ (last row).

Now, comparing Row 1 (Bayes estimator) of Fig. 1 with Row 3 (optimal choice of γ for cooperative agents) quantifies the agents loss due to history independence and use of exponentially moving averaging. Their loss cannot be lower than that obtained by the Bayes estimator, and they are close to this upper bound. I.e., the Bayes estimator has a loss of about $O(\frac{1}{T})$ and the exponential weight moving average has a loss of $O(\frac{\ln T}{T})$.

Update process	Worst case Mean Square Error: $\max_b \text{MSE}(b, P_T)$	Expected Mean Square Error: $E_B \text{MSE}(b, P_T)$ B uniform in $[0,1]$	Guaranteed accuracy for confidence $1-\delta$: $\epsilon(\delta) = \arg\min_\epsilon \max_b$ $\text{Prob}(|P_T - b| < \epsilon) \geq 1-\delta$
Bayes (Known History)	$\frac{1}{4(T+2)}$	$\frac{1}{6(T+2)}$	$\frac{1}{T+2} + \sqrt{\frac{\ln\frac{1}{\delta}}{2T}}$
Individually Optimal $\gamma = \frac{1}{\sqrt{2T}}$	$\frac{1}{4\sqrt{2T}} + e^{-\sqrt{2T}}$	$\frac{1}{12\sqrt{2T}-6} + O\left(e^{-\sqrt{2T}}\right)$	$O\left(\sqrt{\frac{\log(T)\log(1/\delta)}{\sqrt{T}}}\right)$
Socially Optimal $\gamma = \frac{\ln T}{2T}$	$\frac{\ln T}{8T} + \frac{1}{T}$	$\frac{\ln T}{24T} - \frac{1}{12T} + O\left(\frac{1}{T^2}\right)$	$O\left(\sqrt{\frac{\log(T)\log(1/\delta)}{T}}\right)$
Arbitrary γ	$\frac{\gamma}{4} + (1-\gamma)^{2T}$	$\frac{\gamma}{6(2-\gamma)} + (1-\gamma)^{2T}\frac{(2-3\gamma)}{12(2-\gamma)}$	$\sqrt{\frac{\gamma}{2}\ln\frac{1}{\delta}} + (1-\gamma)^T$

Fig. 1. Performance metrics for Exponential Moving Average (EMA) versus Bayes optimal benchmark for estimating (P_T) the unknown bias b of a coin. T is the total number of updates and γ is the averaging constant of EMA. The mean squared error (MSE) in the second column is over realizations of the signals $V_1, \ldots, V_T$. The expectation in the third column is assuming a uniform prior of $b \in [0,1]$ and initial prediction $P_0 = \frac{1}{2}$.

Also, comparing Row 2 (Individually Optimal) of Fig. 1 with Row 1 (Bayes estimator) shows that the mean square error increases by a $\Theta(\sqrt{T})$ factor and that the error probability for a given confidence increases by a $\Theta(T^{-\frac{1}{4}})$ factor. Note that the mean square error still vanishes at a polynomial rate (in T), and that, for any constant accuracy, the error probability remains exponentially small.

1.3 Related Models

Our setting can be cast in the model of partial information presented by [2].[1] The issues studied in [2,8,9] are how communication leads agents to revise their posteriors until they converge, given that the agents have common priors. Other work discussing aspects of information aggregation among agents having private information differ by the nature of the information to get aggregated.

The study of information aggregation by word-of-mouth information flow has been previously studied in various settings, e.g., [1,3,4,7] and many others. The settings considered are quite different than ours, and, notably, cascading effects are observed in several of these models. Cascades (see, e.g., [6]) do not develop in our setting because opinions are continuous and not discrete, one can always modify one opinion sufficiently slightly.

---

[1] In a somewhat non-standard use of the Aumann's model, because there are aspects of the state of the world that are not interesting in and of themselves, whereas in our setting agents are only interested in the underlying probability of the event occurring.

There is a strong connection between minimizing the expected mean error and the use of the quadratic scoring rule as the basis for a market making mechanism in information markets. The convergence of such markets to the true answer has been studied [10], however — this assumes that the history of trades is public knowledge, in contrast we assume that the agents are unaware of the history and even unaware of their position.

2 Model and Preliminaries

A social learning process is established to learn the unknown probability of some future event $F \in \{0, 1\}$. The unknown event F is governed by a Bernoulli random variable with a bias b and there is a uniform prior over it, i.e., $b \sim B$ where B is uniformly distributed in $[0, 1]$. The social learning process has T agents, each agent $t \in [1, T]$ receives a private signal $v_t \sim V_t$, where V_t is a Bernoulli random variable with bias b (the same bias as F) and the r.v. V_t are i.i.d. Each agent t outputs an estimate p_t, which is based on its realized private signal v_t and the estimate of the preceding agent p_{t-1}. (Agent 1 observes $p_0 = \frac{1}{2}$.) Once the event is realized, each agent suffers a quadratic loss, i.e., agent t has a loss $\ell_t(F, p_t) = (F - p_t)^2$.

We denote a series of T i.i.d. such random variables by $\boldsymbol{V}_{[1,T]} = (V_1, \ldots, V_T)$ and their respective realizations by $\boldsymbol{v}_{[1,T]} = (v_1, \ldots, v_T)$. We denote by $P_t(\boldsymbol{V})$ the random variable giving the distribution of the prediction of agent t, i.e., $p_t \sim P_t$. P_t depends on $\boldsymbol{V}_{[1,t]}$ and how agents $1, \ldots, t-1$ compute their prediction. Now, each of the T agents knows T but not its own position $t \in \{1, \ldots, T\}$.

We view the final posted estimate p_T, an aggregation of the T agent's private signals, as an estimator for b, the unknown probability of F occurring.

2.1 Estimation of the Unknown Bias

Given $\boldsymbol{v}_{[1,T]}$, an *estimator* $\theta(\cdot) : \{0,1\}^T \to [0,1]$ for the unknown bias b is a function that maps $\boldsymbol{v}_{[1,T]}$ to some estimated bias in $[0, 1]$. Two such estimators are presented next. The **Bayes estimator** $\widehat{\theta}(\cdot)$ for b is

$$\widehat{\theta}(\boldsymbol{v}_{[1,T]}) = \frac{\sum_{t=1}^{T} v_t + 1}{T + 2}. \tag{1}$$

Equivalently, the Bayes estimator can be computed iteratively as follows for $t = 1, \ldots, T$.[2]

$$\widehat{\theta}(\boldsymbol{v}_{[1,t]}) = \left(1 - \frac{1}{t+2}\right)\widehat{\theta}(\boldsymbol{v}_{[t-1]}) + \frac{1}{t+2}v_t. \tag{2}$$

The **Exponential Moving Average (EMA)** estimator $\theta_\gamma(\cdot)$ is parameterized by a predefined constant γ:

$$\theta_\gamma(\boldsymbol{v}_{[1,T]}) = (1 - \gamma)^T \theta_\gamma(\emptyset) + \sum_{t=1}^{T}(1 - \gamma)^{T-t}\gamma v_t, \tag{3}$$

---

[2] Note that $\widehat{\theta}(\emptyset) = \frac{1}{2}$ which is consistent with $B \sim U[0,1]$.

where $\theta_\gamma(\emptyset) = \frac{1}{2}$. The interpretation of γ as the importance associated to recent signals is evident by the EMA's equivalent iterative form

$$\theta_\gamma(v_{[1,t]}) = (1 - \gamma)\theta_\gamma(v_{[1,t-1]}) + \gamma v_t. \tag{4}$$

Note the resemblance of $\theta_\gamma(\cdot)$ to the iterative form of the Bayes estimator $\hat{\theta}(\cdot)$. The difference being that $\frac{1}{t+2}$ in (2) is replaced with a fixed γ in (4).

Estimator Performance Metrics. A common metric for assessing the performance of an estimator $\hat{\theta}(\cdot) : X \to \Theta$ is the *Mean Squared Error* (MSE):

$$\text{MSE}(\theta, \hat{\theta}) \triangleq E_{\hat{\theta}}[(\theta - \hat{\theta})^2]. \tag{5}$$

Note that θ above is fixed, and the expectation is over realizations of the random variable $\hat{\theta}$ that depends on θ through the observations X. The *Bayes Risk* of an estimator is its expected risk (over a prior distribution of θ).

We therefore consider in our analysis the following three properties of an estimator $\hat{\theta}(\cdot)$

- $\max_\theta \text{MSE}(\theta, \hat{\theta})$: Quantifying the worst case MSE, over all possible values of θ.
- $\max_\theta \Pr(|\theta - \hat{\theta}| > \epsilon)$: The worst case confidence $\delta(\epsilon)$ of a desired ϵ-accuracy.
- $E[\text{MSE}(\theta, \hat{\theta})]$: This is the *Bayes Risk*, the expected MSE over a prior probability distribution for θ.

Bayes Optimal Estimation. The *Bayes Optimal* estimator given observation X is

$$\hat{\theta}_{\text{Bayes}}(X) = E[\theta|X], \tag{6}$$

that is, the expected posterior, minimizes the Bayes Risk. [3] Finally, considering our unknown-bias estimation setting and assuming a uniform prior $b \sim U[0,1]$, the *Bayes Estimator* (Eq. (1)) satisfies $\hat{\theta}(v_{[1,T]}) = E[b|v_{[1,T]}]$ and is therefore *Bayes Optimal*, justifying its use as our key benchmark in assessing the performance of EMA estimators $\theta_\gamma(\cdot)$ for $\gamma \in [0,1]$.

2.2 History-Independent Distributed Learning

Recall that each of the T agents knows T but not its position $t \in \{1, \ldots, T\}$. If agents knew their position in the sequence, and all agents were rational, then they could update the estimate for b using the iterative Bayes update of Eq. (2), since this would minimize the quadratic loss. We consider the alternative setting where agents are history independent, and that the order of updates is a random permutation of the agents. In this case, agents cannot update the estimate p_t for b

---

[3] This actually holds (see [5]) also for a more general definition of Bayes Risk, where a *Bregman loss* is used to generalize the MSE (5).

using the Bayes estimator — the tth agent does not know how many updates have been done, i.e., $t-1$, and therefore can't use Eq. (2). For reasons of symmetry, in this history independent setting, the update strategies of all agents should be identical. It is thus natural to consider exponential moving average updates (4).

In what follows, we consider the limited strategy space using EMA with $\gamma \in [0,1]$ available to the agents in the history-independent setting, and analyze the individually optimal strategy (in equilibrium) and the socially optimal strategy. We then compare the performance of the corresponding estimator of each strategy to that of the Bayes optimal estimator.

3 Strategies: Socially Optimal, Individually Optimal

Socially Optimal Strategy. The socially optimal strategy is the value $\gamma^*_{max} \in [0,1]$ for which the score of the resulting computation is maximal:

$$\gamma^*_{max} \triangleq \arg\min_{\gamma \in [0,1]} E[(F - P_T)^2].$$

Note that the expectation is over the realization of F and $P_T = \theta_\gamma(V_{[1,T]})$.

Theorem 1. *For the EMA estimator $P_T = \theta_\gamma(V_{[1,T]})$,*

$$\gamma^*_{max} = \frac{\ln T}{2T} + \phi, \quad \text{where } |\phi| \leq \frac{2}{T}.$$

Proof (Sketch). We first note that for the quadratic loss

$$\arg\min_{\gamma \in [0,1]} E[(F - P_T)^2] = \arg\max_{\gamma \in [0,1]} E[2bP_T - P_T^2].$$

Next we derive the following closed form for the above optimization target:

$$E[2bP_T - P_T^2] = \frac{4 - 3\gamma}{6(2-\gamma)} - (1-\gamma)^{2T}\left(P_0^2 - P_0 + \frac{4-3\gamma}{6(2-\gamma)}\right),$$

and show that the unique optimum is as required. □

It is worthwhile to compare the resulting γ^*_{max}, minimizing agent loss using the EMA estimator with that of the Bayesian estimator. In the Bayesian estimator the agent is aware of the history and knows his location in the permutation, and when his location is t he updates using $\gamma_{\text{Bayes}} = \frac{1}{t+2}$. If we average over all the locations we have that the average update magnitude is $\frac{1}{T}\sum_{t=1}^{T}\frac{1}{t+2} \approx \frac{\ln T}{T}$. Note that this is only a factor of 2 larger than the resulting update minimizing total agents' loss using EMA.

Individually Optimal Strategy. We now consider the social learning process with agents using the EMA estimator, $\theta_\gamma(V)$, where agents are strategic. I.e., we seek a value of γ such that, for all t, given that agents $1, \ldots, t-1$ compute their prediction using EMA with parameter γ, then it is a best response for agent t to do likewise. Such a choice of γ gives a symmetric equilibrium.

To find such a value of γ, let $\lambda(\gamma)$ be the best response of an agent, assuming that all other agents use update factor γ. (For brevity we will use λ, when clear from the context.) An agent arriving at time t will update the outstanding P_{t-1} as follows

$$P_t(\gamma, \lambda) = (1 - \lambda)P_{t-1}(\gamma) + \lambda V_t,$$

where $P_{t-1}(\gamma) = P_{t-1}(\gamma, \gamma)$ assumes that the first $t-1$ agents update using θ_γ. Since the agent does not know her location, her expected loss is,

$$\ell(\gamma, \lambda) = \frac{1}{T} \sum_{t=1}^{T} E_{B, V}[(F - P_t(\gamma, \lambda))^2].$$

Therefore, an agent minimizes her expected loss, given that all other agents update using γ, by choosing the best response $\lambda^*(\gamma) = \arg\min_\lambda \ell(\gamma, \lambda)$ and in equilibrium $\lambda^*(\gamma) = \gamma$.

An update parameter that achieves equilibrium is denoted γ_{eq}^*, i.e., $\lambda^*(\gamma_{eq}^*) = \gamma_{eq}^*$. Note that we are assuming that the total number of agents T is known by the agents. Hence the utilities and updates defined above may all depend on T (e.g., $\gamma_{eq}^*(T)$) which is omitted from the notations for clarity when not needed explicitly. In Sect. 5.1 we discuss the extension to the case where the agents have only a prior distribution over the number of agents.

Theorem 2. *For the EMA estimator* $P_T = \theta_\gamma(V_{[1,T]})$, *then,*

$$\gamma_{eq}^* = \sqrt{\frac{1}{2T}} - \beta, \quad and\, \beta \in [0, \frac{6}{T}],$$

is the unique symmetric equilibrium.

Proof (Sketch). We first derive a closed form for the best response

$$\lambda^*(\gamma) = \frac{\gamma^2 T + (1 - \gamma)^2(1 - (1 - \gamma)^{2T})a(\gamma)}{2\gamma T + (1 - \gamma)^2(1 - (1 - \gamma)^{2T})a(\gamma)}$$

where $a(\gamma) = \frac{1}{2}(1 - \frac{2\gamma}{2 - \gamma})$, and then prove that γ_{eq}^* as stated in the theorem is the unique solution to $\lambda^*(\gamma) = \gamma$. $\square$

4   Estimators Performance

In this section we study the basic performance measures for the Bayes estimator and EMA as a function of T, the total number of agents. We first compute the worse case MSE and high probability deviation for the Bayes estimator and the

EMA estimator (for a general γ). We then compute the performance for the specific γ value of the symmetric equilibrium and compare to the performance of the optimal (Bayes) estimator. The main goal is to show that the loss due to the restriction of the agents to use EMA is rather minimal (assuming a non-strategic behavior of the agents) and to quantify the effect of strategic behavior on the resulting estimator's performance.

We will need the following lemma, a bound for the worse case probability (over possible values of b, the expected value of each of the signals $\{V_t\}_{t=1}^T$) of ϵ-deviation of an estimator $P_T = \widehat{\theta}(V)$ from b.

Lemma 1. *For any estimator $P_T = \widehat{\theta}(V)$ and any $b \in [0, 1]$, define*

$$\beta \triangleq \max_b |E(P_T) - b| \ , then$$

$$\max_b Pr(|P_T - b| > \epsilon) \leq \max_b Pr(|P_T - E(P_T)| > \epsilon - \beta).$$

We start with the analysis of the Bayes estimator.

Theorem 3. *For the Bayes estimator $P_T = \widehat{\theta}(V)$,*
(1) $\forall b \in [0, 1], \mathrm{MSE}(b, P_T) \leq \frac{1}{4(T+2)}$,

(2) $\forall b \in [0, 1],,$ *With probability at least $1 - \delta$ we have $|P_T - b| \leq \frac{1}{T+2} + \sqrt{\frac{\ln \frac{1}{\delta}}{2T}}$,*
or equivalently, $Pr(|P_T - b| > \epsilon) \leq \exp(-2(\epsilon - \frac{1}{T+2})^2 T)$, and
(3) $E_{b \sim B}[\mathrm{MSE}(b, P_T)] = \frac{1}{6(T+2)}$.

Proof (Sketch). The first claim (1) follows by a direct computation. For the high probability bound (2), we first use McDiarmid's inequality as follows. Recall that

$$P_T = \widehat{\theta}(V) = \frac{1}{T+2} + \sum_{t=1}^T \frac{1}{T+2} V_t \ .$$

This implies that the influence of V_t on P_T is abounded by $c_t = \frac{1}{T+2}$. Therefore, $\sum_t c_t^2 = \frac{T}{(T+2)^2}$ and we get:

$$Pr(|P_T - E(P_T)| > \epsilon) \leq e^{\frac{-2\epsilon^2}{\sum_t c_t^2}} = e^{\frac{-2\epsilon^2(T+2)^2}{T}} \leq e^{-2\epsilon^2 T}$$

Plugging into Lemma 1, we get

$$\max_b Pr(|P_T - b| > \epsilon) \ \leq \max_b Pr(|P_T - E(P_T)| > \epsilon - \frac{1}{T+2}) \leq e^{-2(\epsilon - \frac{1}{T+2})^2 T}$$

As required. The equivalent formulation is achieved by setting $\delta = e^{-2(\epsilon - \frac{1}{T+2})^2 T}$ and solving for ϵ. For the third claim (3), we first show by a direct computation that for the Bayes estimator $E[2bP_T - P_T^2] = \frac{1}{3} - \frac{1}{6(T+2)}$, where the averaging is over the realization of P_T and also over $b \sim U[0, 1]$. $\qquad\square$

Similar techniques are used to establish bounds for the performance of the Exponential Moving Average (EMA) as a function of the parameter γ and the number of agents T.

Theorem 4. *For the EMA estimator $P_T = \theta_\gamma(\boldsymbol{V})$,*

(1) $\forall b \in [0,1]$, $MSE(b, P_T) \leq \frac{\gamma}{4} + (1-\gamma)^{2T}$,

(2) $\forall b \in [0,1]$, *with probability at least $1 - \delta$, we have $|P_T - b| \leq (1-\gamma)^T + \sqrt{\frac{\gamma}{2}\ln\frac{1}{\delta}}$, or equivalently, $Pr(|P_T - b| > \epsilon) \leq \exp(-2\frac{(\epsilon - (1-\gamma)^T)^2}{\gamma})$, and*

(3) $E_{b \sim B}[MSE(b, P_T)] = \frac{\gamma}{6(2-\gamma)} + (1-\gamma)^{2T}\frac{(2-3\gamma)}{12(2-\gamma)}$.

Now, based on Theorem 4, we derive the following performance of the EMA estimator $\theta_{\gamma^*_{\max}}(\boldsymbol{V})$:

Corollary 1. *For the EMA estimator $P_T = \theta_{\gamma^*_{\max}}(\boldsymbol{V})$,*

(1) $\forall b \in [0,1]$, $MSE(b, P_T) = O(\frac{\ln T}{T})$,

(2) $\forall b \in [0,1]$, *with probability at least $1 - \delta$, we have $|P_T - b| = O(\sqrt{\log(T)\log(1/\delta)/T})$*

(3) $E_{b \sim B}[MSE(b, P_T)] = \frac{\ln T}{24T} - \frac{1}{12T} + O(\frac{1}{T^2})$.

Proof (Sketch). Note that $(1 - \gamma^*_{\max})^{2T} \approx \frac{1}{T}$, and that for $\gamma = \frac{\ln T}{2T}$ we have $(1-\gamma)^{2T} = \frac{1}{T}$. Also, for $\gamma = \frac{1}{\sqrt{2T}}$ we have $(1-\gamma)^{2T} = e^{-\sqrt{T}}$. Finally, the term $\frac{\gamma}{6(2-\gamma)}$ approaches $\frac{\gamma}{12}$ and the term $\frac{2-3\gamma}{12(2-\gamma)}$ approaches $\frac{1}{12}$. Plugging the above in Theorem 4 yields the corollary. $\square$

We can contrast the bounds with those of the Bayes estimator $P_T = \widehat{\theta}(\boldsymbol{V})$. The MSE bound increased by a logarithmic factor $O(\ln T)$ (from $O(\frac{1}{T})$ to $O(\frac{\ln T}{T})$) and the high probability bound increases only by a factor of $O(\sqrt{\log T})$. This logarithmic increases show that the impact of restricting the updates to EMA is rather limited.

We now revisit the EMA estimator $P_T = \theta_\gamma(\boldsymbol{V})$ performance for the equilibrium update $\gamma^*_{eq} = \frac{1}{\sqrt{2T}}$. As in the case for non-strategic agents, note that $(1 - \gamma^*_{eq})^T \approx e^{-\sqrt{T/2}}$, and calculating similarly to the proof of Corollary 1, we derive the following corollary of Theorem 4 for the case $P_T = \theta_{\gamma^*_{eq}}(\boldsymbol{V})$.

Corollary 2. *For the EMA estimator $P_T = \theta_{\gamma^*_{eq}}(\boldsymbol{V})$,*

(1) $\forall b \in [0,1]$, $MSE(b, P_T) = O(\frac{1}{\sqrt{T}})$,

(2) $\forall b \in [0,1]$, *with probability at least $1 - \delta$, we have $|P_T - b| = O(T^{-1/4}\sqrt{\log(T)\log(1/\delta)})$*

(3) $E_{b \sim B}[MSE(b, P_T)] = \frac{1}{12\sqrt{2T}-6} + O(e^{-\sqrt{2T}})$.

Comparing the bounds above with those of the Bayes estimator $\widehat{\theta}(\boldsymbol{V})$ and with the exponential moving average $\theta_{\gamma^*_{\max}}(\boldsymbol{V})$. Both $\widehat{\theta}(\boldsymbol{V})$ and $\theta_{\gamma^*_{\max}}(\boldsymbol{V})$ achieve a mean square error of $\tilde{O}(\frac{1}{T})$ vs. $O(\frac{1}{\sqrt{T}})$ for the symmetric equilibrium. For the high probability bound (2) the gap is between $\tilde{O}(\frac{1}{\sqrt{T}})$ and $\tilde{O}(\frac{1}{T^{\frac{1}{4}}})$.

This is both good news and bad news. The good news is the process converges to the true probabilities even when agents are unaware of the trading history (and use EMA updates). The bad news is that the convergence rate deteriorates due to selfish strategic behavior.

5   Extensions

5.1   Distribution over the Number of Agents

An interesting extension is to assume further uncertainty, where even the total number of agents, T, is unknown. It may be unrealistic to forecast the number of agents. A more reasonable assumption may be a common prior over the number of agents. The obvious question is how this additional uncertainty impacts our results. We therefore compute the symmetric equilibrium in this setting.

Theorem 5. *For strategic agents that know neither their position in line, nor the total number of agents, but share a prior on the total number of agents with* $E[\frac{1}{T}] \leq \frac{1}{8}$, *the equilibrium update is* $\gamma_{dist}^* = \Theta(\sqrt{E(\frac{1}{T})})$.

Note that the resulting equilibrium update parameters is $\Theta(\sqrt{E(\frac{1}{T})})$, which is different from $\Theta(E(\frac{1}{\sqrt{T}}))$. Conceptually, this is very good news. Recall that the Bayes update would have mean square error equal $\Theta(E(1/T))$. This implies that the EMA equilibrium update γ_{dist}^*, which is only square-root of that quantity, has a mean square error of $\Theta(\sqrt{E(\frac{1}{T})})$, assuming that for $\alpha = \sqrt{E(\frac{1}{T})}$ we have $E[Te^{-\alpha T}] = O(\sqrt{E(\frac{1}{T})})$. This establishes the following corollary to Theorem 4 part (1), for $\gamma_{dist}^* = \Theta(\sqrt{E(\frac{1}{T})})$.

Corollary 3. *If the Bayes estimator MSE is bounded by* ϵ *then for agents in equiliria, the mean square error is at most* $O(\sqrt{\epsilon} + E[Te^{-\epsilon T}])$.

5.2   Single Unaware Agent

Assume all agents do the correct (fully informed, Bayesian) update $\hat{\theta}(\cdot)$, except for one agent which is not aware of the history and his location. Such a setting assesses the penalty of an agent not knowing its location. Alternatively, this measures the maximum price that such an agent would be willing to pay to gain the information, in the extreme case that all other agents know their location. One can view the unaware agent as a late adaptor of a technology that determines an agent's location, and we compute the penalty associated with this late adoption. Technically, this implies that when the unaware agent arrives at the process, the price is set by the Bayesian update $\hat{\theta}(v)$. We now compute the γ that maximizes the agent's score. Let t be the unaware agent, then we have,

$$P_t^\gamma = (1 - \gamma)P_{t-1}^B + \gamma V_t \quad \text{and} \quad P_{t-1}^B = \frac{1 + \sum_{i=1}^{t-1} V_i}{t + 1}.$$

The expected score of the unaware agent, assuming a uniform distribution over his arrival $t \in \{1, \cdots, T\}$, is,

$$\frac{1}{T} \sum_{t=1}^{T} E_{B,V}[(F - P_t^\gamma)^2].$$

The following theorem establishes the optimal update parameter.

Theorem 6. *The optimal update parameter γ_1^* for a single unaware agent model, when $P_0 = \frac{1}{2}$, is*

$$\gamma_1^* = \frac{\ln T}{T + \ln T}.$$

Recall that when none of the agents are informed, the utility maximizing update parameter is $\frac{\ln T}{2T}$ while if all the agents are informed and use Bayes updates then the average update parameter is $\frac{\ln T}{T}$. The update parameter above is a small step from the update of all informed Bayesian agents to all uninformed EMA agents.

5.3   Single Aware Agent

This setting can be seen as the flip-side of the previous setting. Here we consider the case that only a single agent is informed regarding his location. This models the benefit that an agent can gain by being able to access his location. One way of gaining the information is through buying it exclusively, the utility gain bounds the price the agent would be willing to pay for such an information.

Technically, assume a single agent doing the correct (fully informed, Bayesian) update $\hat{\theta}(\cdot)$, and all other agents (not aware of their location) are restricted to use EMA θ_γ strategy and are either unaware or ignore the fact that a single agent is using a different strategy. We define

$$P_{t-1}^\gamma = (1 - \gamma)P_{t-2}^\gamma + \gamma V_{t-1} \quad \text{and} \quad P_t^B = \frac{1 + \sum_{i=1}^{t} V_i}{t + 2}.$$

Now, for an agent using a strategy $P_t(\cdot)$, we define the average expected loss as follows,

$$\ell_T(P_T) \triangleq \frac{1}{T} \sum_{t=1}^{T} E_{B,V}[(F - P_t)^2] = \frac{1}{T} \sum_{t=1}^{T} E_{b \sim B} MSE(b, P_t).$$

We consider the difference between the average expected loss of an unaware agent and the average expected loss of a single aware agent,

$$u_T(\gamma, \text{Bayes}) \triangleq \ell_T(P_T^\gamma) - \ell_T(P_T^B)$$

Theorem 7. *For $\gamma = \frac{\ln T}{2T}$ we have $u_T(\frac{\ln T}{2T}, \text{Bayes}) = \Theta(\frac{\ln^2 T}{T})$ and for $\gamma = \frac{1}{\sqrt{2T}}$ we have $u_T(\frac{1}{\sqrt{2T}}, \text{Bayes}) = \Theta(\frac{1}{\sqrt{T}})$.*

When all agents are symmetric then the loss of an individual agent is $\Theta(\frac{1}{T})$, since the total loss of all the agents is constant. It follows from the theorem above that the loss to a single aware agent is significantly higher. Thus, the value of knowledge is (about) $\frac{1}{\sqrt{T}}$.

6 Closing Remarks and Future Work

Our analysis of history-independent social learning settings assumed quadratic loss. This raises the question regarding the socially optimal and individually optimal updates resulting for other loss functions.

Finally, the choice of strategy space for the agents is usually key in the analysis of equilibrium in game-like scenarios. Future research allowing for strategies beyond the linear updates assumed herein are a natural next research challenge.

References

1. Abernethy, J., Frongillo,R.M.: A collaborative mechanism for crowdsourcing prediction problems (2011). CoRR, abs/1111.2664
2. Aumann, R.J.: Agreeing to disagree. Ann. Stat. **4**(6), 1236–1239 (1976)
3. Bala, V., Goyal, S.: Learning from neighbours. Rev. Econ. Stud. **65**(3), 595–621 (1998)
4. Banerjee, A., Fudenberg, D.: Word-of-mouth learning. Games Econ. Behav. **46**(1), 1–22 (2004)
5. Banerjee, A., Guo, X., Wang, H.: On the optimality of conditional expectation as a Bregman predictor. IEEE Trans. Inf. Theory **51**(7), 2664–2669 (2005)
6. David, E., Jon, K.: Networks, Crowds, and Markets: Reasoning About a Highly Connected World. Cambridge University Press, New York (2010)
7. Ellison, G., Fudenberg, D.: Word-of-mouth communication and social learning. Q. J. Econ. **110**(1), 93–125 (1995)
8. Geanakoplos, J., Polemarchakis, H.M.: We can't disagree forever. Cowles Foundation Discussion Papers 639, Cowles Foundation for Research in Economics, Yale University, July 1982
9. McKelvey, R.D., Page, T.: Common knowledge, consensus, and aggregate information. Econometrica **54**(1), 109–127 (1986)
10. Ostrovsky, M.: Information aggregation in dynamic markets with strategic traders. Research Papers 2053, Stanford University, Graduate School of Business, March 2009

Congestion Games and Networks

On the Robustness of the Approximate Price of Anarchy in Generalized Congestion Games

Vittorio Bilò[✉]

Department of Mathematics and Physics "Ennio De Giorgi",
University of Salento, Provinciale Lecce-Arnesano, P.O. Box 193, 73100 Lecce, Italy
vittorio.bilo@unisalento.it

Abstract. One of the main results shown through Roughgarden's notions of smooth games and robust price of anarchy is that, for any sum-bounded utilitarian social function, the worst-case price of anarchy of coarse correlated equilibria coincides with that of pure Nash equilibria in the class of weighted congestion games with non-negative and non-decreasing latency functions and that such a value can always be derived through the, so called, smoothness argument. We significantly extend this result by proving that, for a variety of (even non-sum-bounded) utilitarian and egalitarian social functions and for a broad generalization of the class of weighted congestion games with non-negative (and possibly decreasing) latency functions, the worst-case price of anarchy of ϵ-approximate coarse correlated equilibria still coincides with that of ϵ-approximate pure Nash equilibria, for any $\epsilon \geq 0$. As a byproduct of our proof, it also follows that such a value can always be determined by making use of the primal-dual method we introduced in a previous work.

1 Introduction

The celebrated notion of robust price of anarchy introduced by Roughgarden in [22] has lately given rise to much interest in the determination of inefficiency bounds for pure Nash equilibria which may automatically extend to some of their appealing generalizations, such as mixed Nash equilibria, correlated equilibria and coarse correlated equilibria. These three types of solutions have a particular flavor since, differently from pure Nash equilibria, they are always guaranteed to exist by Nash's Theorem [19]; moreover, the last two ones can also be efficiently computed and even easily learned when a game is repeatedly played over time.

To this aim, Roughgarden [22] identifies a class of games, called *smooth games*, for which a simple three-line proof, called *smoothness argument*, shows significant upper bounds on the price of anarchy of pure Nash equilibria as long as the social function measuring the quality of any strategy profile in the game is *sum-bounded*, that is, upper bounded by the sum of the players' costs. He then

This work was partially supported by the PRIN 2010–2011 research project ARS TechnoMedia: "Algorithmics for Social Technological Networks" funded by the Italian Ministry of University.

M. Gairing and R. Savani (Eds.): SAGT 2016, LNCS 9928, pp. 93–104, 2016.
DOI: 10.1007/978-3-662-53354-3_8

defines the *robust price of anarchy* of a smooth game as the best-possible (i.e., the lowest) upper bound which can be derived by making use of this argument and provides an *extension theorem* which shows that, still for sum-bounded social functions, the price of anarchy of coarse correlated equilibria of any smooth game is upper bounded by its robust price of anarchy. Finally, he shows that several games considered in the literature happen to be smooth and that the class of (unweighted) congestion games with non-negative and **non-decreasing latency functions** is *tight* for the **utilitarian social function** (that is, the social function defined as the sum of the players' costs), in the sense that, in this class of games, the worst-case price of anarchy of pure Nash equilibria exactly matches the robust price of anarchy. This last result has been subsequently extended to the class of weighted congestion games by Bhawalkar, Gairing and Roughgarden in [4].

Our Contribution and Significance. In this work, we generalize the tightness result by Bhawalkar, Gairing and Roughgarden along the following four directions (see Sect. 2 for formal definitions):

1. the class of games we consider is a broad generalization of that of weighted congestion games. In particular, we focus on *generalized weighted congestion games*, that is, games in which each player's *perceived cost* is defined as a certain linear combination of all the players' *individual costs* originally experienced in some underlying weighted congestion game.
2. the families of social functions we consider are generalizations of both the utilitarian and the egalitarian social functions (where the egalitarian social function is defined as the maximum of the players' costs). In particular, a family of utilitarian social functions is obtained by summing up a certain contribution from each player, whereas a family of egalitarian social functions is obtained by taking the maximum contribution among the players, where each player's contribution is given by a conic combination of the players' individual costs. We stress that such a combination may significantly differ from the one used to define the players' perceived costs, so that there exist social functions in both families that may not be sum-bounded;
3. the latency functions we consider in the definition of the players' individual costs are selected from a family of allowable non-negative functions with no additional restrictions. This permits us to encompass also latency functions not considered so far in the previous tightness results known in the literature, such as, for instance, the widely used fair cost sharing rule induced by the Shapley value [23];
4. the solution concepts we consider are the approximate versions of all the four types of equilibria named so far. In particular, for any real value $\epsilon \geq 0$, we focus on either ϵ-approximate pure Nash equilibria and ϵ-approximate coarse correlated equilibria.

More precisely, but still informally speaking, we prove the following result (Theorem 1 in Sect. 3): *for a variety of utilitarian and egalitarian social functions and for any real value $\epsilon \geq 0$, the worst-case price of anarchy of ϵ-approximate pure Nash equilibria coincides with that of ϵ-approximate coarse* *correlated equilibria in the class generalized weighted congestion games with non-negative latency functions.*

As it can be appreciated, the above tightness result generalizes the previous one by Bhawalkar, Gairing and Roughgarden along all four directions simultaneously. The technique we use to prove the theorem is the primal-dual method that we introduced in [5]. In fact, as a byproduct of our proof, it also follows that, in the above considered scenario of investigation, *the worst-case price of anarchy of ε-approximate pure Nash equilibria can always be determined through the primal-dual method.*

Related Work. The notion of price of anarchy as a measure of the inefficiency caused by selfish behavior in non-cooperative games has been introduced in a seminal paper by Koutsoupias and Papadimitriou [17] in 1999. Since then, several classes of games have been studied under this perspective. Among these classes, congestion games introduced by Rosenthal in [21] and their weighted variants [18] occupy a preeminent role.

Tight bounds for the worst-case price of anarchy of pure Nash equilibria in congestion games with polynomial latency functions under the utilitarian social function have been given by Awerbuch, Azar and Epstein [3], Christodoulou and Koutsoupias [14] and Aland et al. [1] which have been subsequently generalized to approximate pure Nash equilibria by Christodoulou, Koutsoupias and Spirakis [16]. Christodoulou and Koutsoupias [15] show that the worst-case price of anarchy of correlated equilibria is the same as that for pure Nash equilibria in weighted and unweighted congestion games when considering affine latency functions. As already said, such an equivalence has been further extended to coarse correlated equilibria and to any class of non-negative and non-decreasing latency functions by Roughgarden [22] in the unweighted case and by Bhawalkar, Gairing and Roughgarden [4] in the weighted case, by making use of the smoothness argument and the robust price of anarchy.

Robust bounds on the worst-case price of anarchy have been lately achieved via extensions (even to non-sum-bounded social functions) of the smoothness argument in some generalizations of (unweighted) congestion games. In particular, de Anagnostopoulos et al. [2] and Rahn and Schäfer [20] consider the altruistic extension of congestion games in which, similarly to our model of generalized congestion games, the perceived cost of each player is defined as a linear combination of the individual costs of all the players in the game. However, they restrict their analysis to the case in which the social function is the sum of the players' individual costs.

Much less attention has been devoted in the literature to the egalitarian social function, for which Christodoulou and Koutsoupias [14] give an asymptotically tight bound on the worst-case price of anarchy in unweighted congestion games with affine latency functions.

We introduced the primal-dual method in [5] as a tool for obtaining tight bounds on the inefficiencies caused by selfish behavior in weighted congestion games and their possible generalizations for a variety of solutions concepts. Since then, the method has been fruitfully exploited in [6,8–13].

Paper Organization. The paper is organized as follows. In the next section, we give all necessary definitions and notation and provide also some preliminary remarks. Section 3 contains the technical contribution of the paper, with the proof of our main theorem. Due to space limitations, some material has been omitted, see [7] for a complete version.

2   Definitions, Notation and Preliminaries

A *weighted congestion game* is a tuple $\mathsf{CG} = \left([n], (w_i)_{i\in[n]}, E, (\Sigma_i)_{i\in[n]}, (\ell_e)_{e\in E}\right)$ such that $[n] = \{1, 2, \ldots, n\}$ is a set of $n \geq 2$ players, $w_i > 0$ is the *weight* of player i, E is a non-empty set of *resources*, $\Sigma_i \subseteq 2^E \setminus \{\emptyset\}$ is a non-empty *set of strategies* for player i and $\ell_e : \mathbb{R}_{\geq 0} \to \mathbb{R}_{\geq 0}$ is the *latency function* of resource $e \in E$. Denote as $\Sigma = \prod_{i\in[n]} \Sigma_i$ the set of all strategy profiles of CG, that is, the set of outcomes which can be realized when each player $i \in [n]$ chooses a strategy in Σ_i. A *strategy profile* $\boldsymbol{\sigma} = (\sigma_1, \ldots, \sigma_n)$ is then a vector of strategies, where, for each $i \in [n]$, $\sigma_i \in \Sigma_i$ denotes the choice of player i in $\boldsymbol{\sigma}$. For a strategy profile $\boldsymbol{\sigma}$ and a resource $e \in E$, the value $n_e(\boldsymbol{\sigma}) = \sum_{i\in[n]:e\in\sigma_i} w_i$ denotes the *congestion* of resource e in $\boldsymbol{\sigma}$, that is, the sum of the weights of all the players choosing e in $\boldsymbol{\sigma}$. The *individual cost* of player i in $\boldsymbol{\sigma}$ is defined as $c_i(\boldsymbol{\sigma}) = w_i \sum_{e\in\sigma_i} \ell_e(n_e(\boldsymbol{\sigma}))$.

Given a finite space of functions $\mathcal{F} \subseteq \{f : \mathbb{R}_{\geq 0} \to \mathbb{R}_{\geq 0}\}$, let $\mathcal{B}(\mathcal{F}) = \{f_k : \mathbb{R}_{\geq 0} \to \mathbb{R}_{\geq 0} \mid k \in [r]\}$ be a basis for $\mathcal{F}$ of cardinality r, whose elements (functions) are numbered from 1 to r. We say that CG is defined over $\mathcal{F}$ if, for each $e \in E$, it holds that $\ell_e = \sum_{k\in[r]} v_k^e f_k$, where $v_k^e \in \mathbb{R}$ is a scalar. Throughout the paper, we will impose only **minimal assumptions** on $\mathcal{F}$; in particular, we will assume that any $f \in \mathcal{F}$ is non-negative with $f(x) = 0$ if and only if $x = 0$.

For any n-dimensional vector of (positive) weights $\boldsymbol{w} = (w_1, \ldots, w_n)$, we denote with $\mathcal{C}_{\boldsymbol{w}}(\mathcal{F})$ the class of all the weighted congestion games with players' weights induced by $\boldsymbol{w}$ and defined over $\mathcal{F}$. Moreover, for a fixed quadruple $\mathsf{T}_{\boldsymbol{w}} = ([n], \boldsymbol{w}, E, (\Sigma_i)_{i\in[n]})$, called a *congestion model*, the set $\mathcal{C}_{\mathsf{T}_{\boldsymbol{w}}}(\mathcal{F}) = \{\mathsf{CG} \in \mathcal{C}_{\boldsymbol{w}}(\mathcal{F}) \mid \mathsf{CG} = (\mathsf{T}_{\boldsymbol{w}}, (\ell_e)_{e\in E})\}$ is the set of all the weighted congestion games induced by $\mathsf{T}_{\boldsymbol{w}}$ and defined over $\mathcal{F}$. Note that, since for each game $\mathsf{CG} \in \mathcal{C}_{\mathsf{T}_{\boldsymbol{w}}}(\mathcal{F})$ and $e \in E$ there exist r numbers $v_1^e, \ldots, v_r^e$ such that $\ell_e = \sum_{k\in[r]} v_k^e f_k$, it follows that CG can be specified by the pair $(\mathsf{T}_{\boldsymbol{w}}, (v_k^e)_{e\in E, k\in[r]})$. Moreover, it holds that $\mathcal{C}_{\boldsymbol{w}}(\mathcal{F}) = \bigcup_{\mathsf{T}_{\boldsymbol{w}}} \mathcal{C}_{\mathsf{T}_{\boldsymbol{w}}}(\mathcal{F})$. Finally, we denote with $\Sigma(\mathsf{T}_{\boldsymbol{w}})$ the set of strategy profiles induced by the congestion model $\mathsf{T}_{\boldsymbol{w}}$.

A *generalized weighted congestion game* is a pair (CG, α) where $\mathsf{CG} = ([n], (w_i)_{i\in[n]}, E, (\Sigma_i)_{i\in[n]}, (\ell_e)_{e\in E})$ is a weighted congestion game and $\alpha \in \mathbb{R}^{n\times n}$ is an n-dimensional square matrix. Game (CG, α) has the same set of players and strategies of CG, but the *perceived* cost of player i in the strategy profile $\boldsymbol{\sigma}$ is defined as $\widehat{c}_i(\boldsymbol{\sigma}) = \sum_{j\in[n]} \alpha_{ij} c_j(\boldsymbol{\sigma}) = \sum_{e\in E} \sum_{k\in[r]} v_k^e f_k(n_e(\boldsymbol{\sigma})) \sum_{j\in[n]:e\in\sigma_j} \alpha_{ij} w_j$, where $c_i(\boldsymbol{\sigma})$ is the individual cost that player i experiences in $\boldsymbol{\sigma}$ in the underlying weighted congestion game CG.

Note that, when α is the identity matrix, (CG, α) coincides with CG, while, in all the other cases, (CG, α) may not be isomorphic to any weighted congestion game, so that the set of generalized weighted congestion games expands that of weighted congestion games.

Given a strategy profile σ, a player $i \in [n]$ and a strategy $x \in \Sigma_i$, we denote with (σ_{-i}, x) the strategy profile obtained from σ when player i changes her strategy from σ_i to x, while the strategies of all the other players are kept fixed. In particular, for any $\epsilon \geq 0$, the perceived cost suffered by player i in σ minus $1+\epsilon$ times the perceived cost suffered by player i in (σ_{-i}, x) in a generalized weighted congestion game can be expressed as follows (for a strategy profile σ, a player i, a resource e, an index k and a value ϵ, we set $A_e^i(\sigma) = \sum_{j \in [n]: e \in \sigma_j} \alpha_{ij} w_j$, $T_{e,i}^+(\sigma, \epsilon, k) = f_k(n_e(\sigma)) A_e^i(\sigma) - (1+\epsilon) f_k(n_e(\sigma) + w_i)\left(A_e^i(\sigma) + \alpha_{ii} w_i\right)$ and $T_{e,i}^-(\sigma, \epsilon, k) = f_k(n_e(\sigma)) A_e^i(\sigma) - (1+\epsilon) f_k(n_e(\sigma) - w_i)\left(A_e^i(\sigma) - \alpha_{ii} w_i\right)$):

$$
\widehat{c}_i(\sigma) - (1+\epsilon) \cdot \widehat{c}_i(\sigma_{-i}, x)
$$

$$
= \sum_{j \in [n]} \alpha_{ij} c_j(\sigma) - (1+\epsilon) \sum_{j \in [n]} \alpha_{ij} c_j(\sigma_{-i}, x)
$$

$$
= A_e^i(\sigma) \left(\sum_{e \in \sigma_i} \ell_e(n_e(\sigma)) - (1+\epsilon) \sum_{e \in x} \ell_e(n_e(\sigma_{-i}, x)) \right)
$$

$$
= \sum_{e \in \sigma_i \setminus x} \sum_{k \in [r]} v_k^e T_{i,e}^-(\sigma, \epsilon, k)
$$

$$
+ \sum_{e \in x \setminus \sigma_i} \sum_{k \in [r]} v_k^e T_{i,e}^+(\sigma, \epsilon, k) - \epsilon \sum_{e \in \sigma_i \cap x} \sum_{k \in [r]} v_k^e f_k(n_e(\sigma)) A_e^i(\sigma). \tag{1}
$$

Next two definitions formalize the two concepts of approximate equilibria that we will consider throughout the paper.

Definition 1. *For any $\epsilon \geq 0$, an ϵ-approximate coarse correlated equilibrium is a probability distribution p defined over Σ such that, for any player $i \in [n]$ and strategy $x \in \Sigma_i$, it holds that $\sum_{\sigma \in \Sigma} p_\sigma \cdot \widehat{c}_i(\sigma) \leq (1+\epsilon) \sum_{\sigma \in \Sigma} p_\sigma \cdot \widehat{c}_i(\sigma_{-i}, x)$, where, for each $\sigma \in \Sigma$, p_σ is the probability assigned to σ by p.*

Definition 2. *For any $\epsilon \geq 0$, an ϵ-approximate pure Nash equilibrium is a strategy profile σ such that, for any player $i \in [n]$ and strategy $x \in \Sigma_i$, it holds that $\widehat{c}_i(\sigma) \leq (1+\epsilon) \cdot \widehat{c}_i(\sigma_{-i}, x)$.*

Denote as $\mathsf{PNE}_\epsilon(CG, \alpha)$ and $\mathsf{CCE}_\epsilon(CG, \alpha)$, respectively, the set of ϵ-approximate pure Nash equilibria and ϵ-approximate coarse correlated equilibria of the generalized weighted congestion game (CG, α). It is easy to see that, for any $\epsilon \geq 0$, an ϵ-approximate pure Nash equilibrium σ is an ϵ-approximate coarse correlated equilibrium p such that $p_\sigma = 1$ and $p_\tau = 0$ for any $\tau \in \Sigma \setminus \{\sigma\}$. So, $\mathsf{PNE}_\epsilon(CG, \alpha) \subseteq \mathsf{CCE}_\epsilon(CG, \alpha)$. Moreover, the sets $\mathsf{PNE}_0(CG, \alpha)$ and $\mathsf{CCE}_0(CG, \alpha)$ coincide with the sets of pure Nash equilibria and coarse correlated equilibria of (CG, α), respectively.

For an n-dimensional *non-null* square matrix $\beta \in \mathbb{R}_{\geq 0}^{n \times n}$ and a player $i \in [n]$, let $\beta\text{-}cost_i : \Sigma \to \mathbb{R}_{>0}$ be the contribution of player i to the definition of the social function which is defined as follows:

$$\beta\text{-}cost_i(\boldsymbol{\sigma}) = \sum_{j \in [n]} \beta_{ij} c_j(\boldsymbol{\sigma}) = \sum_{e \in E} \sum_{k \in [r]} v_k^e f_k(n_e(\boldsymbol{\sigma})) \sum_{j \in [n]: e \in \sigma_j} \beta_{ij} w_j.$$

Let $\Delta(\Sigma)$ be the set of all the probability distributions defined over Σ. For a $\boldsymbol{p} \in \Delta(\Sigma)$, the β-*utilitarian social function* is a function $\beta\text{-}\mathsf{SUM} : \Delta(\Sigma) \to \mathbb{R}_{>0}$ such that

$$\beta\text{-}\mathsf{SUM}(\boldsymbol{p}) = \sum_{i \in [n]} \mathbb{E}_{\boldsymbol{\sigma} \sim \boldsymbol{p}} \left[\beta\text{-}cost_i(\boldsymbol{\sigma}) \right] = \mathbb{E}_{\boldsymbol{\sigma} \sim \boldsymbol{p}} \left[\sum_{i \in [n]} \beta\text{-}cost_i(\boldsymbol{\sigma}) \right]$$

$$= \sum_{\boldsymbol{\sigma} \in \Sigma} p_{\boldsymbol{\sigma}} \left(\sum_{e \in E} \sum_{k \in [r]} v_k^e f_k(n_e(\boldsymbol{\sigma})) \sum_{i \in [n]} \sum_{j \in [n]: e \in \sigma_j} \beta_{ij} w_j \right)$$

and the β-*egalitarian social function* is a function $\beta\text{-}\mathsf{MAX} : \Delta(\Sigma) \to \mathbb{R}_{>0}$ such that

$$\beta\text{-}\mathsf{MAX}(\boldsymbol{p}) = \max_{i \in [n]} \left\{ \mathbb{E}_{\boldsymbol{\sigma} \sim \boldsymbol{p}} \left[\beta\text{-}cost_i(\boldsymbol{\sigma}) \right] \right\}$$

$$= \max_{i \in [n]} \left\{ \sum_{\boldsymbol{\sigma} \in \Sigma} p_{\boldsymbol{\sigma}} \sum_{e \in E} \sum_{k \in [r]} v_k^e f_k(n_e(\boldsymbol{\sigma})) \sum_{j \in [n]: e \in \sigma_j} \beta_{ij} w_j \right\}.$$

We remark that there is also another possible (and indeed more traditional) definition for the β-egalitarian social function, obtained by setting $\beta\text{-}\mathsf{MAX}(\boldsymbol{p}) = \mathbb{E}_{\boldsymbol{\sigma} \sim \boldsymbol{p}} \left[\max_{i \in [n]} \{ \beta\text{-}cost_i(\boldsymbol{\sigma}) \} \right]$. In such a case, however, the application of the primal-dual method seems to be not so natural, so that the study of this social function remains an interesting open problem at the moment.

Consider the case in which $\boldsymbol{p} \in \Delta(\Sigma)$ is indeed a strategy profile $\boldsymbol{\sigma} \in \Sigma$. When β is the identity matrix, $\beta\text{-}\mathsf{SUM}$ (resp. $\beta\text{-}\mathsf{MAX}$) coincides with the sum (resp. the maximum) of the players' individual costs in the underlying weighted congestion game CG, while, when $\beta = \alpha$, $\beta\text{-}\mathsf{SUM}$ (resp. $\beta\text{-}\mathsf{MAX}$) coincides with the sum (resp. the maximum) of the players' perceived costs in (CG, α). In general, an infinite variety of social functions can be defined by tuning the choice of matrix β. For a function $\mathsf{SF} \in \{\mathsf{SUM}, \mathsf{MAX}\}$, we denote with $\boldsymbol{o}$ the *social optimum*, that is, any strategy profile minimizing $\beta\text{-}\mathsf{SF}$. Note that, by the properties of the latency functions and the definition of β, it follows that $\beta\text{-}\mathsf{SF}(\boldsymbol{o}) > 0$. The ϵ-approximate coarse correlated price of anarchy of (CG, α) under the social function $\beta\text{-}\mathsf{SF}$ is defined as $\mathsf{CCPoA}_\epsilon(\beta\text{-}\mathsf{SF}, \mathsf{CG}, \alpha) = \max_{\boldsymbol{p} \in \mathsf{CCE}_\epsilon(\mathsf{CG}, \alpha)} \frac{\beta\text{-}\mathsf{SF}(\boldsymbol{p})}{\beta\text{-}\mathsf{SF}(\boldsymbol{o})}$, while the ϵ-approximate pure price of anarchy of (CG, α) under the social function $\beta\text{-}\mathsf{SF}$ is defined as $\mathsf{PPoA}_\epsilon(\beta\text{-}\mathsf{SF}, \mathsf{CG}, \alpha) = \max_{\boldsymbol{\sigma} \in \mathsf{PNE}_\epsilon(\mathsf{CG}, \alpha)} \frac{\beta\text{-}\mathsf{SF}(\boldsymbol{\sigma})}{\beta\text{-}\mathsf{SF}(\boldsymbol{o})}$.

For an n-dimensional vector of weights $\boldsymbol{w} = (w_1, \ldots, w_n)$ and a matrix $\alpha \in \mathbb{R}^{n \times n}$, we denote with $\mathcal{C}_{\boldsymbol{w}}(\mathcal{F}, \alpha) = \{(\mathsf{CG}, \alpha) : \mathsf{CG} \in \mathcal{C}_{\boldsymbol{w}}(\mathcal{F})\}$ the set of all the generalized weighted congestion games induced by w and α and defined over $\mathcal{F}$. Similarly, for any congestion model T_w, one defines the class $\mathcal{C}_{\mathsf{T}_w}(\mathcal{F}, \alpha)$, so as to obtain $\mathcal{C}_w(\mathcal{F}, \alpha) = \bigcup_{\mathsf{T}_w} \mathcal{C}_{\mathsf{T}_w}(\mathcal{F}, \alpha)$. The worst-case ϵ-approximate coarse correlated price of anarchy of the class $\mathcal{C}_w(\mathcal{F}, \alpha)$ under the social function β-SF is defined as $\mathsf{CCPoA}_\epsilon(\beta\text{-SF}, \mathcal{C}_w(\mathcal{F}, \alpha)) = \sup_{(\mathsf{CG}, \alpha) \in \mathcal{C}_w(\mathcal{F}, \alpha)} \mathsf{CCPoA}_\epsilon(\beta\text{-SF}, \mathsf{CG}, \alpha)$. Similarly, one defines the worst-case ϵ-approximate pure price of anarchy of the class $\mathcal{C}_w(\mathcal{F}, \alpha)$ under the social function β-SF.

By $\mathsf{PNE}_\epsilon(\mathsf{CG}, \alpha) \subseteq \mathsf{CCE}_\epsilon(\mathsf{CG}, \alpha)$, it follows that $\mathsf{PPoA}_\epsilon(\beta\text{-SF}, \mathcal{C}_w(\mathcal{F}, \alpha)) \leq \mathsf{CCPoA}_\epsilon(\beta\text{-SF}, \mathcal{C}_w(\mathcal{F}, \alpha))$ for any real value $\epsilon \geq 0$, n-dimensional vector of weights w, finite space of function $\mathcal{F}$, pair of matrices $\alpha \in \mathbb{R}^{n \times n}$ and $\beta \in \mathbb{R}^{n \times n}_{\geq 0}$ and function $\mathsf{SF} \in \{\mathsf{SUM}, \mathsf{MAX}\}$. Throughout the paper, we will also refer to the worst-case ϵ-approximate pure price of anarchy and to the worst-case ϵ-approximate coarse correlated price of anarchy of subsets of $\mathcal{C}_w(\mathcal{F}, \alpha)$ which are naturally defined by restriction.

We conclude this section with an easy, although crucial result, stating that, independently of which is the adopted social function, both the worst-case ϵ-approximate pure price of anarchy and the worst-case ϵ-approximate coarse correlated price of anarchy of a class of generalized weighted congestion games remain the same even if one restricts to only those games in the given class whose social optimum has social value equal to one. To this aim, for any function $\mathsf{SF} \in \{\mathsf{SUM}, \mathsf{MAX}\}$ and matrix $\beta \in \mathbb{R}^{n \times n}_{\geq 0}$, let $\overline{\mathcal{C}}_w(\mathcal{F}, \alpha) \subset \mathcal{C}_w(\mathcal{F}, \alpha)$ be the subset of all the generalized weighted congestion games induced by w and α and defined over $\mathcal{F}$ such that the social optimum o satisfies $\beta\text{-SF}(o) = 1$. Similarly, for any congestion model T_w, one defines the class $\overline{\mathcal{C}}_{\mathsf{T}_w}(\mathcal{F}, \alpha)$, so as to obtain $\overline{\mathcal{C}}_w(\mathcal{F}, \alpha) = \bigcup_{\mathsf{T}_w} \overline{\mathcal{C}}_{\mathsf{T}_w}(\mathcal{F}, \alpha)$.

Lemma 1. *For any $\epsilon \geq 0$, n-dimensional vector of weights w, finite space of functions $\mathcal{F}$, pair of matrices $\alpha \in \mathbb{R}^{n \times n}$ and $\beta \in \mathbb{R}^{n \times n}_{\geq 0}$ and function $\mathsf{SF} \in \{\mathsf{SUM}, \mathsf{MAX}\}$, it holds that $\mathsf{PPoA}_\epsilon(\beta\text{-SF}, \mathcal{C}_w(\mathcal{F}, \alpha)) = \mathsf{PPoA}_\epsilon(\beta\text{-SF}, \overline{\mathcal{C}}_w(\mathcal{F}, \alpha))$ and $\mathsf{CCPoA}_\epsilon(\beta\text{-SF}, \mathcal{C}_w(\mathcal{F}, \alpha)) = \mathsf{CCPoA}_\epsilon(\beta\text{-SF}, \overline{\mathcal{C}}_w(\mathcal{F}, \alpha))$.*

3   The Main Result

Our main result is the proof of the following general theorem.

Theorem 1. *For any real value $\epsilon \geq 0$, n-dimensional vector of weights w, finite space of functions $\mathcal{F}$, pair of matrices $\alpha \in \mathbb{R}^{n \times n}$ and $\beta \in \mathbb{R}^{n \times n}_{\geq 0}$ and function $\mathsf{SF} \in \{\mathsf{SUM}, \mathsf{MAX}\}$, it holds that $\mathsf{PPoA}_\epsilon(\beta\text{-SF}, \mathcal{C}_w(\mathcal{F}, \alpha)) = \mathsf{CCPoA}_\epsilon(\beta\text{-SF}, \mathcal{C}_w(\mathcal{F}, \alpha))$. Moreover, the value $\mathsf{PPoA}_\epsilon(\beta\text{-SF}, \mathcal{C}_w(\mathcal{F}, \alpha))$ can always be determined via the primal-dual method.*

Proof. Fix a real value $\epsilon \geq 0$, an n-dimensional vector of weights w, a finite space of functions $\mathcal{F}$, a pair of matrices $\alpha \in \mathbb{R}^{n \times n}$ and $\beta \in \mathbb{R}^{n \times n}_{\geq 0}$ and a function $\mathsf{SF} \in \{\mathsf{SUM}, \mathsf{MAX}\}$. We prove the claim in four steps.

Step (1). Definition of the *representative* congestion model T_w^*.

Let $\mathsf{T}_w^* = ([n], w, E^*, (\Sigma_i^*)_{i \in [n]})$ be a congestion model such that

1. $\Sigma_i^* = \{\sigma_i^*, o_i^*\}$ for each $i \in [n]$, i.e., each player $i \in [n]$ has exactly two strategies denoted as σ_i^* and o_i^*;
2. the set of resources E^* and the strategies σ_i^* and o_i^* for each $i \in [n]$ are properly defined in such a way that, for each $P, Q \subseteq [n]$, there exists exactly one resource $e(P,Q) \in E^*$ for which it holds that $\{i \in [n] \mid e(P,Q) \in \sigma_i^*\} = P$ and $\{i \in [n] \mid e(P,Q) \in o_i^*\} = Q$. Hence, $|E^*| = 2^n \cdot 2^n = 4^n$.

Intuitively, the representative congestion model T_w^* is defined in such a way that the pair of strategy profiles $\boldsymbol{\sigma}^* = (\sigma_1^*, \ldots, \sigma_n^*)$ and $\boldsymbol{o}^* = (o_1^*, \ldots, o_n^*)$ is able to encompass all possible configurations of congestions that may arise in any pair of strategy profiles and for any congestion model induced by w. In particular, the following fundamental property holds.

Property 1. For any congestion model $\mathsf{T}_w = ([n], w, E, (\Sigma_i)_{i\in[n]})$, resource $e \in E$ and pair of profiles $\boldsymbol{\sigma}', \boldsymbol{\sigma}'' \in \Sigma(\mathsf{T}_w)$, there exists a resource $\overline{e} \in E^*$ such that $\{i \in [n] \mid e \in \sigma_i'\} = \{i \in [n] \mid \overline{e} \in \sigma_i^*\}$ and $\{i \in [n] \mid e \in \sigma_i''\} = \{i \in [n] \mid \overline{e} \in o_i^*\}$.

Proof. Fix a congestion model $\mathsf{T}_w = ([n], w, E, (\Sigma_i)_{i\in[n]})$, a resource $e \in E$ and pair of profiles $\boldsymbol{\sigma}', \boldsymbol{\sigma}'' \in \Sigma(\mathsf{T}_w)$. Let $\{i \in [n] \mid e \in \sigma_i'\} := P$ and $\{i \in [n] \mid e \in \sigma_i''\} := Q$. To prove the claim, it suffices choosing $\overline{e} = e(P,Q)$. □

Step (2). Definition of a primal-dual formulation for $\mathsf{PPoA}_\epsilon(\beta\text{-}\mathsf{SF}, \overline{\mathcal{C}}_w(\mathcal{F}, \alpha))$.

Fix a function $\mathsf{SF} \in \{\mathsf{SUM}, \mathsf{MAX}\}$. Our aim is to use the optimal solution of a linear program $\mathsf{PP}_{\mathsf{PNE}}(\mathsf{SF}, \mathsf{T}_w^*, \boldsymbol{\sigma}^*, \boldsymbol{o}^*)$ to achieve an upper bound on the worst-case ϵ-approximate pure price of anarchy of any game in $\overline{\mathcal{C}}_{\mathsf{T}_w^*}(\mathcal{F}, \alpha)$ under the restriction that the latency functions are suitably tuned so as to make $\boldsymbol{\sigma}^*$ the worst ϵ-approximate pure Nash equilibrium and $\boldsymbol{o}^*$ a social optimum (of social value 1). The linear program $\mathsf{PP}_{\mathsf{PNE}}(\mathsf{SUM}, \mathsf{T}_w^*, \boldsymbol{\sigma}^*, \boldsymbol{o}^*)$ for the β-utilitarian social function is defined as follows.

$$maximize \sum_{e\in E^*} \sum_{k\in[r]} v_k^e f_k(n_e(\boldsymbol{\sigma}^*)) \sum_{i\in[n]} \sum_{j\in[n]: e\in\sigma_j^*} \beta_{ij} w_j$$

$$subject\ to$$
$$\sum_{e\in\sigma_i^*\setminus o_i^*} \sum_{k\in[r]} v_k^e T_{e,i}^-(\boldsymbol{\sigma}^*, \epsilon, k) + \sum_{e\in o_i^*\setminus\sigma_i^*} \sum_{k\in[r]} v_k^e T_{e,i}^+(\boldsymbol{\sigma}^*, \epsilon, k)$$
$$-\epsilon \sum_{e\in o_i^*\cap\sigma_i^*} \sum_{k\in[r]} v_k^e f_k(s_e) A_e^i(\boldsymbol{\sigma}^*) \le 0, \qquad \forall i \in [n]$$
$$\sum_{e\in E^*} \sum_{k\in[r]} v_k^e f_k(n_e(\boldsymbol{o}^*)) \sum_{i\in[n]} \sum_{j\in[n]: e\in o_j^*} \beta_{ij} w_j \le 1,$$
$$v_k^e \ge 0, \qquad \forall e \in E^*, k \in [r]$$

The first n constraints guarantee that no player can lower her perceived cost of a factor more than $1 + \epsilon$ by switching to the strategy she uses in the social optimum $\boldsymbol{o}^*$ (see Eq. (1)), while the last constraint normalizes to at most 1 the value $\beta\text{-}\mathsf{SUM}(\boldsymbol{o}^*)$.

The dual program $\mathsf{DP}_{\mathsf{PNE}}(\mathsf{SUM}, \mathsf{T}_w^*, \boldsymbol{\sigma}^*, \boldsymbol{o}^*)$ is the following (we associate a variable y_i with the ith constraint of the first n ones and a variable γ with the normalizing constraint).

minimize γ
subject to

$$\sum_{i\in[n]:e\in\sigma_i^*\setminus o_i^*} y_i T_{e,i}^-(\boldsymbol{\sigma}^*,\epsilon,k) + \sum_{i\in[n]:e\in o_i^*\setminus\sigma_i^*} y_i T_{e,i}^+(\boldsymbol{\sigma}^*,\epsilon,k)$$

$$-\epsilon \sum_{i\in[n]:e\in o_i^*\cap\sigma_i^*} y_i f_k(n_e(\boldsymbol{\sigma}^*)) A_e^i(\boldsymbol{\sigma}^*)$$

$$+\gamma f_k(n_e(\boldsymbol{o}^*)) \sum_{i\in[n]}\sum_{j\in[n]:e\in o_j^*} \beta_{ij}w_j \geq f_k(n_e(\boldsymbol{\sigma}^*)) \sum_{i\in[n]}\sum_{j\in[n]:e\in\sigma_j^*} \beta_{ij}w_j, \quad \forall e \in E^*, k \in [r]$$

$$y_i \geq 0, \qquad\qquad\qquad\qquad\qquad\qquad\qquad\qquad\qquad\qquad\qquad \forall i \in [n]$$
$$\gamma \geq 0$$

For the primal-dual formulations for the β-egalitarian social function, we refer the reader to [7].

We stress that, being all the values ϵ, $(w_i)_{i\in[n]}$, $(\alpha_{ij},\beta_{ij})_{i,j\in[n]}$, $n_e(\boldsymbol{\sigma}^*)$ and $n_e(\boldsymbol{o}^*)$ fixed constants in the proposed formulations, $\mathsf{PP}_{\mathsf{PNE}}(\mathsf{SUM},\mathsf{T}_{\boldsymbol{w}}^*,\boldsymbol{\sigma}^*,\boldsymbol{o}^*)$ is a linear program defined over the variables $(v_k^e)_{e\in E^*,k\in[r]}$ and $\mathsf{PP}_{\mathsf{PNE}}(\mathsf{MAX},$ $\mathsf{T}_{\boldsymbol{w}}^*,\boldsymbol{\sigma}^*,\boldsymbol{o}^*)$ is a linear program defined over the variables $(v_k^e)_{e\in E^*,k\in[r]}$ and t, as needed. Note that, for $\mathsf{SF} \in \{\mathsf{SUM},\mathsf{MAX}\}$, $\mathsf{PP}_{\mathsf{PNE}}(\mathsf{SF},\mathsf{T}_{\boldsymbol{w}}^*,\boldsymbol{\sigma}^*,\boldsymbol{o}^*)$ is, in general, under-constrained. In fact, in order to assure that $\boldsymbol{\sigma}^*$ and $\boldsymbol{o}^*$ are the worst ϵ-approximate pure Nash equilibrium and the social optimum, respectively, one should guarantee β-$\mathsf{SF}(\boldsymbol{\sigma}^*) \geq \beta$-$\mathsf{SF}(\boldsymbol{\sigma})$ for each other ϵ-approximate pure Nash equilibrium $\boldsymbol{\sigma} \in \Sigma^*$, if any, and β-$\mathsf{SF}(\boldsymbol{o}^*) \leq \beta$-$\mathsf{SF}(\boldsymbol{\sigma})$ for each $\boldsymbol{\sigma} \in \Sigma^*$. Moreover, the normalizing constraints have also been relaxed so as to assure β-$\mathsf{SF}(\boldsymbol{o}^*) \leq 1$ rather than β-$\mathsf{SF}(\boldsymbol{o}^*) = 1$. However, as we will discuss in the proof of Lemma 2, either removing or relaxing these constraints can only worsen the resulting upper bounds.

The significance of the previously defined pairs of primal-dual formulations is witnessed by the following lemma which states that the value of an optimal solution to $\mathsf{PP}_{\mathsf{PNE}}(\mathsf{SF},\mathsf{T}_{\boldsymbol{w}}^*,\boldsymbol{\sigma}^*,\boldsymbol{o}^*)$ provides an upper bound on $\mathsf{PPoA}_\epsilon(\beta$-$\mathsf{SF},\overline{\mathcal{C}}_{\boldsymbol{w}}(\mathcal{F},\alpha))$.

Lemma 2. *For a fixed* $\mathsf{SF} \in \{\mathsf{SUM},\mathsf{MAX}\}$, *let* $\overline{x}$ *be the value of an optimal solution to* $\mathsf{PP}_{\mathsf{PNE}}(\mathsf{SF},\mathsf{T}_{\boldsymbol{w}}^*,\boldsymbol{\sigma}^*,\boldsymbol{o}^*)$ *when this linear problem is not unlimited, otherwise let* $\overline{x} = \infty$. *Then* $\mathsf{PPoA}_\epsilon(\beta$-$\mathsf{SF},\overline{\mathcal{C}}_{\boldsymbol{w}}(\mathcal{F},\alpha)) \leq \overline{x}$.

Step (3). Proof of existence of a game $(\mathsf{CG},\alpha) \in \overline{\mathcal{C}}_{\boldsymbol{w}}(\mathcal{F},\alpha)$ such that $\mathsf{PPoA}_\epsilon(\beta$-$\mathsf{SF},\mathsf{CG},\alpha) = \overline{x}$.

Lemma 3. *For a fixed* $\mathsf{SF} \in \{\mathsf{SUM},\mathsf{MAX}\}$, *let* $\overline{x}$ *be the value of an optimal solution to* $\mathsf{PP}_{\mathsf{PNE}}(\mathsf{SF},\mathsf{T}_{\boldsymbol{w}}^*,\boldsymbol{\sigma}^*,\boldsymbol{o}^*)$ *when this linear problem is not unlimited, otherwise let* $\overline{x} = \infty$. *Then* $\mathsf{PPoA}_\epsilon(\beta$-$\mathsf{SF},\overline{\mathcal{C}}_{\boldsymbol{w}}(\mathcal{F},\alpha)) = \overline{x}$.

Step (4). Definition of a primal-dual formulation for $\mathsf{CCPoA}_\epsilon(\beta$-$\mathsf{SF},\overline{\mathcal{C}}_{\boldsymbol{w}}(\mathcal{F},\alpha))$ and proof of the "Extension Lemma".

Fix a congestion model $\mathsf{T}_{\boldsymbol{w}} = ([n],\boldsymbol{w},E,(\Sigma_i)_{i\in[n]})$, a probability distribution $\boldsymbol{p} \in \Delta(\Sigma(\mathsf{T}_{\boldsymbol{w}}))$ and a strategy profile $\boldsymbol{o} \in \Sigma(\mathsf{T}_{\boldsymbol{w}})$. We define the following primal program $\mathsf{PP_{CCE}}(\mathsf{SUM}, \mathsf{T}_w, \boldsymbol{p}, \boldsymbol{o})$ for the β-utilitarian social function.

$$maximize \sum_{\sigma \in \Sigma} p_\sigma \sum_{e \in E} \sum_{k \in [r]} v_k^e f_k(n_e(\sigma)) \sum_{i \in [n]} \sum_{j \in [n]:e \in \sigma_j} \beta_{ij} w_j$$

subject to

$$\sum_{\sigma \in \Sigma} p_\sigma \sum_{e \in \sigma_i \setminus o_i} \sum_{k \in [r]} v_k^e T_{e,i}^-(\sigma, \epsilon, k) + \sum_{\sigma \in \Sigma} p_\sigma \sum_{e \in o_i^* \setminus \sigma_i} \sum_{k \in [r]} v_k^e T_{e,i}^+(\sigma, \epsilon, k)$$

$$-\epsilon \sum_{\sigma \in \Sigma} p_\sigma \sum_{e \in o_i^* \cap \sigma_i} \sum_{k \in [r]} v_k^e f_k(n_e(\sigma)) A_e^i(\sigma) \leq 0, \qquad\qquad \forall i \in [n]$$

$$\sum_{e \in E} \sum_{k \in [r]} v_k^e f_k(n_e(o)) \sum_{i \in [n]} \sum_{j \in [n]:e \in o_j} \beta_{ij} w_j \leq 1,$$

$$v_k^e \geq 0, \qquad\qquad \forall e \in E, k \in [r]$$

The dual program $\mathsf{DP_{CCE}}(\mathsf{SUM}, \mathsf{T}_w, \boldsymbol{p}, \boldsymbol{o})$ is the following (again, we associate a variable y_i with the ith constraint of the first n ones and a variable γ with the normalizing constraint).

$$minimize \; \gamma$$
$$subject \; to$$
$$\sum_{\sigma \in \Sigma} p_\sigma \sum_{i \in [n]:e \in \sigma_i^* \setminus o_i^*} y_i T_{e,i}^-(\sigma, \epsilon, k) + \sum_{\sigma \in \Sigma} p_\sigma \sum_{i \in [n]:e \in o_i^* \setminus \sigma_i^*} y_i T_{e,i}^+(\sigma, \epsilon, k)$$
$$-\epsilon \sum_{\sigma \in \Sigma} p_\sigma \sum_{i \in [n]:e \in o_i^* \cap \sigma_i^*} y_i f_k(n_e(\sigma)) A_e^i(\sigma)$$
$$+\gamma f_k(n_e(o)) \sum_{i \in [n]} \sum_{j \in [n]:e \in o_j} \beta_{ij} w_j \geq \sum_{\sigma \in \Sigma} p_\sigma f_k(n_e(\sigma)) \sum_{i \in [n]} \sum_{j \in [n]:e \in \sigma_j} \beta_{ij} w_j, \quad \forall e \in E, k \in [r]$$
$$y_i \geq 0, \qquad\qquad \forall i \in [n]$$
$$\gamma \geq 0$$

For the primal-dual formulations for the β-egalitarian social function, we refer the reader to [7].

Again, even though both $\mathsf{PP_{CCE}}(\mathsf{SUM}, \mathsf{T}_w, \boldsymbol{p}, \boldsymbol{o})$ and $\mathsf{PP_{CCE}}(\mathsf{MAX}, \mathsf{T}_w, \boldsymbol{p}, \boldsymbol{o})$ may be, in general, under-constrained, by the same arguments used in the discussion of the pairs of primal-dual formulations used for bounding the worst-case ϵ-approximate pure price of anarchy, it follows that, for each function $\mathsf{SF} \in \{\mathsf{SUM}, \mathsf{MAX}\}$, the optimal solution to $\mathsf{PP_{CCE}}(\mathsf{SF}, \mathsf{T}_w, \boldsymbol{p}, \boldsymbol{o})$ yields an upper bound on the worst-case ϵ-approximate coarse correlated price of anarchy of the class $\overline{\mathcal{C}}_{\mathsf{T}_w}(\mathcal{F}, \alpha)$ attainable when $\boldsymbol{p}$ is taken for the worst ϵ-approximate coarse correlated equilibrium and $\boldsymbol{o}$ for the social optimum (of social value 1). Let us denote such a class with $\widehat{\mathcal{C}}_{\mathsf{T}_w}(\mathcal{F}, \alpha)$.

The following lemma shows that any upper bound on $\mathsf{PPoA}_\epsilon(\beta\text{-}\mathsf{SF}, \mathcal{C}_{\mathsf{T}_w^*}(\mathcal{F}, \alpha))$ proved via the primal-dual method automatically extends to $\mathsf{CCPoA}_\epsilon(\beta\text{-}\mathsf{SF}, \widehat{\mathcal{C}}_{\mathsf{T}_w}(\mathcal{F}, \alpha))$.

Lemma 4 (Extension Lemma). *For any function* $\mathsf{SF} \in \{\mathsf{SUM}, \mathsf{MAX}\}$, *congestion model* $\mathsf{T}_w = ([n], \boldsymbol{w}, E, (\Sigma_i)_{i \in [n]})$, *probability distribution* $\boldsymbol{p} \in \Delta(\Sigma(\mathsf{T}_w))$ *and strategy profile* $\boldsymbol{o} \in \Sigma(\mathsf{T}_w)$, *it holds that any feasible solution to* $\mathsf{DP_{PNE}}(\mathsf{SF}, \mathsf{T}_w^*, \boldsymbol{\sigma}, \boldsymbol{o})$ *is also a feasible solution to* $\mathsf{DP_{CCE}}(\mathsf{SF}, \mathsf{T}_w, \boldsymbol{p}, \boldsymbol{o})$.

We now have all the ingredients needed to conclude the proof of the theorem.

Fix a function $\mathsf{SF} \in \{\mathsf{SUM}, \mathsf{MAX}\}$. Assume, first, that $\mathsf{PP_{PNE}}(\mathsf{SF}, \mathsf{T}_w^*, \boldsymbol{\sigma}^*, \boldsymbol{o}^*)$ is unlimited. Then, by Lemma 3, it holds that $\mathsf{PPoA}_\epsilon(\beta\text{-}\mathsf{SF}, \overline{\mathcal{C}}_w(\mathcal{F}, \alpha)) = \infty$ which, together with $\mathsf{PPoA}_\epsilon(\beta\text{-}\mathsf{SF}, \overline{\mathcal{C}}_w(\mathcal{F}, \alpha)) \leq \mathsf{CCPoA}_\epsilon(\beta\text{-}\mathsf{SF}, \overline{\mathcal{C}}_w(\mathcal{F}, \alpha))$, immediately implies that $\mathsf{PPoA}_\epsilon(\beta\text{-SF}, \overline{\mathcal{C}}_w(\mathcal{F}, \alpha)) = \mathsf{CCPoA}_\epsilon(\beta\text{-SF}, \overline{\mathcal{C}}_w(\mathcal{F}, \alpha))$. By applying Lemma 1, we obtain $\mathsf{PPoA}_\epsilon(\beta\text{-SF}, \mathcal{C}_w(\mathcal{F}, \alpha)) = \mathsf{CCPoA}_\epsilon(\beta\text{-SF}, \mathcal{C}_w(\mathcal{F}, \alpha))$.

In the case in which $\mathsf{PP}_{\mathsf{PNE}}(\mathsf{SF}, \mathsf{T}_w^*, \sigma^*, o^*)$ admits an optimal solution of value $\overline{x}$, by Lemma 3, it holds that $\mathsf{PPoA}_\epsilon(\beta\text{-SF}, \overline{\mathcal{C}}_w(\mathcal{F}, \alpha)) = \overline{x}$. Moreover, by the Strong Duality Theorem, there exists a feasible solution (y^*, γ^*) to $\mathsf{DP}_{\mathsf{PNE}}(\mathsf{SF}, \mathsf{T}_w^*, \sigma, o)$ of value $\gamma^* = \overline{x}$. Choose an arbitrary game $(\mathsf{CG}, \alpha) \in \overline{\mathcal{C}}_w(\mathcal{F}, \alpha)$ such that $\mathsf{CCPoA}_\epsilon(\beta\text{-SF}, \overline{\mathcal{C}}_w(\mathcal{F}, \alpha)) = \mathsf{CCPoA}_\epsilon(\beta\text{-SF}, \mathsf{CG}, \alpha)$ and let T_w be the congestion model defining CG, p be the worst ϵ-approximate coarse correlated equilibrium of (CG, α) and o be the social optimum (of social value 1). By the definition of T_w, p and o, it follows that the optimal solution to $\mathsf{PP}_{\mathsf{CCE}}(\mathsf{SF}, \mathsf{T}_w, p, o)$ has a value of at least $\mathsf{CCPoA}_\epsilon(\beta\text{-SF}, \overline{\mathcal{C}}_w(\mathcal{F}, \alpha))$, which, by the Weak Duality Theorem, implies in turn that any feasible solution to $\mathsf{DP}_{\mathsf{CCE}}(\mathsf{SF}, \mathsf{T}_w, p, o)$ has a value of at least $\mathsf{CCPoA}_\epsilon(\beta\text{-SF}, \overline{\mathcal{C}}_w(\mathcal{F}, \alpha))$. By Lemma 4, it follows that (y^*, γ^*) is also a feasible solution to $\mathsf{DP}_{\mathsf{CCE}}(\mathsf{SF}, \mathsf{T}_w, p, o)$. This implies that $\mathsf{CCPoA}_\epsilon(\beta\text{-SF}, \overline{\mathcal{C}}_w(\mathcal{F}, \alpha)) \leq \gamma^* = \overline{x} = \mathsf{PPoA}_\epsilon(\beta\text{-SF}, \overline{\mathcal{C}}_w(\mathcal{F}, \alpha))$. Again, by applying Lemma 1, we obtain that $\mathsf{PPoA}_\epsilon(\beta\text{-SF}, \mathcal{C}_w(\mathcal{F}, \alpha)) = \mathsf{CCPoA}_\epsilon(\beta\text{-SF}, \mathcal{C}_w(\mathcal{F}, \alpha))$.

It is clear from our discussion that the value $\mathsf{PPoA}_\epsilon(\beta\text{-SF}, \overline{\mathcal{C}}_w(\mathcal{F}, \alpha)) = \mathsf{PPoA}_\epsilon(\beta\text{-SF}, \mathcal{C}_w(\mathcal{F}, \alpha))$ can always be (theoretically) determined via the primal-dual method, that is, by computing the value of the optimal solution of either the primal program $\mathsf{PP}_{\mathsf{PNE}}(\mathsf{SF}, \mathsf{T}_w^*, \sigma, o)$ or the dual one $\mathsf{DP}_{\mathsf{PNE}}(\mathsf{SF}, \mathsf{T}_w^*, \sigma, o)$ for each function $\mathsf{SF} \in \{\mathsf{SUM}, \mathsf{MAX}\}$, and this concludes the proof (solving the dual program, in particular, requires to determine the minimum value γ^* for which all the $r \cdot 4^n$ possible constraints induced by the $|E^*| = 4^n$ pairs of values yielded by the representative congestion model T_w^* on each of the r components of the latency functions are satisfied). $\square$

References

1. Aland, S., Dumrauf, D., Gairing, M., Monien, B., Schoppmann, F.: Exact price of anarchy for polynomial congestion games. SIAM J. Comput. **40**(5), 1211–1233 (2011)
2. Anagnostopoulos, A., Becchetti, L., de Keijzer, B., Schäfer, G.: Inefficiency of games with social context. Theory Comput. Syst. **57**(3), 782–804 (2015)
3. Awerbuch, B., Azar, Y., Epstein, L.: The price of routing unsplittable flow. In: Proceedings of the 37th Annual ACM Symposium on Theory of Computing (STOC), pp. 57–66. ACM Press (2005)
4. Bhawalkar, K., Gairing, M., Roughgarden, T.: Weighted congestion games: price of anarchy, universal worst-case examples, and tightness. ACM Trans. Econ. Comput. **2**(4), 1–23 (2014)
5. Bilò, V.: A unifying tool for bounding the quality of non-cooperative solutions in weighted congestion games. In: Erlebach, T., Persiano, G. (eds.) WAOA 2012. LNCS, vol. 7846, pp. 215–228. Springer, Heidelberg (2013)
6. Bilò, V.: On linear congestion games with altruistic social context. In: Cai, Z., Zelikovsky, A., Bourgeois, A. (eds.) COCOON 2014. LNCS, vol. 8591, pp. 547–558. Springer, Heidelberg (2014)

7. Bilò, V.: On the robustness of the approximate price of anarchy in generalized congestion games. CoRR abs/1412.0845 (2014)
8. Bilò, V., Fanelli, A., Moscardelli, L.: On lookahead equilibria in linear congestion games. Math. Struct. Comput. Sci. (to appear)
9. Bilò, V., Flammini, M., Gallotti, V.: On bidimensional congestion games. In: Even, G., Halldórsson, M.M. (eds.) SIROCCO 2012. LNCS, vol. 7355, pp. 147–158. Springer, Heidelberg (2012)
10. Bilò, V., Flammini, M., Monaco, G., Moscardelli, L.: Some anomalies of farsighted strategic behavior. Theory Comput. Syst. 56(1), 156–180 (2015)
11. Bilò, V., Paladini, M.: On the performance of mildly greedy players in cut games. J. Comb. Optim. (to appear)
12. Bilò, V., Vinci, C.: On stackelberg strategies in affine congestion games. In: Markakis, E., Schäfer, G. (eds.) WINE 2015. LNCS, vol. 9470, pp. 132–145. Springer, Heidelberg (2015). doi:10.1007/978-3-662-48995-6_10
13. Bilò, V., Vinci, C.: Dynamic taxes for polynomial congestion games. In: Proceedings of the 17th ACM Conference on Economics and Computation (EC) (to appear)
14. Christodoulou, G., Koutsoupias, E.: The price of anarchy of finite congestion games. In: Proceedings of the 37th Annual ACM Symposium on Theory of Computing (STOC), pp. 67–73. ACM Press (2005)
15. Christodoulou, G., Koutsoupias, E.: On the price of anarchy and stability of correlated equilibria of linear congestion games. In: Brodal, G.S., Leonardi, S. (eds.) ESA 2005. LNCS, vol. 3669, pp. 59–70. Springer, Heidelberg (2005)
16. Christodoulou, G., Koutsoupias, E., Spirakis, P.G.: On the performance of approximate equilibria in congestion games. Algorithmica 61(1), 116–140 (2011)
17. Koutsoupias, E., Papadimitriou, C.: Worst-case equilibria. In: Meinel, C., Tison, S. (eds.) STACS 1999. LNCS, vol. 1563, pp. 404–413. Springer, Heidelberg (1999)
18. Monderer, D., Shapley, L.S.: Potential games. Games Econ. Behav. 14(1), 124–143 (1996)
19. Nash, J.F.: Equilibrium points in n-person games. Proc. Nat. Acad. Sci. 36(1), 48–49 (1950)
20. Rahn, M., Schäfer, G.: Bounding the inefficiency of altruism through social contribution games. In: Chen, Y., Immorlica, N. (eds.) WINE 2013. LNCS, vol. 8289, pp. 391–404. Springer, Heidelberg (2013)
21. Rosenthal, R.W.: A class of games possessing pure-strategy Nash equilibria. Int. J. Game Theory 2(1), 65–67 (1973)
22. Roughgarden, T.: Intrinsic robustness of the price of anarchy. Commun. ACM 55(7), 116–123 (2012)
23. Shapley, L.S.: The value of n-person games. In: Contributions to the Theory of Games, pp. 31–40. Princeton University Press, Princeton (1953)

Efficiency of Equilibria
in Uniform Matroid Congestion Games

Jasper de Jong[1], Max Klimm[2], and Marc Uetz[1(✉)]

[1] Universiteit Twente, Enschede, The Netherlands
{j.dejong-3,m.uetz}@utwente.nl
[2] Technische Universität Berlin, Berlin, Germany
max.klimm@tu-berlin.de

Abstract. Network routing games, and more generally congestion games play a central role in algorithmic game theory, comparable to the role of the traveling salesman problem in combinatorial optimization. It is known that the price of anarchy is independent of the network topology for non-atomic congestion games. In other words, it is independent of the structure of the strategy spaces of the players, and for affine cost functions it equals 4/3. In this paper, we show that the situation is considerably more intricate for atomic congestion games. More specifically, we consider congestion games with affine cost functions where the players' strategy spaces are symmetric and equal to the set of bases of a k-uniform matroid. In this setting, we show that the price of anarchy is strictly larger than the price of anarchy for singleton strategy spaces where it is 4/3. As our main result we show that the price of anarchy can be bounded from above by $28/13 \approx 2.15$. This constitutes a substantial improvement over the price of anarchy bound 5/2, which is known to be tight for network routing games with affine cost functions.

1 Introduction

Understanding the impact of selfish behavior on the performance of a system is an important question in algorithmic game theory. One of the cornerstones of the substantial literature on this topic is the famous result of Roughgarden and Tardos [27]. They considered the traffic model of Wardrop [30] in a network with affine flow-dependent congestion cost functions on the edges. Given a set of commodities, each specified by a source node, a target node, and a flow demand, a Wardrop equilibrium is a multicommodity flow with the property that every commodity uses only paths that minimize the cost. For this setting, Roughgarden and Tardos proved that the total cost of an equilibrium flow is not worse than 4/3 times that of a system optimum. This ratio was coined the *price of anarchy* by Koutsoupias and Papadimitriou [19] who introduced it as a measure of a system's performance degradation due to selfish behavior. A surprising consequence of the result of Roughgarden and Tardos is that the worst case price of anarchy in congested networks is attained for very simple single-commodity networks already considered a century ago by Pigou [23]. Pigou-style networks consist

© Springer-Verlag Berlin Heidelberg 2016
M. Gairing and R. Savani (Eds.): SAGT 2016, LNCS 9928, pp. 105–116, 2016.
DOI: 10.1007/978-3-662-53354-3_9

of only two nodes connected by two parallel links. In fact, Roughgarden [26] proved that for *any* set of cost functions, the price of anarchy is independent of the network topology as it is always attained for such simple Pigou-style networks where a feasible strategy of each commodity is to choose exactly one out of the two links.

A model related to Wardrop's model is that of a congestion game with unsplittable (i.e., *atomic*) players. In such a game, there is a finite set of players and a strategy of each player is to choose a set of resources allowable to her. Without any restrictions on the strategy spaces, the price of anarchy for affine cost functions is 5/2 as shown by Christodoulou and Koutsoupias [9] and Awerbuch et al. [5]. As a contrast, for simple Pigou-style instances with symmetric and singleton strategies, Lücking et al. [20] showed that the price of anarchy is only 4/3. These results imply that for atomic congestion games the price of anarchy *does* depend on the combinatorial structure of players' strategies.

In this work, we shed new light on the impact of the combinatorial structure of strategy spaces on the inefficiency of equilibria in atomic congestion games. Specifically, we focus on the minimum combinatorial structure that one may think of, namely symmetric k-uniform congestion games where the strategy set of each player consists of all k-elementary subsets of resources. These games are a natural generalization of the singleton case, and we consider it interesting because it constitutes a first step into the direction where strategies are bases of a (general) matroid. As for potential applications, one may think of, e.g., load balancing games where each player controls the same amount of jobs; see also Abed et al. [1] for a related model in the context of coordination mechanisms.

Our Results. We prove that the price of anarchy in congestion games with affine cost functions is at most 28/13 when strategy spaces are symmetric and bases of a k-uniform matroid. The proof uses in its core several combinatorial arguments on the amount and cost of resources that are over- respectively under-demanded in any given Nash equilibrium as opposed to an optimal solution. It also exploits the affinity of the cost functions, along the lines of earlier arguments of Fotakis [12] for the singleton case. The main point of the technical side of the paper is the insight that the combinatorial structure of strategy spaces, here of the simplest possible form, allows to furnish combinatorial arguments that yield improved results on the price of anarchy. We are not aware of earlier attempts in this direction, and believe this opens new possibilities for our understanding of a "classical" showcase problem in algorithmic game theory. We also show that the price of anarchy for the k-uniform matroid case cannot be the same as for singleton congestion games, as we bound it away from 4/3: For affine cost functions, and k large enough, the price of anarchy is at least 1.343. For $k = 5$ it is at least $47/35 \approx 1.3428$.

Related Work. Since the early works of Pigou [23], Beckman et al. [6], and Braess [7] it is well known that user equilibria in congested networks may be suboptimal for the overall performance of the system. In order to quantify this inefficiency, Koutsoupias and Papadimitriou [19] proposed to study the ratio of the total cost of an equilibrium and the total cost of an optimal solution. This ratio is now known as the *price of anarchy*. Roughgarden and Tardos [27] showed that the price of anarchy for non-atomic games with affine costs is $4/3$. The worst case is attained for simple networks of two parallel links previously studied by Pigou [23]. Roughgarden [26] gave a closed form expression for the price of anarchy for arbitrary cost functions which is again attained for Pigou-style networks, e.g., for polynomials with positive coefficient and maximum degree d the price of anarchy is of order $\Theta(d/\ln(d))$.

Awerbuch et al. [5] and Christodoulou and Koutsoupias [9] considered the related model with atomic players that was introduced by Rosenthal [24]. They showed that for affine cost functions the price of anarchy is $5/2$. Aland et al. [4] gave tight bounds on the price of anarchy for polynomial cost functions with maximum degree d which behaves asymptotically as $\Theta((d/\ln d)^{d+1})$. It is interesting to note that these worst-case bounds are not attained for simple Pigou-style networks with symmetric and singleton strategies as in the non-atomic case. Based on previous work of Suri et al. [28], Caragiannis et al. [8] showed that for affine costs, the worst case is attained for *asymmetric* singleton strategies. For a similar result for polynomial costs, see Gairing and Schoppmann [15]. In fact, for singleton games with symmetric strategies, the price of anarchy is considerably better than in the general case. In fact, Fotakis [12] showed that the price of anarchy of symmetric singleton atomic games is equal to the price of anarchy of non-atomic games. This improves and generalizes previous bounds by Lücking et al. [20] and Gairing et al. [14].

The class of k-uniform games that we consider in this paper is also related to the class of integer-splittable congestion games introduced by Rosenthal [25] and the classes of k-splittable and integer k-splittable congestion games studied by Meyers [21]. In contrast to our model, the models above allow that a player uses a resource with multiple units of demand at the same time. It turns out that allowing for this kind of self-congestion has a severe impact on the existence of pure Nash equilibria [11,25] but for networks of parallel links it is known that pure Nash equilibria are guaranteed to exist [17,29].

The impact of combinatorial structure on the existence and computability of pure Nash equilibria has been studied for many variants of congestion games. Ackermann et al. [2] proved that for atomic games with unweighted players all sequences of best replies converge in polynomial time to a pure Nash equilibrium if the set of strategies of each player corresponds to the set of bases of a matroid. For weighted congestion games, the matroid property guarantees the existence of a pure Nash equilibrium [3] while without that property a pure Nash equilibrium may fail to exist [16]. Similarly, congestion games with player-specific costs and matroid strategies have a pure Nash equilibrium which can be computed efficiently [3] which is in contrast to the general case [22]. For similar results in the context of resource buying games, see also Harks and Peis [18].

To the best of our knowledge, the impact of matroid structures on the efficiency of Nash equilibria has not been considered before. The only result in this direction is a yet unpublished work of Fujishige et al. [13]. They showed that Braess' paradox cannot occur in non-atomic games with matroid strategies, i.e., the quality of the user equilibrium cannot deteriorate when removing a resource. This result, however, has no consequences for the inefficiency of equilibria in non-atomic games since the worst case is attained for Pigou-style networks where the strategies are symmetric and 2-uniform matroids.

2   Preliminaries

Let $N = \{1, \ldots, n\}$ be a finite set of players and let R be a finite set of resources. Each player i is associated with a set of subsets of resources $S_i \subseteq 2^R$ allowable to her. A strategy of a player is to choose a subset $s_i \in S_i$ from this set. A strategy vector $s = (s_i)_{i \in N}$ consists of n strategies, one for each player. Every resource r is endowed with a cost function $c_r : \mathbb{N} \to \mathbb{R}$ that maps the total number of its users $x_r = |\{i \in N : r \in s_i\}|$ to a cost value $c_r(x_r)$. The cost functions c_r are called affine if $c_r(x_r) = \alpha_r + \beta_r x_r$, for constants $\alpha_r, \beta_r \geq 0$, $r \in R$. The private cost of player i in strategy vector s is then defined as

$$\pi_i(s) = \sum_{r \in s_i} c_r(x_r).$$

We use standard game theory notation; for a strategy vector $s \in S = S_1 \times \cdots \times S_n$, a player i and an alternative strategy $s_i' \in S_i$, we denote by (s_i', s_{-i}) the strategy vector in which all players play as in s except for i who plays s_i'. A strategy vector s is a Nash equilibrium if,

$$\pi_i(s) \leq \pi_i(s_i', s_{-i}) \quad \text{for all } i \in N \text{ and } s_i' \in S_i.$$

Given an instance of a game $I = (N, R, S, (c_r)_{r \in R})$, we denote the set of Nash equilibria of I by $S^{\mathsf{NE}}(I)$.

We are interested in how restrictions on the set of strategies of each player influence the inefficiency of equilibria. We measure the efficiency of a strategy vector $s \in S$ in terms of the social costs $C(s)$ defined as

$$C(s) = \sum_{i \in N} \pi_i(s).$$

We denote by $S^{\mathsf{OPT}}(I)$ the set of strategy vectors s that minimize $C(s)$. For an instance I of a game, the price of anarchy is defined as

$$\mathsf{PoA}(I) = \max_{s^{\mathsf{NE}} \in S^{\mathsf{NE}}(I)} \frac{C(s^{\mathsf{NE}})}{C(s^{\mathsf{OPT}})},$$

where $s^{\mathsf{OPT}} \in S^{\mathsf{OPT}}(I)$ is a strategy vector minimizing C. For a class $\mathcal{G}$ of games, the price of anarchy is defined as $\mathsf{PoA}(\mathcal{G}) = \sup_{I \in \mathcal{G}} \mathsf{PoA}(I)$. We drop $\mathcal{G}$ whenever it is clear from context. We are specifically interested in singleton and k-uniform matroid strategy spaces. A game is said to be a singleton game, if $|s_i| = 1$ for all $s_i \in S_i$ and $i \in N$. A game is called k-uniform game if for each player, there is a subset $R_i \subseteq R$ such that $S_i = \{R' \subseteq R_i : |R'| = k\}$. A game is called symmetric, if $S_i = S_j$ for all $i, j \in N$.

3   Symmetric k-uniform Games

The main result of this paper is the following.

Theorem 1. *The price of anarchy of symmetric k-uniform congestion games with affine cost functions is at most $\frac{28}{13} \approx 2.15$.*

For the proof of Theorem 1, we are going to prove that $C(s^{NE}) \leq \frac{28}{13} C(s^{OPT})$ for any given worst-case Nash equilibrium s^{NE} and optimal solution s^{OPT}, of an arbitrary instance I of a symmetric k-uniform congestion game. For the remainder of this section, fix an instance I, a worst-case Nash equilibrium s^{NE} and a system optimal solution s^{OPT}.

To gain some intuition on congestion games with k-uniform matroid strategies, let us first consider the following example of a k-uniform congestion game that will serve as a running example throughout this section. Even though it has only a moderate price of anarchy of $16/14$, it showcases the crucial structures that we exploit later in this section when proving Theorem 1.

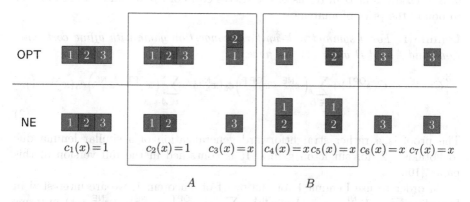

Fig. 1. A symmetric 4-uniform congestion game with seven resources. The height of the stack of each resource corresponds to its cost, e.g., resource 1 is used by players 1, 2 and 3 both in s^{NE} and s^{OPT} and the corresponding cost is 1; resource 3 is used by players 1 and 2 in s^{OPT} and the corresponding cost is 2.

Example 1. Consider the symmetric 4-uniform congestion game in Fig. 1. There are seven resources $R = \{1, \ldots, 7\}$. The first two resources have constant cost functions $c_1(x) = c_2(x) = 1$ for all $x \in \mathbb{N}$. The cost function of the other five resources is the identity, i.e., $c_r(x) = x$ for all $r \in \{3, \ldots, 7\}$. There are three players whose strategy is to choose exactly 4 resources, i.e., $S_i = \{R' \subset R : |R'| = 4\}$ for all $i \in \{1, 2, 3\}$. In the system optimum, the two resources with constant costs are used by all players and each player chooses two of the remaining five resources, see the upper profile in Fig. 1. One of the resources with non-constant costs has to be used by two players leading to overall costs of 14. However, there is a Nash equilibrium, in which not all of the resources with constant costs are used by all players, see the lower profile in Fig. 1. This Nash equilibrium has a total cost of 16. The price of anarchy of this instance is $16/14 \approx 1.14$.   ◁

In order to derive bounds on the price of anarchy for the proof of Theorem 1, we bound the excess costs of the resources that are chosen by more players in the Nash equilibrium than in the system optimum in terms of the excess costs of the resources that are chosen by more players in the system optimum than in the Nash equilibrium. To this end, we denote by A the set of resources chosen by more players in s^{OPT} than in s^{NE}, and by B the set of resources chosen by more players in s^{NE} than in s^{OPT}, i.e.,

$$A = \left\{ r \in R : x_r^{\mathsf{OPT}} > x_r^{\mathsf{NE}} \right\} \quad \text{and} \quad B = \left\{ r \in R : x_r^{\mathsf{OPT}} < x_r^{\mathsf{NE}} \right\}. \quad (1)$$

Henceforth, we term the resources in A underloaded and the resources in B overloaded. For an illustration, see also Fig. 1 where the set of underloaded resources is $A = \{2, 3\}$ and the set of overloaded resources is $B = \{4, 5\}$.

As we show in the following lemma, it is sufficient to bound the excess costs of the resources in B in terms of the excess costs of the resources in A in order to bound the price of anarchy.

Lemma 1. *For a symmetric k-uniform congestion game with affine cost functions and A and B as in* (1), *we have*

$$\frac{3}{4}C(s^{\mathsf{NE}}) \leq C(s^{\mathsf{OPT}}) + \sum_{b \in B} \left(x_b^{\mathsf{NE}} - x_b^{\mathsf{OPT}} \right) c_b(x_b^{\mathsf{NE}}) - \sum_{a \in A} \left(x_a^{\mathsf{OPT}} - x_a^{\mathsf{NE}} \right) c_a(x_a^{\mathsf{NE}} + 1).$$
$$(2)$$

The proof is a rather straightforward generalization of a similar lemma due to Fotakis [12] for singleton games. It is contained in the full version of this paper [10].

In order to use Lemma 1 for the proof of Theorem 1, we are interested in bounding $\sum_{b \in B}(x_b^{\mathsf{NE}} - x_b^{\mathsf{OPT}})c_b(x_b^{\mathsf{NE}}) - \sum_{a \in A}(x_a^{\mathsf{OPT}} - x_a^{\mathsf{NE}})c_a(x_a^{\mathsf{NE}} + 1)$ in terms of $C(s^{\mathsf{NE}})$. It is interesting to note that for symmetric singleton games, it holds that

$$c_r(x_r^{\mathsf{NE}}) \leq c_{r'}(x_{r'}^{\mathsf{NE}} + 1) \quad (3)$$

for all $r, r' \in R$ by the Nash inequality. This implies in particular that

$$\sum_{b \in B}(x_b^{\mathsf{NE}} - x_b^{\mathsf{OPT}})c_b(x_b^{\mathsf{NE}}) \leq \sum_{a \in A}(x_a^{\mathsf{OPT}} - x_a^{\mathsf{NE}})c_a(x_a^{\mathsf{NE}} + 1), \quad (4)$$

which together with Lemma 1 implies an upper bound on the price of anarchy of 4/3. This is the road taken by Fotakis [12] in order to derive this bound.

However, neither inequality (3) nor inequality (4) hold in k-uniform congestion games due to the more complicated strategy spaces. E.g., for the Nash equilibrium s^{NE} and system optimum s^{OPT} in Fig. 1 we have

$$c_4(x_4^{\mathsf{NE}}) = 2 > c_2(x_2^{\mathsf{NE}} + 1) = 1,$$

contradicting (3), as well as

$$c_4(x_4^{NE}) + c_5(x_5^{NE}) = 4 > 3 = c_2(x_2^{NE} + 1) + c_3(x_2^{NE} + 1),$$

contradicting (4). More generally speaking, inequality (3) does not necessarily hold if all players choosing r in s^{NE} also choose r'. The main technical work in our proof of Theorem 1 is to derive an alternative upper bound for the right hand side in (2). Specifically, we will work towards showing that for k-uniform congestion games, we have

$$\sum_{b \in B}(x_b^{NE} - x_b^{OPT})c_b(x_b^{NE}) - \sum_{a \in A}(x_a^{OPT} - x_a^{NE})c_a(x_a^{NE} + 1) \leq \frac{2}{7}C(s^{NE}). \quad (5)$$

In order to show inequality (5), some further notation is necessary. A natural way of decomposing the cost of a strategy vector s is to consider the tuples (i, r) with the property that player i uses resource r in strategy s. One may think of such a tuple as a single unit of demand that player i places on resource r under strategy vector s. The cost of a unit of demand is equal to the cost of the corresponding resource under that strategy profile, and the cost of strategy profile is then equal to the sum of the costs of the units of demand. Let

$$P_A \subseteq \{(i, a) : a \in A, a \in s_i^{NE}\}$$

be a subset of the units of demand placed in s^{NE} on the resources in A such that $|\{(i, a) \in P_A\}| = x_a^{OPT} - x_a^{NE}$ for all $a \in A$, i.e., for each resource $a \in A$, P_A contains exactly as many units of demand as there are more on these resources in s^{OPT} than in s^{NE}. Similarly, let

$$P_B \subseteq \{(i, b) : b \in B, b \in s_i^{OPT}\}$$

be such that $|\{(i, b) \in P_B\}| = x_b^{NE} - x_b^{OPT}$ for all $b \in B$.

Given these definitions, we want to bound the total costs of the units in P_B with respect to the total costs of the units in P_A. We first identify a subset of these units, for which a simple bound can be obtained, i.e., we identify units of demand $(i, a) \in P_A$ and $(j, b) \in P_B$ such that $c_b(x_b^{NE}) \leq c_a(x_a^{NE} + 1)$. For our purposes, it is sufficient to do this iteratively in a greedy way, see the greedy cancelling process in Algorithm 1.

Intuitively, this algorithm maps all units of demand in P_B whose cost are bounded by the cost of another unit in P_A and removes both units from the sets P_A and P_B. In the following, we denote by $P_A' \subseteq P_A$ and $P_B' \subseteq P_B$ the set of units that survives this elimination. We denote by $x_a'^{OPT}$ and $x_b'^{NE}$ the number of units of demand that survive this elimination on each underloaded and overloaded resource respectively. Note that by definition of P_A and P_B, we have that $x_b'^{NE} \geq x_b^{OPT}$ for $b \in B$, and $x_a'^{OPT} \geq x_a^{NE}$ for $a \in A$ after the cancelling.

Also note that during the course of the algorithm there may be different pairs $(i, a) \in P_A$ and $(j, b) \in P_B$ for which the condition in the if-loop is satisfied. For

$P'_A \leftarrow P_A,\ P'_B \leftarrow P_B;$
$x'^{OPT}_a \leftarrow x^{OPT}_a, \forall a \in A;$
$x'^{NE}_b \leftarrow x^{NE}_b, \forall b \in B;$
while true do
    if *there are* $(i,a) \in P_A$ *and* $(j,b) \in P_B$ *with* $c_b(x^{NE}_b) \le c_a(x^{NE}_a + 1)$ **then**
        $P'_A \leftarrow P'_A \setminus \{(i,a)\};$
        $P'_B \leftarrow P'_B \setminus \{(j,b)\};$
        $x'^{OPT}_a \leftarrow x'^{OPT}_a - 1;$
        $x'^{NE}_b \leftarrow x'^{NE}_b - 1;$
    else
        return $P'_A, P'_B, x'^{OPT}_a, \forall a \in A, x'^{NE}_b, \forall b \in B;$
    end
end

Algorithm 1. Cancelling process

our following arguments it is irrelevant, which of these is removed from P_A and P_B. Let

$$A' = \{a \in A : \text{ there is } (i,a) \in P'_A \text{ for some } i \in N\}, \tag{6a}$$
$$B' = \{b \in B : \text{ there is } (i,b) \in P'_B \text{ for some } i \in N\} \tag{6b}$$

be the resources that remain over- respectively underloaded in s^{NE} as opposed to s^{OPT} after the cancelling process. The following lemma then follows directly by definition of the above cancelling process and states that the cost of cancelled packets on B with respect to $c_b(x^{NE}_b)$ is bounded by the cost of the cancelled packets on A with respect to $c_a(x^{NE}_a + 1)$.

Lemma 2. *For a symmetric k-uniform congestion game with affine cost functions, A and B as in (1), and A' and B' as in (6), we have*

$$\sum_{b \in B}(x^{NE}_b - x'^{NE}_b)c_b(x^{NE}_b) - \sum_{a \in A}(x^{OPT}_a - x'^{OPT}_a)c_a(x^{NE}_a + 1) \le 0.$$

For the following arguments, it may be helpful to consult Fig. 2 that shows the outcome of the cancelling process and the resulting sets A' and B' for the congestion game introduced in Example 1. Let us define

$$P = \{(i,r) : r \in R, r \in s^{NE}_i\} \tag{7}$$

as the set of all units of demand in s^{NE}. The next lemma is the first, crucial ingredient that allows us to obtain improved bounds on the price of anarchy. It states that for each "overloaded" unit of demand on a resource in P'_B, there are "enough" other units on other resources. Subsequently, we also bound the cost of these other units. The proof of the lemma is deferred to the full version [10].

Lemma 3. *For a symmetric k-uniform congestion game with affine cost functions, let P be as in (7) and let (P'_A, P'_B) be the output of Algorithm 1. Then, $|P \setminus P'_B| \ge 3|P'_B|$.*

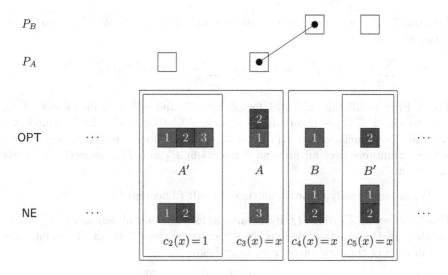

Fig. 2. Underloaded resources $A = \{2,3\}$ and overloaded resources $B = \{4,5\}$ for the game considered in Example 1. In the cancelling process one unit of demand of resource 4 cancelled out with a unit of demand of resource 3. After the cancelling process only resource 2 is underloaded and only resource 5 is overloaded, i.e., $A' = \{2\}$ and $B' = \{5\}$.

Before we proceed, we provide two structural lemmas which restrict the space of instances with worst-case price of anarchy.

Lemma 4. *The worst-case price of anarchy of symmetric k-uniform congestion games is attained on games that have the property that no resource is chosen by all players both in an optimal strategy vector and a worst-case Nash equilibrium.*

The proof is by contradiction and can be found in the full version. The next lemma is a technical lemma specifically about the structure of worst-case instances with $\mathsf{PoA}(I) \geq 4/3$. Again, the proof is given in the full version [10].

Lemma 5. *For any instance I of a symmetric k-uniform congestion game with affine cost functions and $\mathsf{PoA}(I) \geq 4/3$ and a resource $r \in R \setminus B'$, chosen by all players in s^{NE}, there exists an instance $\tilde{I}$, with resource r removed, such that $\mathsf{PoA}(\tilde{I}) \geq \mathsf{PoA}(I)$.*

The restrictions on the structure of worst-case instances obtained in Lemma 5 will be used later in the proof of Theorem 1. Before we can do that, however, we proceed to bound the costs of the resources in A' with the following two lemmas.

Lemma 6. *For a symmetric k-uniform congestion game with affine cost functions, we have $c_r(x_r^{\mathsf{NE}}) \leq 2c_{r'}(x_{r'}^{\mathsf{NE}}+1)$ for any two resources r, r', where $x_r^{\mathsf{NE}} \geq 1$ and $x_{r'}^{\mathsf{NE}} < n$.*

The proof is contained in the full version [10].

Lemma 7. *For a symmetric k-uniform congestion game with affine cost functions, we have*

$$\sum_{b \in B'} (x_b'^{\mathsf{NE}} - x_b^{\mathsf{OPT}}) c_b(x_b^{\mathsf{NE}}) \leq 2 \sum_{a \in A'} (x_a'^{\mathsf{OPT}} - x_a^{\mathsf{NE}}) c_a(x_a^{\mathsf{NE}} + 1).$$

Proof. First recall that $x_b^{\mathsf{NE}} \geq 1$ for all $b \in B'$, and $x_a^{\mathsf{NE}} < n$ for all $a \in A'$ as resources in $A' \subseteq A$ are chosen more often in s^{OPT} than in s^{NE}. By Lemma 6, we can therefore conclude that $c_b(x_b^{\mathsf{NE}}) \leq 2c_a(x_a^{\mathsf{NE}} + 1)$ for all resources $b \in B'$ and $a \in A'$. Summing over all units of demands in P_A' and P_B', respectively, yields the result. □

We are now ready to prove our main result (Theorem 1).

Proof. (Proof of Theorem 1). By Lemma 3, for each unit of demand in P_B', there are three distinct units of demand in s^{NE} on resources $r \in R \setminus B'$. We bound the cost of each of these resource units from below by

$$c_r(x_r^{\mathsf{NE}}) \geq \frac{c_r(x_r^{\mathsf{NE}} + 1)}{2} \geq \frac{c_b(x_b^{\mathsf{NE}})}{4}, \tag{8}$$

for any $b \in B$. Here the first inequality follows directly from the fact that the cost functions are affine. The second inequality follows from Lemma 6 for resources r with $x_r^{\mathsf{NE}} < n$. However, by Lemma 5, it is without loss of generality to assume that no resource $r \in R \setminus B'$ is chosen by all players in s^{NE}, unless the price of anarchy is not larger than $4/3$. Therefore, we finally get

$$C(s^{\mathsf{NE}}) \geq \sum_{b \in B'} (x_b'^{\mathsf{NE}} - x_b^{\mathsf{OPT}}) c_b(x_b^{\mathsf{NE}}) + \sum_{b \in B'} x_b^{\mathsf{OPT}} c_b(x_b^{\mathsf{NE}}) + \sum_{r \in R \setminus B'} x_r^{\mathsf{NE}} c_r(x_r^{\mathsf{NE}})$$

$$\geq \sum_{b \in B'} (x_b'^{\mathsf{NE}} - x_b^{\mathsf{OPT}}) c_b(x_b^{\mathsf{NE}}) + \sum_{r \in R \setminus B'} x_r^{\mathsf{NE}} c_r(x_r^{\mathsf{NE}})$$

$$\geq \frac{7}{4} \sum_{b \in B'} (x_b'^{\mathsf{NE}} - x_b^{\mathsf{OPT}}) c_b(x_b^{\mathsf{NE}}). \tag{9}$$

Here, the first inequality uses $x_b'^{\mathsf{NE}} \leq x_b^{\mathsf{NE}}$ for any resource $b \in B$, which follows from the cancelling process. The last inequality uses that $\sum_{b \in B'} (x_b'^{\mathsf{NE}} - x_b^{\mathsf{OPT}}) = |P_B'|$, and by Lemma 3, $\sum_{r \in R \setminus B'} x_r^{\mathsf{NE}} \geq |P \setminus P_B'| \geq 3|P_B'| = 3 \sum_{b \in B'} (x_b'^{\mathsf{NE}} - x_b^{\mathsf{OPT}})$, and each of these resource units has cost at least $c_b(x_b^{\mathsf{NE}})/4$, for all $b \in B$ by (8).

Combining (9) with Lemmas 2 and 7 yields

$$\sum_{b \in B} (x_b^{\mathsf{NE}} - x_b^{\mathsf{OPT}}) c_b(x_b^{\mathsf{NE}}) - \sum_{a \in A} (x_a^{\mathsf{OPT}} - x_a^{\mathsf{NE}}) c_a(x_a^{\mathsf{NE}} + 1)$$

$$\leq \sum_{b \in B'} (x_b'^{\mathsf{NE}} - x_b^{\mathsf{OPT}}) c_b(x_b^{\mathsf{NE}}) - \sum_{a \in A'} (x_a'^{\mathsf{OPT}} - x_a^{\mathsf{NE}}) c_a(x_a^{\mathsf{NE}} + 1)$$

$$\leq \frac{1}{2} \sum_{b \in B'} (x_b'^{\mathsf{NE}} - x_b^{\mathsf{OPT}}) c_b(x_b^{\mathsf{NE}}) \leq \frac{2}{7} C(s^{\mathsf{NE}}),$$

where the first inequality is by Lemma 2, the second by Lemma 7, and the third by (9). Finally, plugging this into (2) proves Theorem 1.   □

4   Lower Bound

We show that generalizing the strategy spaces from singletons to k-uniform matroids increases the price of anarchy of congestion games. The proof is by a parametric set of instances and contained in the full version [10].

Theorem 2. *The price of anarchy of symmetric k-uniform congestion games with affine cost functions is at least $7 - 4\sqrt{2} \approx 1.343$ for large enough k.*

5   Conclusions

The most interesting open problem, next to improving lower and upper bounds for the k-uniform matroid case we study here, is to analyze the price of anarchy for the generalized problem with arbitrary matroid strategy spaces. We also note that larger lower bounds on the price of anarchy can be achieved for more general settings such as matroid strategy spaces, or non-affine cost functions. They are not included in this paper, however.

References

1. Abed, F., Correa, J.R., Huang, C.-C.: Optimal coordination mechanisms for multi-job scheduling games. In: Schulz, A.S., Wagner, D. (eds.) ESA 2014. LNCS, vol. 8737, pp. 13–24. Springer, Heidelberg (2014)
2. Ackermann, H., Röglin, H., Vöcking, B.: On the impact of combinatorial structure on congestion games. J. ACM **55**(6), 1–22 (2008)
3. Ackermann, H., Röglin, H., Vöcking, B.: Pure Nash equilibria in player-specific and weighted congestion games. Theoret. Comput. Sci. **410**(17), 1552–1563 (2009)
4. Aland, S., Dumrauf, D., Gairing, M., Monien, B., Schoppmann, F.: Exact price of anarchy for polynomial congestion games. SIAM J. Comput. **40**(5), 1211–1233 (2011)
5. Awerbuch, B., Azar, Y., Epstein, A.: The price of routing unsplittable flow. In: Proceedings of the 37th Annual ACM Symposium Theory Computing, pp. 57–66 (2005)
6. Beckmann, M., McGuire, C.B., Winsten, C.B.: Studies in the Economics and Transportation. Yale University Press, New Haven (1956)
7. Braess, D.: Über ein Paradoxon aus der Verkehrsplanung. Unternehmensforschung **12**, 258–0268 (1968). (German)
8. Caragiannis, I., Flammini, M., Kaklamanis, C., Kanellopoulos, P., Moscardelli, L.: Tight bounds for selfish and greedy load balancing. In: Bugliesi, M., Preneel, B., Sassone, V., Wegener, I. (eds.) ICALP 2006. LNCS, vol. 4051, pp. 311–322. Springer, Heidelberg (2006)
9. Christodoulou, G., Koutsoupias, E.: The price of anarchy of finite congestion games. In: Proceedings of the 37th Annual ACM Symposium Theory Computing, pp. 67–73 (2005)

10. de Jong, J., Klimm, M., Uetz, M.: Efficiency of equilibria in uniform matroid congestion games. CTIT Technical report TR-CTIT-16-04, University of Twente (2016). http://eprints.eemcs.utwente.nl/26855/
11. Dunkel, J., Schulz, A.S.: On the complexity of pure-strategy Nash equilibria in congestion and local-effect games. Math. Oper. Res. **33**(4), 851–868 (2008)
12. Fotakis, D.: Stackelberg strategies for atomic congestion games. Theory Comput. Syst. **47**, 218–249 (2010)
13. Fujishige, S., Goemans, M.X., Harks, T., Peis, B., Zenklusen, R.: Matroids are immune to Braess paradox. arXiv:1504.07545 (2015)
14. Gairing, M., Lücking, T., Mavronicolas, M., Monien, B., Rode, M.: Nash equilibria in discrete routing games with convex latency functions. In: Díaz, J., Karhumäki, J., Lepistö, A., Sannella, D. (eds.) ICALP 2004. LNCS, vol. 3142, pp. 645–657. Springer, Heidelberg (2004)
15. Gairing, M., Schoppmann, F.: Total latency in singleton congestion games. In: Deng, X., Graham, F.C. (eds.) WINE 2007. LNCS, vol. 4858, pp. 381–387. Springer, Heidelberg (2007)
16. Goemans, M.X., Mirrokni, V.S., Vetta, A.: Sink equilibria and convergence. In: Proceedings of the 46th Annual IEEE Symposium Foundations of Computer Science, pp. 142–154 (2005)
17. Harks, T., Klimm, M., Peis, B.: Resource competition on integral polymatroids. In: Liu, T.-Y., Qi, Q., Ye, Y. (eds.) WINE 2014. LNCS, vol. 8877, pp. 189–202. Springer, Heidelberg (2014)
18. Harks, T., Peis, B.: Resource buying games. Algorithmica **70**(3), 493–512 (2014)
19. Koutsoupias, E., Papadimitriou, C.: Worst-case equilibria. In: Meinel, C., Tison, S. (eds.) STACS 1999. LNCS, vol. 1563, pp. 404–413. Springer, Heidelberg (1999)
20. Lücking, T., Mavronicolas, M., Monien, B., Rode, M.: A new model for selfish routing. Theoret. Comput. Sci. **406**(3), 187–206 (2008)
21. Meyers, C., Problems, N.F., Games, C.: Complexity and approximation results. Ph.D. thesis, MIT, Operations Research Center (2006)
22. Milchtaich, I.: Congestion games with player-specific payoff functions. Games Econom. Behav. **13**(1), 111–124 (1996)
23. Pigou, A.C.: The Economics of Welfare. Macmillan, London (1920)
24. Rosenthal, R.W.: A class of games possessing pure-strategy Nash equilibria. Internat. J. Game Theory **2**(1), 65–67 (1973)
25. Rosenthal, R.W.: The network equilibrium problem in integers. Networks **3**, 53–59 (1973)
26. Roughgarden, T.: The price of anarchy is independent of the network topology. J. Comput. System Sci. **67**, 341–364 (2002)
27. Roughgarden, T., Tardos, É.: How bad is selfish routing? J. ACM **49**(2), 236–259 (2002)
28. Suri, S., Tóth, C.D., Zhou, Y.: Selfish load balancing and atomic congestion games. Algorithmica **47**(1), 79–96 (2007)
29. Tran-Thanh, L., Polukarov, M., Chapman, A., Rogers, A., Jennings, N.R.: On the existence of pure strategy nash equilibria in integer–splittable weighted congestion games. In: Persiano, G. (ed.) SAGT 2011. LNCS, vol. 6982, pp. 236–253. Springer, Heidelberg (2011)
30. Wardrop, J.G.: Some theoretical aspects of road traffic research. Proc. Inst. Civ. Eng. **1**(3), 325–362 (1952)

On the Price of Anarchy of Highly Congested Nonatomic Network Games

Riccardo Colini-Baldeschi[1], Roberto Cominetti[2], and Marco Scarsini[1(✉)]

[1] Dipartimento di Economia e Finanza, LUISS, Viale Romania 32, 00197 Rome, Italy
{rcolini,marco.scarsini}@luiss.it
[2] Facultad de Ingeniería y Ciencias, Universidad Adolfo Ibáñez, Santiago, Chile
roberto.cominetti@uai.cl

Abstract. We consider nonatomic network games with one source and one destination. We examine the asymptotic behavior of the price of anarchy as the inflow increases. In accordance with some empirical observations, we show that, under suitable conditions, the price of anarchy is asymptotic to one. We show with some counterexamples that this is not always the case. The counterexamples occur in simple parallel graphs.

1 Introduction

The study of network routing costs and their efficiency goes back to Pigou [23] who, in the first edition of his book, introduces his famous two-road model. Wardrop [31] develops a model where many players (vehicles on the road) choose a road in order to minimize their cost (travel time) and the influence of each one of them, singularly taken, is negligible. He introduces a concept of equilibrium that has become the standard in the literature on nonatomic network games.

When travelers minimize their travel time without considering the negative externalities that their behavior has on other travelers, the collective outcome of the choices of all travelers is typically inefficient, i.e., it is worse than the outcome that a benevolent planner would have achieved. Various measures have been proposed to quantify this inefficiency. Among them the price of anarchy has been the most successful. Introduced by Koutsoupias and Papadimitriou [14] and given this name by Papadimitriou [21], it is the ratio of the worst social equilibrium cost and the optimum cost. The price of anarchy has been studied by several authors and interesting bounds for it have been found under some conditions on the cost functions.

Most of the existing results about the price of anarchy consider worst-case scenarios. They are not necessarily helpful in specific situations. In a nice recent paper O'Hare et al. [19] show, both theoretically and with the aid of simulations, how the price of anarchy is affected by changes in the total inflow of players. They consider data for three cities and they write: *"In each city, it can be seen that there are broadly three identifiably distinct regions of behaviour: an initial region in which the Price of Anarchy is one; an intermediate region of fluctuations; and a final region of decay, which has a similar characteristic shape across all*

© Springer-Verlag Berlin Heidelberg 2016
M. Gairing and R. Savani (Eds.): SAGT 2016, LNCS 9928, pp. 117–128, 2016.
DOI: 10.1007/978-3-662-53354-3_10

three networks. The similarities in this general behaviour across the three cities suggest that there may be common mechanisms that drive this variation."

The core of the paper [19] is an analysis of the intermediate fluctuations. In our paper we will mainly look at the asymptotic behavior of the price of anarchy. We consider nonatomic congestion games with single source and single destination. We show that for a large class of cost functions the price of anarchy is, indeed, asymptotic to one, as the mass of players grows. Nevertheless, we find counterexamples where its lim sup is not 1 and it can even be infinite.

Contribution. The goal of this paper is twofold. On one hand we provide some positive results that show that under some conditions the price of anarchy of nonatomic network games is indeed asymptotic to one. On the other hand, we present counterexamples where the lim sup of the price of anarchy is not one.

We first show that, for any single-source, single-destination graph, the price of anarchy is asymptotic to one whenever the cost of at least one path is bounded. Then we focus on parallel graphs and we show that the price of anarchy is asymptotic to one for a large class of costs that we characterize in terms of regularly varying functions (see [3] for properties of these functions). This class includes affine functions and cost functions that can be bounded by a pair of affine functions with the same slope.

Next, we present counterexamples where the behavior of the price of anarchy is periodic on a logarithmic scale, so that its lim sup is larger than one both as the mass of players grows unbounded and as it goes to zero. In another counterexample the lim sup of the price of anarchy is infinite. A further counterexample shows that the price of anarchy may not converge to one even for convex costs. An interesting point is that all the counterexamples concern a very simple parallel graph with just two edges, so that the bad behavior of the price of anarchy depends solely on the costs and not on the topology of the graph. This is in stark contrast with the results in [19], where the irregular behavior of the price of anarchy in the intermediate region of inflow heavily depends on the structure of the graph.

Related Literature. Wardrop's nonatomic model has been studied by Beckmann et al. [2] and many others. The formal foundation of games with a continuum of players came with Schmeidler [30] and then with Mas Colell [16]. Nonatomic congestion games have been studied, among others, by Milchtaich [17, 18].

Various bounds for the price of anarchy in nonatomic games have been proved, under different conditions. In particular Roughgarden and Tardos [27] prove that, when the cost functions are affine, the price of anarchy in nonatomic games is at most 4/3, irrespective of the topology of the network. The bound is sharp and is attained even in very simple networks. Several authors have extended this bound to larger classes of functions. Roughgarden [25] shows that if the class of cost functions includes the constants, then the worst price of anarchy is achieved on parallel networks with just two edges. In his paper he considers bounds for the price of anarchy when the cost functions are polynomials of degree at most d. Dumrauf and Gairing [8] do the same when the degrees of the polynomials are between s and d. Roughgarden and Tardos [28] provide a unifying result for the class of standard costs, i.e., costs c that are differentiable and such that $xc(x)$ is convex. Correa et al. [5] consider the price of anarchy for networks where edges have a capacity and costs are not necessarily convex, differentiable, or even continuous. In [7] they reinterpret and extend these results using a geometric approach. In [6] they consider the problem of minimizing the maximum latency rather than the average latency and provide results about the price of anarchy in this framework. The reader is referred to [26,29] for a survey.

Some papers show how in real life the price of anarchy may substantially differ from the worst-case scenario, [15,32]. González Vayá et al. [12] deal with a problem of optimal schedule for the electricity demand of a fleet of plug-in electric vehicles. Without using the term, they show that the price of anarchy goes to one as the number of vehicles grows. Cole and Tao [4] study large Walrasian auctions and large Fisher markets and show that in both cases the price of anarchy goes to one as the market size increases. Feldman et al. [10] define a concept of (λ, μ)-smoothness for sequences of games, and show that the price of anarchy in atomic congestion games converges to the price of anarchy of the corresponding nonatomic game, when the number of players grows. Patriksson [22] and Josefsson and Patriksson [13] perform sensitivity analysis of Wardrop equilibrium to some parameters of the model. Closer to the scope of our paper, Englert et al. [9] examine how the equilibrium of a congestion game changes when either the total mass of players is increased by ε or an edge that carries an ε fraction of the mass is removed. For polynomial cost functions they bound the increase of the equilibrium cost when a mass ε of players is added to the system. Other recent papers, such as [20,24], have also raised questions about the practical validity of known results about the price of anarchy.

2   The Model

Consider a finite directed multigraph $\mathscr{G} = (V, E)$, where V is a set of vertices and E is a set of edges. The graph G together with a source s and a destination t, with $s, t \in V$, is called a network. A path P is a set of consecutive edges that go from source to destination. Call $\mathscr{P}$ the set of all paths. Each path P has a flow $x_P \geq 0$ and call $\boldsymbol{x} = (x_P)_{P \in \mathscr{P}}$. The total flow from source to destination is denoted by $M \in \mathbb{R}_+$. A flow $\boldsymbol{x}$ is feasible if $\sum_{P \in \mathscr{P}} x_P = M$. Call $\mathscr{F}_M$ the set of feasible flows. For each edge $e \in E$ there exists a cost function $c_e(\cdot) : \mathbb{R}_+ \to \mathbb{R}_+$, that is assumed (weakly) increasing and continuous. Call $\boldsymbol{c} = (c_e)_{e \in E}$. This defines a nonatomic congestion game $\Gamma_M = (\mathscr{G}, M, \boldsymbol{c})$. The number M can be seen as the mass of players who play the game.

The cost of a path P with respect to a flow $\boldsymbol{x}$ is the sum of the cost of its edges: $c_P(\boldsymbol{x}) = \sum_{e \in P} c_e(x_e)$, where

$$x_e = \sum_{\substack{P \in \mathscr{P}: \\ e \in P}} x_P.$$

A flow $\boldsymbol{x}^*$ is an *equilibrium flow* if for every $P, Q \in \mathscr{P}$ such that $x_P^* > 0$ we have $c_P(\boldsymbol{x}^*) \leq c_Q(\boldsymbol{x}^*)$. Denote $\mathscr{E}(\Gamma_M)$ the set of all such equilibrium flows.

For each flow $\boldsymbol{x}$ define the *social cost* associated to it as

$$C(\boldsymbol{x}) := \sum_{P \in \mathscr{P}} x_P c_P(\boldsymbol{x}) = \sum_{e \in E} x_e c_e(x_e),$$

and let $\mathsf{Opt}(\Gamma_M) = \min_{\boldsymbol{x} \in \mathscr{F}_M} C(\boldsymbol{x})$ be the *optimum cost* of Γ_M. Define also the *worst equilibrium cost* of Γ_M as $\mathsf{WEq}(\Gamma_M) = \max_{\boldsymbol{x} \in \mathscr{E}(\Gamma_M)} C(\boldsymbol{x})$. Actually, in the present setting the cost $C(\boldsymbol{x}^*)$ is the same for every equilibrium $\boldsymbol{x}^*$ (see [11]).

The *price of anarchy* of the game Γ_M is then defined as

$$\mathsf{PoA}(\Gamma_M) := \frac{\mathsf{WEq}(\Gamma_M)}{\mathsf{Opt}(\Gamma_M)}.$$

We will be interested in the price of anarchy of this game, as $M \to \infty$. We will show that, under suitable conditions, it is asymptotic to one. We call *asymptotically well behaved* the congestion games for which this happens.

3   Well Behaved Congestion Games

3.1   General Result

The following general result shows that for any network the price of anarchy is asymptotic to one when at least one path has a bounded cost.

Theorem 1. *For each path $P \in \mathscr{P}$ denote*

$$c_P^\infty = \sum_{e \in P} c_e^\infty \quad \text{with} \quad c_e^\infty = \lim_{z \to \infty} c_e(z)$$

and suppose that $B := \min_{P \in \mathscr{P}} c_P^\infty$ is finite. Then, $\lim_{M \to \infty} \mathsf{PoA}(\Gamma_M) = 1$.

Proof. Let $\boldsymbol{x}^*$ be an equilibrium for Γ_M. Then if $x_P^* > 0$ we have

$$c_P(\boldsymbol{x}^*) = \min_{Q \in \mathscr{P}} c_Q(\boldsymbol{x}^*) \leq \min_{Q \in \mathscr{P}} c_Q^\infty = B$$

and therefore

$$\mathsf{WEq}(\Gamma_M) = \sum_{P \in \mathscr{P}} x_P^* c_P(\boldsymbol{x}^*) \leq \sum_{P \in \mathscr{P}} x_P^* B = MB.$$

It follows that

$$\mathsf{PoA}(\Gamma_M) \leq \frac{MB}{\mathsf{Opt}(\Gamma_M)},$$

so that it suffices to prove that $\mathsf{Opt}(\Gamma_M)/M \to B$. To this end denote $\Delta(\mathscr{P})$ the simplex defined by $\boldsymbol{y} = (y_P)_{P \in \mathscr{P}} \geq 0$ and $\sum_{P \in \mathscr{P}} y_P = 1$, so that

$$\frac{1}{M} \mathsf{Opt}(\Gamma_M) = \min_{\boldsymbol{x} \in \mathscr{F}_M} \sum_{P \in \mathscr{P}} \frac{x_P}{M} c_P(\boldsymbol{x})$$

$$= \min_{\boldsymbol{y} \in \Delta(\mathscr{P})} \sum_{P \in \mathscr{P}} y_P c_P(M\boldsymbol{y}).$$

Denote $\Phi_M(y) = \sum_{P \in \mathscr{P}} y_P c_P(My)$. Since the cost functions $c_e(\cdot)$ are non-decreasing, the family $\Phi_M(\cdot)$ monotonically increases with M towards the limit function

$$\Phi_\infty(y) = \sum_{P \in \mathscr{P}: y_P > 0} y_P c_P^\infty.$$

Now we use the fact that a monotonically increasing family of functions epi-converges (see [1]) and since $\Delta(\mathscr{P})$ is compact it follows that the minimum $\min_{y \in \Delta(\mathscr{P})} \Phi_M(y)$ converges as $M \to \infty$ towards

$$\min_{y \in \Delta(\mathscr{P})} \Phi_\infty(y).$$

Clearly this latter optimal value is B and is attained by setting $y_P > 0$ only on those paths P that attain the smallest value $c_P^\infty = B$, and therefore we conclude

$$\frac{1}{M} \mathsf{Opt}(\Gamma_M) = \min_{y \in \Delta(\mathscr{P})} \Phi_M(y) \to B,$$

as was to be proved.   □

3.2   Parallel Graphs

In this section we examine the asymptotic behavior of the price of anarchy when the game is played on a parallel graph.

Let $\mathscr{G} = (V, E)$ be a parallel graph such that $V = \{s, t\}$ are the vertices and $E = \{e_1, e_2, \ldots, e_n\}$ are the edges. For each edge $e_i \in E$ the function $c_i(\cdot)$ represents the cost function of the edge e_i. Call $\Gamma_M = (\mathscr{G}, M, c)$ the corresponding game. In the whole section we will deal with this graph.

Adding a Constant to Costs. First we prove a preservation result. We show that if the price of anarchy of a game converges to 1, then adding positive constants to each cost does not alter this asymptotic behavior.

Theorem 2. *Given a game $\Gamma_M = (\mathscr{G}, M, c)$ and a vector $a \in [0, \infty)^n$, consider a new game $\Gamma_M^a(\mathscr{G}, M, c^a)$, where*

$$c_i^a(x) = a_i + c_i(x).$$

If $c_i(\cdot)$ is strictly increasing and continuous, $\lim_{x \to \infty} c_i(x) = \infty$ for all $e_i \in E$, and $\lim_{M \to \infty} \mathsf{PoA}(\Gamma_M) = 1$, then $\lim_{M \to \infty} \mathsf{PoA}(\Gamma_M^a) = 1$.

Regularly Varying Functions

Definition 3. *Let $\beta \geq 0$. A function $\Theta : (0, +\infty) \to (0, +\infty)$ is called β-regularly varying if for all $a > 0$*

$$\lim_{x \to \infty} \frac{\Theta(a \cdot x)}{\Theta(x)} = a^\beta \in (0, +\infty).$$

When $\beta = 1$, we just say that the function is regularly varying.

The following theorem shows that asymptotically the price of anarchy goes to 1 for a large class of cost functions.

Theorem 4. *Consider the game Γ_M and suppose that for some $\beta > 0$ there exists a β-regularly varying function $c(\cdot) \in C^1$ such that the function $x \mapsto c(x) + xc'(x)$ is strictly increasing and for all $e_i \in E$ the function $c_i(\cdot)$ is strictly increasing and continuous with*

$$\lim_{x \to \infty} \frac{c^{-1} \circ c_i(x)}{x} = \alpha_i \in (0, +\infty] \tag{1}$$

and that at least one α_i is finite. Then

$$\lim_{M \to \infty} \mathsf{PoA}(\Gamma_M) = 1.$$

Proof. We begin by noting that if some cost $c_i(\cdot)$ is bounded, then the result follows directly from Theorem 1. Suppose now that $c_i(x) \to \infty$ when $x \to \infty$ in all links and consider first the case where all the α_i are finite. In this case the equilibrium flows x_i^* must diverge to ∞ as $M \to \infty$ and the equilibrium is characterized by $c_i(x_i^*) = \lambda$. This allows to derive an upper bound for the cost of the equilibrium. That is, (1) implies that for small $\varepsilon > 0$ we have

$$\frac{c^{-1} \circ c(x_i^*)}{x_i^*} = \frac{c^{-1}(\lambda)}{x_i^*} \in (\alpha_i - \varepsilon, \alpha_i + \varepsilon),$$

provided M is large enough. It then follows that

$$\sum_{i=1}^{n} \frac{c^{-1}(\lambda)}{\alpha_i + \varepsilon} \leq \sum_{i=1}^{n} x_i^* = M,$$

so that, denoting

$$a(\varepsilon) = \left(\sum_{i=1}^{n} \frac{1}{\alpha_i + \varepsilon} \right)^{-1},$$

we get $\lambda \leq c(Ma(\varepsilon))$ and

$$\mathsf{WEq} = M\lambda \leq Mc(Ma(\varepsilon)).$$

Next we derive a lower bound for the optimal cost

$$\mathsf{Opt}(\Gamma_M) = \min_{x \in \mathscr{F}_M} \sum_{i=1}^{n} x_i c_i(x_i).$$

We note that when $M \to \infty$ the optimal solutions are such that $x_i(M) \to \infty$ so that using (1) and the fact that $\alpha_i - \varepsilon > 0$ we get for all M large enough

$$\min_{x \in \mathscr{F}_M} \sum_{i=1}^{n} x_i c_i(x_i) \geq \min_{x \in \mathscr{F}_M} \sum_{i=1}^{n} x_i c((\alpha_i - \varepsilon)x_i).$$

The optimality condition for the latter yields

$$c((\alpha_i - \varepsilon)x_i) + (\alpha_i - \varepsilon)x_i c'((\alpha_i - \varepsilon)x_i) = \mu.$$

For the sake of brevity we denote $\tilde{c}(x) = c(x) + xc'(x)$ and $y_i = (\alpha_i - \varepsilon)x_i$ so that the optimality condition becomes $\tilde{c}(y_i) = \mu$. This yields $y_i = \tilde{c}^{-1}(\mu)$ and therefore

$$M = \sum_{i=1}^{n} x_i = \sum_{i=1}^{n} \frac{\tilde{c}^{-1}(\mu)}{\alpha_i - \varepsilon}.$$

Denoting

$$b(\varepsilon) = \left(\sum_{i=1}^{n} \frac{1}{\alpha_i - \varepsilon} \right)^{-1},$$

we then get $\mu = \tilde{c}(Mb(\varepsilon))$ and we obtain the following lower bound for the optimal cost

$$\mathsf{Opt}(\Gamma_M) \geq \min_{x \in \mathscr{F}_M} \sum_{i=1}^{n} x_i c((\alpha_i - \varepsilon)x_i) = Mc(\tilde{c}^{-1}(\mu)) = Mc(Mb(\varepsilon)).$$

Combining the previous bounds we obtain the following estimate for the price of anarchy

$$\mathsf{PoA}(\Gamma_M) \leq \frac{Mc(Ma(\varepsilon))}{Mc(Mb(\varepsilon))}.$$

Letting $M \to \infty$ and using the fact that c is β-regularly varying we deduce

$$\limsup_{M \to \infty} \mathsf{PoA}(\Gamma_M) \leq \left(\frac{a(\varepsilon)}{b(\varepsilon)} \right)^{\beta}$$

and since $a(\varepsilon)/b(\varepsilon) \to 1$ as $\varepsilon \to 0$ we conclude

$$\limsup_{M \to \infty} \mathsf{PoA}(\Gamma_M) = 1.$$

If some $\alpha_i = \infty$, then call $I_0 := \{i : \alpha_i < \infty\}$. In equilibrium

$$M = \sum_{i=1}^{n} c_i^{-1}(\lambda) \geq \sum_{i \in I_0} c_i^{-1}(\lambda) \geq \sum_{i \in I_0} \frac{1}{\alpha_i + \varepsilon} c^{-1}(\lambda),$$

hence

$$\lambda \leq c \left(M \left(\sum_{i \in I_0} \frac{1}{\alpha_i + \varepsilon} \right)^{-1} \right).$$

In the optimum proceed as before with $\alpha_i' \nearrow \alpha_i$. $\square$

The following results follow easily from Theorem 4.

Corollary 5. *In the game Γ_M if for all $i \in E$ we have $\lim_{x\to\infty} c_i(x)/x = m_i \in (0, +\infty]$ and at least one $m_i < \infty$, then*

$$\lim_{M\to\infty} \mathsf{PoA}(\Gamma_M) = 1.$$

Corollary 6. *In the game Γ_M if for all $i \in E$ we have $\lim_{x\to\infty} c_i'(x) = m_i$ with $m_i \in (0, +\infty]$ and at least one m_i is finite, then*

$$\lim_{M\to\infty} \mathsf{PoA}(\Gamma_M) = 1.$$

Corollary 7. *In the game Γ_M if for all $i \in E$ for some $\beta > 0$ there exists a β-regularly varying function $c(\cdot)$ such that*

$$\lim_{x\to\infty} \frac{c_i(x)}{c(x)} = m_i \in (0, +\infty], \tag{2}$$

and at least one m_i is finite, then

$$\lim_{M\to\infty} \mathsf{PoA}(\Gamma_M) = 1.$$

Corollary 8. *In the game Γ_M if, for all $e_i \in E$, $c_i(x) = a_i + b_i x$, then*

$$\lim_{M\to\infty} \mathsf{PoA}(\Gamma_M) = 1.$$

Costs Bounded by Affine Functions. The next theorem examines the case where each cost function is bounded above and below by two affine functions with the same slope.

Theorem 9. *Consider the game Γ_M and assume that for every $e_i \in E$*

$$\ell_i(x) := a_i + b_i x \leq c_i(x) \leq \alpha_i + b_i x =: L_i(x).$$

Then

$$\lim_{M\to\infty} \mathsf{PoA}(\Gamma_M) = 1.$$

4   Ill Behaved Games

In this section we will consider some examples where the price of anarchy is not asymptotic to one, as the inflow goes to infinity.

Consider a standard Pigou graph and assume that the costs are as follows:

$$c_1(x) = x,$$
$$c_2(x) = a^k \quad \text{for } x \in (a^{k-1}, a^k], \quad k \in \mathbb{Z}, \tag{3}$$

with $a \geq 2$, as in Fig. 1. In this game the cost of one edge is the identity, whereas for the other edge it is a step function that touches the identity at intervals that grow exponentially. The cost function c_2 is not continuous, but a very similar game can be constructed by approximating it with a continuous function.

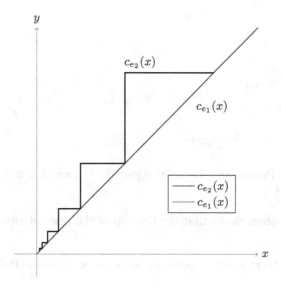

Fig. 1. Step function.

Theorem 10. *Consider the game Γ_M with costs as in (3). We have*

$$\liminf_{M\to\infty} \mathsf{PoA}(\Gamma_M) = 1, \quad \limsup_{M\to\infty} \mathsf{PoA}(\Gamma_M) = \frac{4+4a}{4+3a}.$$

Remark 11. We can immediately see that

$$\limsup_{M\to\infty} \mathsf{PoA}(\Gamma_M) = \frac{6}{5} \quad \text{for } a = 2$$

and

$$\limsup_{M\to\infty} \mathsf{PoA}(\Gamma_M) \to \frac{4}{3} \quad \text{as } a \to \infty.$$

The proof of Theorem 10 shows that there is a periodic behavior of the price of anarchy (on a logarithmic scale). This implies that

$$\liminf_{M\to 0} \mathsf{PoA}(\Gamma_M) = 1, \quad \limsup_{M\to 0} \mathsf{PoA}(\Gamma_M) = \frac{4+4a}{4+3a}.$$

That is, even for very small values of M the price of anarchy is not necessarily close to 1.

Figure 2 plots the price of anarchy for $M \in [2a^k, 2a^{k+1}]$, when $a = 3$.

The next theorem shows that the price of anarchy may fail to be asymptotic to one, even when the cost functions are all convex.

Theorem 12. *There exist congestion games Γ_M where the cost functions are all increasing and convex and both*

$$\limsup_{M\to\infty} \mathsf{PoA}(\Gamma_M) > 1 \quad and \quad \limsup_{M\to 0} \mathsf{PoA}(\Gamma_M) > 1.$$

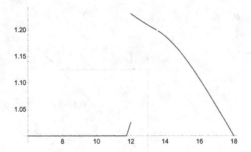

Fig. 2. Price of anarchy for $M \in [2a^k, 2a^{k+1}]$, with $a = 3$, $k = 1$.

The next theorem shows that the lim sup of the price of anarchy may even be infinite.

Theorem 13. *There exist congestion games Γ_M with* $\limsup\limits_{M \to \infty} \mathsf{PoA}(\Gamma_M) = \infty$.

5   Conclusions

The classical result of [27] can be restated as follows. Given a nontrivial single-commodity network, for any fixed total flow M, there exists a vector $\boldsymbol{c}$ of affine costs that depend on M, such that the price of anarchy of the corresponding game is $4/3$.

In this paper we have proved that, given a single-commodity network, for any vector $\boldsymbol{c}$ of costs that is bounded on some path P, there exists a total flow M such that the price of anarchy of the corresponding game is arbitrarily close to 1. Similar results have been obtained under different conditions on the network and the costs. What is relevant is that in our model the order of the quantifiers is reversed with respect to the classical bounds of the price of anarchy, such as [27].

Acknowledgments. Riccardo Colini-Baldeschi is a member of GNCS-INdAM. Roberto Cominetti gratefully acknowledges the support and hospitality of LUISS during a visit in which this research was initiated. His research is also supported by Núcleo Milenio Información y Coordinación en Redes ICM/FIC P10-024F. Marco Scarsini is a member of GNAMPA-INdAM. His work is partially supported by PRIN and MOE2013-T2-1-158.
The authors thank three referees for their insightful comments.

References

1. Attouch, H.: Variational Convergence for Functions and Operators. Pitman, Boston (1984)
2. Beckmann, M.J., McGuire, C., Winsten, C.B.: Studies in the Economics of Transportation. Yale University Press, New Haven (1956)

3. Bingham, N.H., Goldie, C.M., Teugels, J.L.: Regular Variation, Encyclopedia of Mathematics and its Applications, vol. 27. Cambridge University Press, Cambridge (1989)
4. Cole, R., Tao, Y.: The price of anarchy of large Walrasian auctions. Technical report, arXiv:1508.07370v4 (2015). http://arxiv.org/abs/1508.07370
5. Correa, J.R., Schulz, A.S., Stier-Moses, N.E.: Selfish routing in capacitated networks. Math. Oper. Res. **29**(4), 961–976 (2004). http://dx.doi.org/10.1287/moor.1040.0098
6. Correa, J.R., Schulz, A.S., Stier-Moses, N.E.: Fast, fair, and efficient flows in networks. Oper. Res. **55**(2), 215–225 (2007). http://dx.doi.org/10.1287/opre.1070.0383
7. Correa, J.R., Schulz, A.S., Stier-Moses, N.E.: A geometric approach to the price of anarchy in nonatomic congestion games. Games Econom. Behav. **64**(2), 457–469 (2008). http://dx.doi.org/10.1016/j.geb.2008.01.001
8. Dumrauf, D., Gairing, M.: Price of anarchy for polynomial Wardrop games. In: Spirakis, P.G., Mavronicolas, M., Kontogiannis, S.C. (eds.) WINE 2006. LNCS, vol. 4286, pp. 319–330. Springer, Heidelberg (2006). http://dx.doi.org/10.1007/11944874_29
9. Englert, M., Franke, T., Olbrich, L.: Sensitivity of Wardrop equilibria. Theory Comput. Syst. **47**(1), 3–14 (2010). http://dx.doi.org/10.1007/s00224-009-9196-4
10. Feldman, M., Immorlica, N., Lucier, B., Roughgarden, T., Syrgkanis, V.: The price of anarchy in large games. Technical report, arXiv:1503.04755 (2015)
11. Florian, M., Hearn, D.: Network equilibrium and pricing. In: Hall, R.W. (ed.) Handbook of Transportation Science, pp. 373–411. Springer, US, 978-0-306-48058-4 (2003). http://dx.doi.org/10.1007/0-306-48058-1_11
12. González Vayá, M., Grammatico, S., Andersson, G., Lygeros, J.: On the price of being selfish in large populations of plug-in electric vehicles. In: 2015 54th IEEE Conference on Decision and Control (CDC), pp. 6542–6547 (2015)
13. Josefsson, M., Patriksson, M.: Sensitivity analysis of separable traffic equilibrium equilibria with application to bilevel optimization in network design. Transp. Res. Part B: Methodol. **41**(1), 4–31 (2007). http://dx.doi.org/10.1016/j.trb.2005.12.004
14. Koutsoupias, E., Papadimitriou, C.: Worst-case equilibria. In: Meinel, C., Tison, S. (eds.) STACS 1999. LNCS, vol. 1563, pp. 404–413. Springer, Heidelberg (1999). http://dx.doi.org/10.1007/3-540-49116-3_38
15. Law, L.M., Huang, J., Liu, M.: Price of anarchy for congestion games in cognitive radio networks. IEEE Trans. Wireless Commun. **11**(10), 3778–3787 (2012)
16. Mas-Colell, A.: On a theorem of Schmeidler. J. Math. Econom. **13**(3), 201–206 (1984). http://dx.doi.org/10.1016/0304-4068(84)90029-6
17. Milchtaich, I.: Generic uniqueness of equilibrium in large crowding games. Math. Oper. Res. **25**(3), 349–364 (2000). http://dx.doi.org/10.1287/moor.25.3.349.12220
18. Milchtaich, I.: Social optimality and cooperation in nonatomic congestion games. J. Econom. Theory **114**(1), 56–87 (2004). http://dx.doi.org/10.1016/S0022-0531(03)00106-6
19. O'Hare, S.J., Connors, R.D., Watling, D.P.: Mechanisms that govern how the price of anarchy varies with travel demand. Transp. Res. Part B: Methodol. **84**, 55–80 (2016). http://dx.doi.org/10.1016/j.trb.2015.12.005
20. Panageas, I., Piliouras, G.: Approximating the geometry of dynamics in potential games. Technical report, arXiv:1403.3885v5 (2015). https://arxiv.org/abs/1403.3885

21. Papadimitriou, C.: Algorithms, games, and the Internet. In: Proceedings of the Thirty-Third Annual ACM Symposium on Theory of Computing, pp. 749–753. (2001). http://dx.doi.org/10.1145/380752.380883

22. Patriksson, M.: Sensitivity analysis of traffic equilibria. Transp. Sci. **38**(3), 258–281 (2004). http://pubsonline.informs.org/doi/abs/10.1287/trsc.1030.0043

23. Pigou, A.C.: The Economics of Welfare, 1st edn. Macmillan and Co., London (1920)

24. Piliouras, G., Nikolova, E., Shamma, J.S.: Risk sensitivity of price of anarchy under uncertainty. In: Proceedings of the Fourteenth ACM Conference on Electronic Commerce, EC 2013, pp. 715–732. ACM, New York (2013). http://doi.acm.org/10.1145/2482540.2482578

25. Roughgarden, T.: The price of anarchy is independent of the network topology. J. Comput. System Sci. **67**(2), 341–364 (2003). http://dx.doi.org/10.1016/S0022-0000(03)00044-8

26. Roughgarden, T.: Routing games. In: Algorithmic Game Theory, pp. 461–486. Cambridge Univ. Press, Cambridge (2007)

27. Roughgarden, T., Tardos, É.: How bad is selfish routing? J. ACM **49**(2), 236–259 (2002). (electronic) http://dx.doi.org/10.1145/506147.506153

28. Roughgarden, T., Tardos, É.: Bounding the inefficiency of equilibria in nonatomic congestion games. Games Econom. Behav. **47**(2), 389–403 (2004). http://dx.doi.org/10.1016/j.geb.2003.06.004

29. Roughgarden, T., Tardos, É.: Introduction to the inefficiency of equilibria. In: Algorithmic Game Theory, pp. 443–459. Cambridge Univ. Press, Cambridge (2007)

30. Schmeidler, D.: Equilibrium points of nonatomic games. J. Statist. Phys. **7**, 295–300 (1973)

31. Wardrop, J.G.: Some theoretical aspects of road traffic research. In: Proceedings of the Institute of Civil Engineers, Pt. II, vol. 1, pp. 325–378 (1952). http://dx.doi.org/10.1680/ipeds.1952.11362

32. Youn, H., Gastner, M.T., Jeong, H.: Price of anarchy in transportation networks: efficiency and optimality control. Phys. Rev. Lett. **101**, 128701 (2008). http://dx.doi.org/10.1103/PhysRevLett.101.128701

The Impact of Worst-Case Deviations in Non-Atomic Network Routing Games

Pieter Kleer[1] and Guido Schäfer[1,2(✉)]

[1] Centrum Wiskunde & Informatica (CWI),
Networks and Optimization Group, Amsterdam, The Netherlands
{kleer,schaefer}@cwi.nl
[2] Department of Econometrics and Operations Research,
Vrije Universiteit Amsterdam, Amsterdam, The Netherlands

Abstract. We introduce a unifying model to study the impact of worst-case latency deviations in non-atomic selfish routing games. In our model, latencies are subject to (bounded) deviations which are taken into account by the players. The quality deterioration caused by such deviations is assessed by the *Deviation Ratio*, i.e., the worst case ratio of the cost of a Nash flow with respect to deviated latencies and the cost of a Nash flow with respect to the unaltered latencies. This notion is inspired by the *Price of Risk Aversion* recently studied by Nikolova and Stier-Moses [9]. Here we generalize their model and results. In particular, we derive tight bounds on the Deviation Ratio for multi-commodity instances with a common source and arbitrary non-negative and non-decreasing latency functions. These bounds exhibit a linear dependency on the size of the network (besides other parameters). In contrast, we show that for general multi-commodity networks an exponential dependency is inevitable. We also improve recent smoothness results to bound the Price of Risk Aversion.

1 Introduction

In the classical selfish routing game introduced by Wardrop [12], there is an (infinitely) large population of (non-atomic) players who selfishly choose minimum latency paths in a network with flow-dependent latency functions. An assumption that is made in this model is that the latency functions are given deterministically. Although being a meaningful abstraction (which also facilitates the analysis of such games), this assumption is overly simplistic in situations where latencies are subject to deviations which are taken into account by the players.

In this paper, we study how much the quality of a Nash flow deteriorates in the worst case under (bounded) deviations of the latency functions. More precisely, given an instance of the selfish routing game with latency functions $(l_a)_{a \in A}$ on the arcs, we define the *Deviation Ratio (DR)* as the worst case ratio $C(f^\delta)/C(f^0)$ of a Nash flow f^δ with respect to deviated latency functions $(l_a + \delta_a)_{a \in A}$, where $(\delta_a)_{a \in A}$ are arbitrary deviation functions from a feasible set, and a Nash flow f^0 with respect to the unaltered latency functions $(l_a)_{a \in A}$. Here the

© Springer-Verlag Berlin Heidelberg 2016
M. Gairing and R. Savani (Eds.): SAGT 2016, LNCS 9928, pp. 129–140, 2016.
DOI: 10.1007/978-3-662-53354-3_11

social cost function C refers to the total average latency (without the deviations). Our motivation for studying this social cost function is that a central designer usually cares about the long-term performance of the system (accounting for the average latency or pollution). On the other hand, the players typically do not know the exact latencies and use estimates or include "safety margins" in their planning. Similar viewpoints are adopted in [7,9].

In order to model bounded deviations, we extend an idea previously put forward by Bonifaci, Salek and Schäfer [1] in the context of the *restricted network toll problem*: We assume that for every arc $a \in A$ we are given lower and upper bound restrictions $\theta_a^{\min}$ and $\theta_a^{\max}$, respectively, and call a deviation δ_a *feasible* if $\theta_a^{\min}(x) \leq \delta_a(x) \leq \theta_a^{\max}(x)$ for all $x \geq 0$.

Our notion of the Deviation Ratio is inspired by and builds upon the *Price of Risk Aversion (PRA)* recently introduced by Nikolova and Stier-Moses [9]. The authors investigate selfish routing games with uncertain latencies by considering deviations of the form $\delta_a = \gamma v_a$, where $\gamma \geq 0$ is the risk-aversion of the players and v_a is the variance of some random variable with mean zero. They derive upper bounds on the Price of Risk Aversion for single-commodity networks with arbitrary non-negative and non-decreasing latency functions if the *variance-to-mean-ratio* v_a/l_a of every arc $a \in A$ is bounded by some constant $\kappa \geq 0$. It is not hard to see that their model is a special case of our model if we choose $\theta_a^{\min} = 0$ and $\theta_a^{\max} = \gamma \kappa l_a$ (see Sect. 2 for more details).

Our contributions. The main contributions presented in this paper are as follows:

1. Upper bounds: We derive a general upper bound on the Deviation Ratio for multi-commodity networks with a common source and arbitrary non-negative and non-decreasing latency functions (Theorem 3).

In order to prove this upper bound, we first generalize a result by Bonifaci et al. [1] characterizing the inducibility of a fixed flow by δ-deviations to multi-commodity networks with a common source (Theorem 2). This characterization naturally gives rise to the concept of an *alternating path*, which plays a crucial role in the work by Nikolova and Stier-Moses [9] and was first used by Lin, Roughgarden, Tardos and Walkover [6] in the context of the *network design problem*.

We then specialize our bound to the case of so-called (α, β)-*deviations*, where $\theta_a^{\min} = \alpha l_a$ and $\theta_a^{\max} = \beta l_a$ with $-1 < \alpha \leq 0 \leq \beta$. We prove that the Deviation Ratio is at most $1 + (\beta - \alpha)/(1 + \alpha)\lceil (n-1)/2 \rceil r$, where n is the number of nodes of the network and r is the sum of the demands of the commodities (Theorem 3). In particular, this reveals that the Deviation Ratio depends linearly on the size of the underlying network (among other parameters).

By using this result, we obtain a bound on the Price of Risk Aversion (Theorem 6) which generalizes the one in [9] in two ways: (i) it holds for multi-commodity networks with a common source and (ii) it allows for negative risk-aversion parameters (i.e., capturing risk-taking players as well). Further, we show that our result can be used to bound the relative error in social cost incurred by small latency perturbations (Theorem 7), which is of independent interest.

2. *Lower bounds:* We prove that our bound on the Deviation Ratio for (α, β)-deviations is best possible. More specifically, for single-commodity networks we show that our bound is tight in all its parameters. Our lower bound construction holds for arbitrary $n \in \mathbb{N}$ and is based on the *generalized Braess graph* [10] (Example 1). In particular, this complements a recent result by Lianeas, Nikolova and Stier-Moses [5] who show that their bound on the Price of Risk Aversion is tight for single-commodity networks with $n = 2^j$ nodes for all $j \in \mathbb{N}$.

Further, for multi-commodity networks with a common source we show that our bound is tight in all parameters if n is odd, while a small gap remains if n is even (Theorem 4). Finally, for general multi-commodity graphs we establish a lower bound showing that the Deviation Ratio can be exponential in n (Theorem 5). In particular, this shows that there is an exponential gap between the cases of multi-commodity networks with and without a common source. In our proof, we adapt a graph structure used by Lin, Roughgarden, Tardos and Walkover [6] in their lower bound construction for the network design problem on multi-commodity networks (see also [10]).

3. *Smoothness bounds:* We improve (and slightly generalize) recent smoothness bounds on the Price of Risk Aversion given by Meir and Parkes [7] and independently by Lianeas et al. [5]. In particular, we derive tight bounds for the *Biased Price of Anarchy (BPoA)* [7], i.e., the ratio between the cost of a deviated Nash flow and the cost of a social optimum, for *arbitrary* $(0, \beta)$-deviations (Theorem 8).[1] Note that the Biased Price of Anarchy yields an upper bound on the Deviation Ratio/Price of Risk Aversion. We also derive smoothness results for general path deviations (which are not representable by arc deviations). As a result, we obtain bounds on the Price of Risk Aversion (Theorem 9) under the non-linear *mean-std* model [5,9] (see Sect. 2).

It is interesting to note that the smoothness bounds on the Biased Price of Anarchy [7] and the Price of Risk Aversion [5] are independent of the network structure (but dependent on the class of latency functions). In contrast, the bound on the Deviation Ratio depends on certain parameters of the network.[2]

Our results answer a question posed in the work by Nikolova and Stier-Moses [9] regarding possible relations between their Price of Risk Aversion model [9], the restricted network toll problem [1], and the network design problem [10]. In particular, our results also show that the analysis in [9] is not inherent to the used variance function, but rather depends on the restrictions imposed on the feasible deviations.

Related work. The modeling and studying of uncertainties in routing games has received a lot of attention in recent years. An extensive survey on this topic is given by Cominetti [2].

---

[1] We remark that for certain types of $(0, \beta)$-deviations, e.g., *scaled marginal tolls*, better bounds can be obtained (see, e.g., [7]).

[2] For example, there are parallel-arc networks for which the Biased Price of Anarchy is unbounded, whereas the Deviation Ratio is a constant.

As mentioned above, our investigations are inspired by the study of the Price of Risk Aversion by Nikolova and Stier-Moses [9]. They prove that for single-commodity instances with non-negative and non-decreasing latency functions the Price of Risk Aversion is at most $1 + \gamma\kappa\lceil(n-1)/2\rceil$. We elaborate in more detail on the connections to their work in Sect. 2.

There are several papers that study the problem of imposing tolls (which can be viewed as latency deviations) on the arcs of a network to reduce the cost of the resulting Nash flow. Conceptually, our model is related to the *restricted network toll problem* by Bonifaci et al. [1]. The authors study the problem of computing non-negative tolls that have to obey some upper bound restrictions $(\theta_a)_{a\in A}$ such that the cost of the resulting Nash flow is minimized. This is tantamount to computing best-case deviations in our model with $\theta_a^{\min} = 0$ and $\theta_a^{\max} = \theta_a$. In contrast, our focus here is on worst-case deviations. As a side result, we prove that computing such worst-case deviations is NP-hard, even for single-commodity instances with linear latencies (Theorem 1).

Roughgarden [10] studies the *network design problem* of finding a subnetwork that minimizes the latency of all flow-carrying paths of the resulting Nash flow. He proves that the *trivial algorithm* (which simply returns the original network) gives an $\lfloor n/2\rfloor$-approximation algorithm for single-commodity networks and that this is best possible (unless P = NP). Later, Lin et al. [6] show that this algorithm can be exponentially bad for multi-commodity networks. The instances that we use in our lower bound constructions are based on the ones used in [6,10].

Meir and Parkes [7] and independently Lianeas et al. [5] show that for non-atomic network routing games with $(1,\mu)$-*smooth*[3] latency functions it holds that PRA $\leq$ BPoA $\leq (1+\gamma\kappa)/(1-\mu)$. An advantage of such bounds is that they hold for general multi-commodity instances (but depend on the class of latency functions). These bounds stand in contrast to the *topological* bounds obtained here and by Nikolova and Stier-Moses [9] which hold for arbitrary non-negative and non-decreasing latency functions.

2 Preliminaries

Bounded deviation model. Let $\mathcal{I} = (G = (V, A), (l_a)_{a\in A}, (s_i, t_i)_{i\in[k]}, (r_i)_{i\in[k]})$ be an instance of a non-atomic network routing game. Here, $G = (V, A)$ is a directed graph with node set V and arc set $A \subseteq V \times V$, where each arc $a \in A$ has a non-negative, non-decreasing and continuous latency function $l_a : \mathbb{R}_{\geq 0} \to \mathbb{R}_{\geq 0}$. Each commodity $i \in [k]$ is associated with a source-destination pair (s_i, t_i) and has a demand of $r_i \in \mathbb{R}_{>0}$. We assume that $t_i \neq t_j$ if $i \neq j$ for $i, j \in [k]$. If all commodities share a common source node, i.e., $s_i = s_j = s$ for all $i, j \in [k]$, we call $\mathcal{I}$ a *common source multi-commodity instance (with source s)*. We assume without loss of generality that $1 = r_1 \leq r_2 \leq \cdots \leq r_k$ and define $r = \sum_{i\in[k]} r_i$.

---

[3] Meir and Parkes [7] define a function l to be $(1,\mu)$-*smooth* if $xl(y) \leq \mu yl(y) + xl(x)$ for all $x, y \geq 0$ (which is slightly different from Roughgarden's original smoothness definition [11]). Lianeas et al. [5] only require *local smoothness* where y is taken fixed.

We denote by $\mathcal{P}_i$ the set of all simple (s_i, t_i)-paths of commodity $i \in [k]$ in G, and we define $\mathcal{P} = \cup_{i \in [k]} \mathcal{P}_i$. An outcome of the game is a feasible flow $f : \mathcal{P} \to \mathbb{R}_{\geq 0}$, i.e., $\sum_{P \in \mathcal{P}_i} f_P = r_i$ for every $i \in [k]$. Given a flow $f = (f^i)_{i \in [k]}$, we use f_a^i to denote the total flow on arc $a \in A$ of commodity $i \in [k]$, i.e., $f_a^i = \sum_{P \in \mathcal{P}_i : a \in P} f_P$. The total flow on arc $a \in A$ is defined as $f_a = \sum_{i \in [k]} f_a^i$. The latency of a path $P \in \mathcal{P}$ with respect to f is defined as $l_P(f) := \sum_{a \in P} l_a(f_a)$. The *social cost* $C(f)$ of a flow f is given by its total average latency, i.e., $C(f) = \sum_{P \in \mathcal{P}} f_P l_P(f) = \sum_{a \in A} f_a l_a(f_a)$. A flow that minimizes $C(\cdot)$ is called *(socially) optimal*. We use $A_i^+ = \{a \in A : f_a^i > 0\}$ to refer to the support of f^i for commodity $i \in [k]$ and define $A^+ = \cup_{i \in [k]} A_i^+$ as the support of f.

For every arc $a \in A$, we have a continuous function $\delta_a : \mathbb{R}_{\geq 0} \to \mathbb{R}$ modeling the *deviation* on arc a, and we write $\delta = (\delta_a)_{a \in A}$. We define the deviation of a path $P \in \mathcal{P}$ as $\delta_P(f) = \sum_{a \in P} \delta_a(f_a)$. The *deviated latency* on arc $a \in A$ is given by $q_a(f_a) = l_a(f_a) + \delta_a(f_a)$; similarly, the deviated latency on path $P \in \mathcal{P}$ is given by $q_P(f) = l_P(f) + \delta_P(f)$. We say that f is δ-*inducible* if and only if it is a *Wardrop flow* (or *Nash flow*) with respect to $l + \delta$, i.e.,

$$\forall i \in [k], \forall P \in \mathcal{P}_i, f_P > 0 : \qquad q_P(f) \leq q_{P'}(f) \ \forall P' \in \mathcal{P}_i. \tag{1}$$

If f is δ-inducible, we also write $f = f^\delta$. Note that a Nash flow f for the unaltered latencies $(l_a)_{a \in A}$ is 0-inducible, i.e., $f = f^0$.

Let $\theta^{\min} = (\theta_a^{\min})_{a \in A}$ and $\theta^{\max} = (\theta_a^{\max})_{a \in A}$ be given continuous threshold functions satisfying $\theta_a^{\min}(x) \leq 0 \leq \theta_a^{\max}(x)$ for all $x \geq 0$ and $a \in A$, and let $\theta = (\theta^{\min}, \theta^{\max})$. We define $\Delta(\theta) = \{(\delta_a)_{a \in A} \mid \forall a \in A : \theta_a^{\min}(x) \leq \delta_a(x) \leq \theta_a^{\max}(x), \forall x \geq 0\}$ as the set of feasible deviations. Note that $0 \in \Delta(\theta)$ for all threshold functions $\theta^{\min}$ and $\theta^{\max}$. We say that $\delta \in \Delta(\theta)$ is a θ-*deviation*. Furthermore, f is θ-*inducible* if there exists a $\delta \in \Delta(\theta)$ such that f is δ-inducible. For $-1 < \alpha \leq 0 \leq \beta$, we call $\delta \in \Delta(\theta)$ an (α, β)-*deviation* if $\theta^{\min} = \alpha l$ and $\theta^{\max} = \beta l$, and also write $\theta = (\alpha, \beta)$. Throughout the paper, we assume that the deviated latencies are always non-negative, i.e., $l_a(x) + \theta_a^{\min}(x) \geq 0$ for all $x \geq 0$ and $a \in A$.

We (implicitly) assume that only deviations δ are considered for which a Nash flow exists. We briefly elaborate on the existence when $\theta^{\min} = 0$ and $\theta_a^{\max}$ is non-negative, non-decreasing and continuous for all $a \in A$. It is not hard to see that for a deviated Nash flow f^δ there exists some $0 \leq \lambda_a \leq 1$ for every arc $a \in A$ such that $\delta_a(f_a^\delta) = \lambda_a \theta_a^{\max}(f_a^\delta)$. In particular, this means that $\delta' \in \Delta(\theta)$ defined by $\delta'_a = \lambda_a \theta_a^{\max}$ also induces f^δ. Therefore it is sufficient to consider deviations of the form $\delta_a = \lambda_a \theta_a^{\max}$ where $0 \leq \lambda_a \leq 1$ for all $a \in A$. As a consequence, it follows that $q_a = l_a + \delta_a$ is a non-negative, non-decreasing and continuous function for all $a \in A$. It is well-known that for these types of functions, the existence of a Nash flow is guaranteed.

Deviation Ratio. Given an instance $\mathcal{I}$ and threshold functions $\theta = (\theta^{\min}, \theta^{\max})$, we define the *Deviation Ratio* $\mathrm{DR}(\mathcal{I}, \theta) = \sup_{\delta \in \Delta(\theta)} C(f^\delta)/C(f^0)$ as the worst-case ratio of the cost of a θ-inducible flow and the cost of a 0-inducible flow. Intuitively, $\mathrm{DR}(\mathcal{I}, \theta)$ measures the worst-case deterioration of the social cost of a Nash flow due to (feasible) latency deviations.

We emphasize that the social cost function C is defined as above, i.e., with respect to the latencies (not taking into account the deviations). Note that for fixed deviations $\delta \in \Delta(\theta)$, there might be multiple Nash flows that are δ-inducible. In this case, we adopt the convention that $C(f^\delta)$ refers to the social cost of the worst Nash flow that is δ-inducible.

Our main focus in this paper is on establishing (tight) bounds on the Deviation Ratio. As a side-result, we prove that the problem of determining worst-case deviations is NP-hard.

Theorem 1. *It is NP-hard to compute deviations $\delta \in \Delta(\theta)$ such that $C(f^\delta)$ is maximized, even for single-commodity networks with linear latencies.*

Related notions. Nikolova and Stier-Moses [9] (see also [5,8]) consider non-atomic network routing games with uncertain latencies. Here the deviations correspond to variances $(v_a)_{a \in A}$ of some random variable ζ_a (with expectation zero). The *perceived latency* of a path $P \in \mathcal{P}$ with respect to a flow f is then defined as $q_P^\gamma(f) = l_P(f) + \gamma v_P(f)$, where $\gamma \geq 0$ is a parameter representing the *risk-aversion* of the players. They consider two different objectives as to how the deviation $v_P(f)$ of a path P is defined: $v_P(f) = \sum_{a \in P} v_a(f_a)$, called the *mean-var* objective, and $v_P(f) = (\sum_{a \in P} v_a(f_a))^{1/2}$, called the *mean-std* objective. Note that for the mean-var objective there is an equivalent arc-based definition, where the perceived latency of every arc $a \in A$ is defined as $q_a^\gamma(f_a) = l_a(f_a) + \gamma v_a(f_a)$. They define the *Price of Risk Aversion* [9] as the worst-case ratio $C(x)/C(z)$, where x is a *risk-averse* Nash flow with respect to $q^\gamma = l + \gamma v$ and z is a *risk-neutral* Nash flow with respect to l.[4] In their analysis, it is assumed that the *variance-to-mean-ratio* of every arc $a \in A$ under the risk-averse flow x is bounded by some constant $\kappa \geq 0$, i.e., $v_a(x_a) \leq \kappa l_a(x_a)$ for all $a \in A$. Under this assumption, they prove that the Price of Risk Aversion $\text{PRA}(\mathcal{I}, \gamma, \kappa)$ of single-commodity instances $\mathcal{I}$ with non-negative and non-decreasing latency functions is at most $1 + \gamma \kappa \lceil (n-1)/2 \rceil$, where n is the number of nodes.

We now elaborate on the relation to our Deviation Ratio. The main technical difference is that in [9] the variance-to-mean ratio is only considered for the respective flow values x_a. Note however that if we write for every $a \in A$, $v_a(x_a) = \lambda_a l_a(x_a)$ for some $0 \leq \lambda_a \leq \kappa$, then the deviation function $\delta_a(y) = \gamma \lambda_a l_a(y)$ has the property that $x = f^\delta$ is δ-inducible with $\delta \in \Delta(0, \gamma\kappa)$. It follows that for every instance $\mathcal{I}$ and parameters γ, κ, $\text{PRA}(\mathcal{I}, \gamma, \kappa) \leq \text{DR}(\mathcal{I}, (0, \gamma\kappa))$.

Another related notion is the *Biased Price of Anarchy (BPoA)* introduced by Meir and Parkes [7]. Adapted to our setting, given an instance $\mathcal{I}$ and threshold functions θ, the Biased Price of Anarchy is defined as $\text{BPoA}(\mathcal{I}, \theta) = \sup_{\delta \in \Delta(\theta)} C(f^\delta)/C(f^*)$, where f^* is a socially optimal flow. Note that because $C(f^*) \leq C(f)$ for every feasible flow f, we have $\text{DR}(\mathcal{I}, \theta) \leq \text{BPoA}(\mathcal{I}, \theta)$.

Due to space limitations, some material is omitted from this extended abstract and can be found in the full version of the paper (see [4]).

---

[4] The existence of a risk-averse Nash flow is proven in [8].

3   Upper Bounds on the Deviation Ratio

We derive an upper bound on the Deviation Ratio. All results in this section hold for multi-commodity instances with a common source.

We first derive a characterization result for the inducibility of a given flow f. This generalizes the characterization in [1] to common source multi-commodity instances and negative deviations. We define an *auxiliary graph* $\hat{G} = \hat{G}(f) = (V, \hat{A})$ with $\hat{A} = A \cup \bar{A}$, where $\bar{A} = \{(v, u) : a = (u, v) \in A^+\}$. That is, $\hat{A}$ consists of the set of arcs in A, which we call *forward* arcs, and the set $\bar{A}$ of arcs (v, u) with $(u, v) \in A^+$, which we call *reversed* arcs. Further, we define a cost function $c : \hat{A} \rightarrow \mathbb{R}$ as follows:

$$c_a = \begin{cases} l_{(u,v)}(f_a) + \theta^{max}_{(u,v)}(f_a) & \text{for } a = (u, v) \in A \\ -l_{(u,v)}(f_a) - \theta^{min}_{(u,v)}(f_a) & \text{for } a = (v, u) \in \bar{A}. \end{cases} \quad (2)$$

Theorem 2. *Let f be a feasible flow. Then f is θ-inducible if and only if $\hat{G}(f)$ does not contain a cycle of negative cost with respect to c.*

Theorem 2 does not hold for general multi-commodity instances. The proof of Lemma 1 follows directly from Theorem 2.

Lemma 1. *Let x be θ-inducible and let X_i be a flow-carrying (s, t_i)-path for commodity $i \in [k]$ in G. Let χ and ψ be any (s, t_i)-path and (t_i, s)-path in $\hat{G}(x)$, respectively. Then*

$$\sum_{a \in X_i} l_a(x_a) + \theta^{min}_a(x_a) \leq \sum_{a \in \chi \cap A} l_a(x_a) + \theta^{max}_a(x_a) - \sum_{a \in \chi \cap \bar{A}} l_a(x_a) + \theta^{min}_a(x_a)$$

$$\sum_{a \in X_i} l_a(x_a) + \theta^{max}_a(x_a) \geq \sum_{a \in \psi \cap \bar{A}} l_a(x_a) + \theta^{min}_a(x_a) - \sum_{a \in \psi \cap A} l_a(x_a) + \theta^{max}_a(x_a).$$

The following notion of alternating paths turns out to be crucial. It was first introduced by Lin et al. [6] and is also used by Nikolova and Stier-Moses [9].

Definition 1 (Alternating path [6,9]). *Let x and z be feasible flows. We partition $A = X \cup Z$, where $Z = \{a \in A : z_a \geq x_a \text{ and } z_a > 0\}$ and $X = \{a \in A : z_a < x_a \text{ or } z_a = x_a = 0\}$. We say that $\pi_i = (a_1, \ldots, a_r)$ is an alternating s, t_i-path if the arcs in $\pi_i \cap Z$ are oriented in the direction of t_i, and the arcs in $\pi_i \cap X$ are oriented in the direction of s.*

Without loss of generality we may remove all arcs with $z_a = x_a = 0$ (as they do not contribute to the social cost). Note that if along π_i we reverse the arcs of Z then the resulting path is a directed (t_i, s)-path in $\hat{G}(z)$ (which we call the *s-oriented version of π_i*); similarly, if we reverse the arcs of X then the resulting path is an (s, t_i)-path in $\hat{G}(x)$ (which we call the *t_i-oriented version of π_i*).

The following lemma proves the existence of an *alternating path tree*, i.e., a spanning tree of alternating paths, rooted at the common source node s. It is a direct generalization of Lemma 4.6 in [6] and Lemma 4.5 in [9].

Lemma 2. *Let z and x be feasible flows and let Z and X be a partition of A as in Definition 1. Then there exists an alternating path tree.*

We now have all the ingredients to prove the following main result.

Theorem 3. *Let x be θ-inducible and let z be 0-inducible. Further, let $A = X \cup Z$ be a partition as in Definition 1. Let π be an alternating path tree, where π_i denotes the alternating s, t_i-path in π.*

(i) *Suppose $\theta = (\theta^{\min}, \theta^{\max})$. Let X_i be a flow-carrying path of commodity $i \in [k]$ maximizing $l_P(x)$ over all $P \in \mathcal{P}_i$.[5] Then*

$$C(x) \leq C(z) + \sum_{i \in [k]} r_i \left(\sum_{a \in Z \cap \pi_i} \theta_a^{\max}(z_a) - \sum_{a \in X \cap \pi_i} \theta_a^{\min}(z_a) - \sum_{a \in X_i} \theta_a^{\min}(x_a) \right).$$

(ii) *Suppose $\theta = (\alpha, \beta)$ with $-1 < \alpha \leq 0 \leq \beta$. Let η_i is the number of disjoint segments of consecutive arcs in Z on the alternating s, t_i-path π_i for $i \in [k]$.[6] Then*

$$\frac{C(x)}{C(z)} \leq 1 + \frac{\beta - \alpha}{1 + \alpha} \cdot \sum_{i \in [k]} r_i \eta_i \leq 1 + \frac{\beta - \alpha}{1 + \alpha} \cdot \left\lceil \frac{n-1}{2} \right\rceil \cdot r.$$

Proof (i). We have $C(x) = \sum_i \sum_{P \in \mathcal{P}_i} x_P^i l_P(x) \leq \sum_i r_i \sum_{a \in X_i} l_a(x_a)$ by the choice of X_i. By applying the first inequality of Lemma 1 to the flow x in the graph $\hat{G}(x)$, where we choose χ to be the t_i-oriented version of π_i, we obtain

$$\sum_{a \in X_i} l_a(x_a) + \theta_a^{\min}(x_a) \leq \sum_{a \in Z \cap \pi_i} l_a(x_a) + \theta_a^{\max}(x_a) - \sum_{a \in X \cap \pi_i} l_a(x_a) + \theta_a^{\min}(x_a).$$

Let Z_i be an arbitrary flow-carrying path of commodity $i \in [k]$ with respect to z. By applying the second inequality of Lemma 1 to the flow z in the graph $\hat{G}(z)$ with $\theta^{\max} = \theta^{\min} = 0$, where we choose ψ to be the s-oriented version of π_i, we obtain

$$\sum_{a \in Z_i} l_a(z_a) \geq \sum_{a \in Z \cap \pi_i} l_a(z_a) - \sum_{a \in X \cap \pi_i} l_a(z_a).$$

Combining these inequalities and exploiting the definition of X and Z, we obtain

$$\sum_{a \in X_i} l_a(x_a) + \theta_a^{\min}(x_a) \leq \sum_{a \in Z \cap \pi_i} l_a(x_a) + \theta_a^{\max}(x_a) - \sum_{a \in X \cap \pi_i} l_a(x_a) + \theta_a^{\min}(x_a)$$

$$\leq \sum_{a \in Z \cap \pi_i} l_a(z_a) + \theta_a^{\max}(z_a) - \sum_{a \in X \cap \pi_i} l_a(z_a) + \theta_a^{\min}(z_a)$$

$$\leq \sum_{a \in Z_i} l_a(z_a) + \sum_{a \in Z \cap \pi_i} \theta_a^{\max}(z_a) - \sum_{a \in X \cap \pi_i} \theta_a^{\min}(z_a).$$

The claim now follows by multiplying the above inequality with r_i and summing over all commodities $i \in [k]$. Note that $C(z) = \sum_i r_i \sum_{a \in Z_i} l_a(z_a)$. □

---

[5] Note that the values $l_P(x) + \delta_P(x)$ are the same for all flow-carrying paths, but this is not necessarily true for the values $l_P(x)$.
[6] Note that $\eta_i \leq \lceil (n-1)/2 \rceil$.

4   Lower Bounds for (α, β)-deviations

We show that the bound in Theorem 3 is tight in all its parameters for (α, β)-deviations. We start with single-commodity instances.

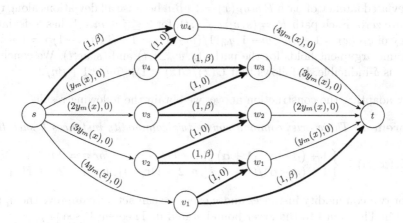

Fig. 1. The fifth Braess graph with (l_a^5, δ_a^5) on the arcs as defined in Example 1. The bold arcs indicate the alternating path π_1.

Our instance is based on the generalized Braess graph [10]. The *m-th Braess graph* $G^m = (V^m, A^m)$ is defined by $V^m = \{s, v_1, \ldots, v_{m-1}, w_1, \ldots, w_{m-1}, t\}$ and A^m as the union of three sets: $E_1^m = \{(s, v_j), (v_j, w_j), (w_j, t) : 1 \leq j \leq m - 1\}$, $E_2^m = \{(v_j, w_{j-1}) : 2 \leq j \leq m\}$ and $E_3^m = \{(v_1, t) \cup \{(s, w_{m-1})\}\}$.

Example 1. We can assume without loss of generality that $\alpha = 0$ (see [4]). Let $\beta \geq 0$ be a fixed constant and let $n = 2m \geq 4 \in \mathbb{N}.$[7] Let G^m be the m-th Braess graph. Furthermore, let $y_m : \mathbb{R}_{\geq 0} \to \mathbb{R}_{\geq 0}$ be a non-decreasing, continuous function[8] with $y_m(1/m) = 0$ and $y_m(1/(m-1)) = \beta$. We define

$$l_a^m(g) = \begin{cases} (m - j) \cdot y_m(g) & \text{for } a \in \{(s, v_j) : 1 \leq j \leq m - 1\} \\ j \cdot y_m(g) & \text{for } a \in \{(w_j, t) : 1 \leq j \leq m - 1\} \\ 1 & \text{otherwise.} \end{cases}$$

Furthermore, we define $\delta_a^m(g) = \beta$ for $a \in E_2^m$, and $\delta_a^m(g) = 0$ otherwise. Note that $0 \leq \delta_a^m(g) \leq \beta l_a^m(g)$ for all $a \in A$ and $g \geq 0$ (see Fig. 1).

A Nash flow $z = f^0$ is given by routing $1/m$ units of flow over the paths $(s, w_{m-1}, t), (s, v_1, t)$ and the paths in $\{(s, v_j, w_{j-1}, t) : 2 \leq j \leq m-1\}$. Note that

---

[7] Note that the value $\lceil (n-1)/2 \rceil$ is the same for $n \in \{2m, 2m+1\}$ with $m \in \mathbb{N}$. The example shows tightness for $n = 2m$. The tightness for $n = 2m + 1$ then follows trivially by adding a dummy node.

[8] For example $y_m(g) = m(m-1)\beta \max\{0, (g - \frac{1}{m})\}$. That is, we define y_m to be zero for $0 \leq g \leq 1/m$ and we let it increase with constant rate to β in $1/(m-1)$.

all these paths have latency one, and the path (s, v_j, w_j, t), for some $1 \leq m \leq j$, also has latency one. We conclude that $C(z) = 1$.

A Nash flow $x = f^\delta$, with δ as defined above, is given by routing $1/(m-1)$ units of flow over the paths in $\{(s, v_j, w_j, t) : 1 \leq j \leq m-1\}$. Each such path P then has a latency of $l_P(x) = 1 + \beta m$. It follows that $C(x) = 1 + \beta m$. Note that the deviated latency of path P is $q_P(x) = 1 + \beta m$ because all deviations along this path are zero. Each path $P' = (s, v_j, w_{j-1}, t)$, for $2 \leq j \leq m-1$, has a deviated latency of $q_{P'}(x) = 1 + \beta + (m-1)y_m(1/(m-1)) = 1 + \beta + (m-1)\beta = 1 + \beta m$. The same argument holds for the paths (s, w_{m-1}, t) and (s, v_1, t). We conclude that x is δ-inducible. It follows that $C(x)/C(z) = 1 + \beta m = 1 + \beta n/2$. $\square$

By adapting the construction above, we obtain the following result.

Theorem 4. *There exist common source two-commodity instances $\mathcal{I}$ such that*

$$DR(\mathcal{I}, (\alpha, \beta)) \geq \begin{cases} 1 + (\beta - \alpha)/(1 + \alpha) \cdot (n-1)/2 \cdot r & \text{for } n = 2m+1 \in \mathbb{N}_{\geq 5} \\ 1 + (\beta - \alpha)/(1 + \alpha) \cdot [(n/2 - 1)r + 1] & \text{for } n = 2m \in \mathbb{N}_{\geq 4}. \end{cases}$$

For two-commodity instances and n even, we can actually improve the upper bound in Theorem 3 to the lower bound stated in Theorem 4 (see [4]).

For general multi-commodity instances the situation is much worse. In particular, we establish an exponential lower bound on the Deviation Ratio. The instance used in proof of Theorem 5 is similar to the one used by Lin et al. [6].

Theorem 5. *For every $p = 2q+1 \in \mathbb{N}$, there exists a two-commodity instance $\mathcal{I}$ whose size is polynomially bounded in p such that $DR(\mathcal{I}, (\alpha, \beta)) \geq 1 + \beta F_{p+1} \approx 1 + 0.45\beta \cdot \phi^{p+1}$, where F_p is the p-th Fibonacci number and $\phi \approx 1.618$ is the golden ratio.*

5   Applications

By using our bounds on the Deviation Ratio, we obtain the following results.

Price of Risk Aversion

Theorem 6. *The Price of Risk Aversion for a common source multi-commodity instance $\mathcal{I}$ with non-negative and non-decreasing latency functions, variance-to-mean-ratio $\kappa > 0$ and risk-aversion parameter $\gamma \geq -1/\kappa$ is at most*

$$PRA(\mathcal{I}, \gamma, \kappa) \leq \begin{cases} 1 - \gamma\kappa/(1 + \gamma\kappa)\lceil (n-1)/2 \rceil r & \text{for } -1/\kappa < \gamma \leq 0 \\ 1 + \gamma\kappa\lceil (n-1)/2 \rceil r & \text{for } \gamma \geq 0. \end{cases}$$

Moreover, these bounds are tight in all its parameters if $n = 2m+1$ and almost tight if $n = 2m$ (see [4]). In particular, for single-commodity instances we obtain tightness for all $n \in \mathbb{N}$.

Stability of Nash flows under small perturbations

Theorem 7. *Let $\mathcal{I}$ be a common source multi-commodity instance with non-negative and non-decreasing latency functions $(l_a)_{a \in A}$. Let f be a Nash flow with respect to $(l_a)_{a \in A}$ and let $\tilde{f}$ be a Nash flow with respect to slightly perturbed latency functions $(\tilde{l}_a)_{a \in A}$ satisfying $\sup_{a \in A,\, x \geq 0} |(l_a(x) - \tilde{l}_a(x))/l_a(x)| \leq \epsilon$ for some small $\epsilon > 0$. Then the relative error in social cost is $(C(\tilde{f}) - C(f))/C(f) \leq 2\epsilon/(1 - \epsilon)\lceil (n-1)/2 \rceil \cdot r = \mathcal{O}(\epsilon r n)$.*

6   Smoothness Based Approaches

We derive tight smoothness bounds on the Biased Price of Anarchy for $(0, \beta)$-deviations. Our bounds improve upon the bounds of $(1 + \beta)/(1 - \mu)$ recently obtained by Meir and Parkes [7] and Lianeas et al. [5] for $(1, \mu)$-smooth latency functions. As a direct consequence, we also obtain better smoothness bounds on the Price of Risk Aversion. Our approach is a generalization of the framework of Correa, Schulz and Stier-Moses [3] (which we obtain for $\beta = 0$).

Let $\mathcal{L}$ be a given set of latency functions and $\beta \geq 0$ fixed. For $l \in \mathcal{L}$, define

$$\hat{\mu}(l, \beta) = \sup_{x, z \geq 0} \left\{ \frac{z[l(x) - (1 + \beta)l(z)]}{xl(x)} \right\} \quad \text{and} \quad \hat{\mu}(\mathcal{L}, \beta) = \sup_{l \in \mathcal{L}} \hat{\mu}(\mathcal{L}, \beta).$$

Theorem 8. *Let $\mathcal{L}$ be a set of non-negative, non-decreasing and continuous functions. Let $\mathcal{I}$ be a general multi-commodity instance with $(l_a)_{a \in A} \in \mathcal{L}^A$. Let x be δ-inducible for some $(0, \beta)$-deviation δ and let z be an arbitrary feasible flow. Then $C(x)/C(z) \leq (1 + \beta)/(1 - \hat{\mu}(\mathcal{L}, \beta))$ if $\hat{\mu}(\mathcal{L}, \beta) < 1$. Moreover, this bound is tight if $\mathcal{L}$ contains all constant functions and is closed under scalar multiplication, i.e., for every $l \in \mathcal{L}$ and $\gamma \geq 0$, $\gamma l \in \mathcal{L}$.*

For example, for affine latencies $\hat{\mu}(\mathcal{L}, \beta) = 1/(4(1+\beta))$ (see [4]) and we obtain a bound of $(1 + \beta)^2/(\frac{3}{4} + \beta)$ on the Biased Price of Anarchy, which is strictly better than the bound $4(1 + \beta)/3$ obtained in [5,7].

We also provide an upper bound on the absolute gap between the Biased Price of Anarchy and the Deviation Ratio (see [4]).

As a final result we derive smoothness bounds for general path deviations, which are not necessarily decomposable into arc deviations (see [4] for formal definitions). The main motivation for investigating such deviations is that we can apply such bounds to the *mean-std* objective of the Price of Risk Aversion model by Nikolova and Stier-Moses [9] (see Sect. 2).

Theorem 9. *Let $\mathcal{I}$ be a general multi-commodity instance with $(l_a)_{a \in A} \in \mathcal{L}^A$. Let x be δ-inducible with respect to some $(0, \beta)$-path deviation δ and let z an arbitrary feasible flow. If $\hat{\mu}(\mathcal{L}, 0) < 1/(1 + \beta)$, then $C(x)/C(z) \leq (1 + \beta)/(1 - (1 + \beta)\hat{\mu}(\mathcal{L}, 0))$.*

7   Conclusions

We introduced a unifying model to study the impact of (bounded) worst-case latency deviations in non-atomic selfish routing games. We demonstrated that the Deviation Ratio is a useful measure to assess the cost deterioration caused by such deviations. Among potentially other applications, we showed that the Deviation Ratio provides bounds on the Price of Risk Aversion and the relative error in social cost if the latency functions are subject to small perturbations.

Our approach to bound the Deviation Ratio (see Sect. 3) is quite generic and, albeit considering a rather general setting, enables us to obtain tight bounds. We believe that this approach will turn out to be useful to derive bounds on the Deviation Ratio of other games (e.g., network cost sharing games).

In general, studying the impact of (bounded) worst-case deviations of the input data of more general classes of games (e.g., congestion games) is an interesting and challenging direction for future work.

References

1. Bonifaci, V., Salek, M., Schäfer, G.: Efficiency of restricted tolls in non-atomic network routing games. In: Persiano, G. (ed.) SAGT 2011. LNCS, vol. 6982, pp. 302–313. Springer, Heidelberg (2011)
2. Cominetti, R.: Equilibrium routing under uncertainty. Math. Program. **151**(1), 117–151 (2015)
3. Correa, J.R., Schulz, A.S., Stier-Moses, N.E.: A geometric approach to the price of anarchy in nonatomic congestion games. Games Econ. Behav. **64**(2), 457–469 (2008)
4. Kleer, P., Schäfer, G.: The impact of worst-case deviations in non-atomic network routing games (2016). CoRR, abs/1605.01510
5. Lianeas, T., Nikolova, E., Stier-Moses, N.E.: Asymptotically tight bounds for inefficiency in risk-averse selfish routing (2015). CoRR, abs/1510.02067
6. Lin, H., Roughgarden, T., Tardos, É., Walkover, A.: Stronger bounds on Braess's paradox and the maximum latency of selfish routing. SIAM J. Discrete Math. **25**(4), 1667–1686 (2011)
7. Meir, R., Parkes, D.: Playing the wrong game: Smoothness bounds for congestion games with behavioral biases. SIGMETRICS Perform. Eval. Rev. **43**(3), 67–70 (2015)
8. Nikolova, E., Stier-Moses, N.E.: A mean-risk model for the traffic assignment problem with stochastic travel times. Oper. Res. **62**(2), 366–382 (2014)
9. Nikolova, E., Stier-Moses, N.E.: The burden of risk aversion in mean-risk selfish routing. In: Proceedings of the ACM Conference on Economics and Computation, pp. 489–506 (2015)
10. Roughgarden, T.: On the severity of Braess's paradox: Designing networks for selfish users is hard. J. Comput. Syst. Sci. **72**(5), 922–953 (2006)
11. Roughgarden, T.: Intrinsic robustness of the price of anarchy. J. ACM **62**(5), 32 (2015)
12. Wardrop, J.G.: Some theoretical aspects of road traffic research. Proc. Inst. Civil Eng. **1**, 325–378 (1952)

On Selfish Creation of Robust Networks

Ankit Chauhan[1], Pascal Lenzner[1(✉)], Anna Melnichenko[1], and Martin Münn[2]

[1] Algorithm Engineering Group,
Hasso-Plattner-Institute Potsdam, Potsdam, Germany
{ankit.chauhan,pascal.lenzner,anna.melnichenko}@hpi.de
[2] Department of Computer Science, University of Liverpool, Liverpool, UK
M.F.Munn@liverpool.ac.uk

Abstract. Robustness is one of the key properties of nowadays networks. However, robustness cannot be simply enforced by design or regulation since many important networks, most prominently the Internet, are not created and controlled by a central authority. Instead, Internet-like networks emerge from strategic decisions of many selfish agents. Interestingly, although lacking a coordinating authority, such naturally grown networks are surprisingly robust while at the same time having desirable properties like a small diameter. To investigate this phenomenon we present the first simple model for selfish network creation which explicitly incorporates agents striving for a central position in the network while at the same time protecting themselves against random edge-failure. We show that networks in our model are diverse and we prove the versatility of our model by adapting various properties and techniques from the non-robust versions which we then use for establishing bounds on the Price of Anarchy. Moreover, we analyze the computational hardness of finding best possible strategies and investigate the game dynamics of our model.

1 Introduction

Networks are everywhere and we crucially rely on their functionality. Hence it is no surprise that designing networks under various objective functions is a well established research area in the intersection of Operations Research, Computer Science and Economics. However, investigating how to create suitable networks *from scratch* is of limited use for understanding most of nowadays networks. The reason for this is that most of our resource, communication and online social networks have not been created and designed by some central authority. Instead, these critical networks emerged from the interaction of many selfish agents who control and shape parts of the network. This clearly calls for a game-theoretic perspective.

One of the most prominent examples of such a selfishly created network is the Internet, which essentially is a network of sub-networks which are each owned and controlled by Internet service providers (ISP). Each ISP decides selfishly how to connect to other ISPs and thereby balancing the cost for creating links (buying the necessary hardware and/or peering agreement contracts for

© Springer-Verlag Berlin Heidelberg 2016
M. Gairing and R. Savani (Eds.): SAGT 2016, LNCS 9928, pp. 141–152, 2016.
DOI: 10.1007/978-3-662-53354-3_12

routing traffic) and the obtained service quality for its customers. Interestingly, although the Internet is undoubtedly an important and critical infrastructure, there is no central authority which ensures its functionality if parts of the network fail. Despite this fact, the Internet seems to be robust against node or edge failures, which hints that a socially beneficial property like network robustness may emerge from selfish behavior.

Modeling agents with a desire for creating a *robust* network has long been neglected and was started to be investigated only very recently. This paper contributes to this endeavor by proposing and analyzing a model of selfish network creation, which explicitly incorporates agents who strive for occupying a central position in the network while at the same time ensuring that the overall network remains functional even under edge-failure.

1.1 Related Work

Previous research on game-theoretic models for network creation has either focused on *centrality models*, where the agents' service quality in the created network depends on the distances to other agents, or on *reachability models* where agents only care about being connected to many other agents.

Some prominent examples of centrality models for selfish network creation are [2,7–10,13,18]. They all have in common that agents correspond to nodes in a network and that the edge set of the network is determined by the combination of the agents' strategies. The utility function of an agent contains a service quality term which depends on the distances to all other agents. To the best of our knowledge the very recent paper by Meirom et al. [19] is the only centrality model which incorporates edge-failures. The authors consider two types of agents, major-league and minor-league agents, which maintain that the network remains 2-connected while trying to minimize distances, which are a linear combination of the length of a shortest path and the length of a best possible vertex disjoint backup path. Under some specific assumptions, e.g. that there are significantly more minor-league than major-league agents, they prove various structural properties of equilibrium networks and investigate the corresponding game-dynamics. In contrast to this, we will investigate a much simpler model with homogeneous agents which is more suitable for analyzing networks created by equal peers. Our results can be understood as zooming in on the sub-network formed by the major-league agents (i.e. top tier ISPs).

In reachability models, e.g. [3–5,11,12,15], the service quality of an agent is simply defined as the number of reachable agents and distances are ignored completely. For reachability models the works of Kliemann [15,16] and the very recent paper by Goyal et al. [12] explicitly incorporate a notion of network robustness in the utility function of every agent. All models consider an external adversary who strikes after the network is built. In [15,16] the adversary randomly removes a single edge and the agents try to maximize the expected number of reachable nodes post attack. Two versions of the adversary are analyzed: edge removal uniformly at random or removal of the edge which hurts the society of agents most. For the former adversary a constant Price of Anarchy is shown, whereas for the latter adversary this positive result is only true if edges can be created unilaterally. In [12] nodes are attacked (and killed) and this attack spreads virus-like to neighboring nodes unless they are protected by a firewall. Interestingly also this model has a low Price of Anarchy and the authors prove a tight linear bound on the amount of edge-overbuilding due to the adversary.

1.2 Model and Notation

We consider the Network Creation Game (NCG) by Fabrikant et al. [10] augmented with the uniform edge-deletion adversary from Kliemann [15] and we call our model *Adversary NCG* (Adv-NCG). More specifically, in an Adv-NCG there are n selfish agents which correspond to the nodes of an undirected multi-graph $G = (V, E)$ and we will use the terms agent and node interchangeably. A pure strategy S_u of agent $u \in V(G)$ is any multi-set over elements from $V \setminus \{u\}$. If v is contained k times in S_u then this encodes that agent u wants to create k undirected edges to node v. Moreover we say that u is the owner of all edges to the nodes in S_u. We emphasize the edge-ownership in our illustrations by drawing directed edges which point away from their owner - all edges are nonetheless understood to be undirected. Given an n-dimensional vector of pure strategies for all agents, then the union of all the edges encoded in all agents' pure strategies defines the edge set E of the multi-graph G. Since there is a bijection of multi-graphs with edge-ownership information and pure strategy-vectors, we will use networks and strategy-vectors interchangeably, e.g. by saying that a network is in equilibrium.

The agents prepare for an adversarial attack on the network after creation. This attack deletes one edge uniformly at random. Hence, agents try to minimize the attack's impact on themselves by minimizing their *expected cost*. Let $G - e$ denote the network G where edge e is removed. Let $\delta_G(u) = \sum_{v \in V(G)} d_G(u, v)$, where $d_G(u, v)$ is the number of edges of a shortest path from u to v in network G. Let

$$dist_G(u) = \frac{1}{|E|} \sum_{e \in E} \delta_{G-e}(u) = \frac{1}{|E|} \sum_{e \in E} \sum_{v \in V} d_{G-e}(u, v)$$

denote agent u's *expected distance cost* after the adversary has removed some edge uniformly at random from G. The expected cost of an agent u in network $G = (V, E)$ with edge-price α is defined as $cost_u(G, \alpha) = edge_u(G, \alpha) + dist_G(u)$, where $edge_u(G, \alpha) = \alpha \cdot |S_u|$ is the total edge-cost for agent u with strategy S_u in (G, α). Thus, compared to the NCG [10], the distance cost term is replaced by the expected distance cost with respect to uniform edge deletion.

Let (G, α) be any network with edge-ownership information. We call any strategy-change from S_u to S'_u of some agent u a *move*. Specifically, if $|S_u| = |S'_u|$, then such a move is called a *multi-swap*, if $|S_u \cap S'_u| < |S_u| - 2$ and a *swap* if $|S_u \cap S'_u| = |S_u| - 2$. If a move of agent u strictly decreases agent u's cost, then it is called an *improving move*. If no improving move exists, then we say that agent u plays its *best response*. Analogously we call a strategy-change towards a best response a *best response move*. A sequence of best response moves which starts and ends with network (G, α) is called a *best response cycle*. We say that (G, α) is in *Pure Nash Equilibrium* (NE) if all agents play a best response.

We measure the overall quality of a network (G, α) with its *social cost*, which is defined as $cost(G, \alpha) = \sum_{u \in V(G)} cost_u(G, \alpha) = edge(G, \alpha) + dist(G)$, where $edge(G, \alpha) = \sum_{u \in V(G)} edge_u(G, \alpha) = \alpha \cdot |E|$ and $dist(G) = \sum_{u \in V(G)} dist_G(u)$. Let $OPT(n, \alpha)$ be a network on n nodes with edge-price α which minimizes the social cost and we call $OPT(n, \alpha)$ the *optimum network* for n and α. Let $\text{maxNE}(n, \alpha)$ be the maximum social cost of any NE network on n agents with edge-price α and analogously let $\text{minNE}(n, \alpha)$ be the NE having minimum social cost. Then, the *Price of Anarchy* is the maximum over all n and α of the ratio $\frac{\text{maxNE}(n,\alpha)}{\text{OPT}(n,\alpha)}$, whereas the *Price of Stability* is the maximum over all n and α of the ratio $\frac{\text{minNE}(n,\alpha)}{\text{OPT}(n,\alpha)}$.

1.3 Our Contribution

This paper introduces and analyzes an accessible model, the Adv-NCG, for selfish network creation in which agents strive for a central position in the network while protecting against random edge-failures.

We show that optimum networks in the Adv-NCG are much more diverse than without adversary, which also indicates that the same holds true for the Nash equilibria of the game. However, we also show that many techniques and results from the NCG can be adapted to cope with the Adv-NCG, which indicates that the influence of the adversary is limited. In particular, we prove NP-hardness of computing a best possible strategy and W[2]-hardness of computing a best multi-swap. Moreover, we show that the Adv-NCG is not weakly acyclic, which is the strongest possible non-convergence result for any game. On the positive side, we prove that the amount of edge-overbuilding due to the adversary is limited, which is then used for proving that upper bounding the diameter essentially bounds the Price of Anarchy from above. We apply this by adapting two diameter-bounding techniques from the NCG to the adversarial version which then yields an upper bound on the PoA of $\mathcal{O}(1 + \alpha/\sqrt{n})$.

Due to space constraints some proofs and details are omitted. All missing material can be found in the full version [6].

2 Optimal Networks

Clearly, every optimal network must be 2-edge connected. Thus, every optimal network must have at least n edges. We first prove the intuitive fact that if edges get more expensive, then the optimum networks will have fewer edges.

Theorem 1. *Let $(G = (V, E), \alpha)$ and $(G' = (V, E'), \alpha')$ be optimal networks on n nodes in the Adv-NCG for α and α', respectively. If $\alpha' > \alpha$, then $|E| \geq |E'|$.*

In the following, we show that the landscape of optimum networks is much richer in the Adv-NCG, compared to the NCG where the optimum is either a clique

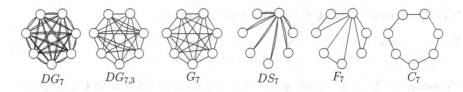

DG_7 $DG_{7,3}$ G_7 DS_7 F_7 C_7

Fig. 1. Different candidates for optimum networks.

or a star, depending on α. In particular, we prove that there are $\Omega(n^2)$ different optimal topologies. We consider the following types of networks: Here DG_n is a clique of n nodes where we have a double edge between all pairs of nodes. Let $DG_{n,k}$ be a n node clique with exactly k pairs of nodes which are connected with double edges. Thus, $DG_{n,0} = G_n$ and $DG_{n,\binom{n}{2}} = DG_n$. Moreover, let F_n denote the fan-graph on n nodes which is a collection of triangles which all share a single node and let DS_n denote a star on n nodes where all connections between the center and the leaves are double edges. Finally, let C_n be a cycle of length n.

Clearly, if $\alpha = 0$, then the optimum network on n nodes must be a DG_n, since in this network no edge deletion of the adversary has any effect, it minimizes the sum of distances of each agent and since edges are for free.

Now consider what happens, if one pair of agents, say u and v, are just connected via a single edge instead of a double edge. The probability that the adversary removes this edge is $\frac{1}{n(n-1)-1}$. The removal would cause an increase in distance cost of 1 between u and v and between v and u. Thus, if $\alpha < \frac{2}{n(n-1)-1}$, then agent u and v would individually be better off buying another edge between each other. Thus, we have the following observation.

Observation 2. *If $0 \leq \alpha \leq \frac{2}{n(n-1)-1}$, then $\mathrm{OPT}(n, \alpha) = DG_n$.*

Lemma 1. *If $\frac{2n(n-1)}{(\binom{n}{2}+k)(\binom{n}{2}+k+1)} \leq \alpha \leq \frac{2n(n-1)}{(\binom{n}{2}+k)(\binom{n}{2}+k-1)}$, for $1 \leq k \leq \binom{n}{2} - 1$, then the network $DG_{n,k}$ is optimal.*

Note, that the proof of the above statement implies that the complete graph G_n is an optimum, if $\frac{4}{\binom{n}{2}+1} \leq \alpha < 2 - \frac{2}{\binom{n}{2}}$.

We also remark that we conjecture that Fig. 1 resembles a snapshot of optimum networks for increasing α from left to right. In fact, extensive simulations indicate that the optimum changes from G_n to DS_n and then, for slightly larger α for F_n. After this the cycles in the fan-graph increase and get fewer in number until, finally, for $\alpha \in \Omega(n^3)$ the cycle appears as optimum.

3   Computing Best Responses and Game Dynamics

In this section we investigate computational aspects of the Adv-NCG. First we analyze the hardness of computing a best response and the hardness of computing a best possible multi-swap. Then we analyze a natural process for finding an equilibrium network by sequentially performing improving moves.

3.1   Hardness of Best Response Computation

We first introduce useful properties for ruling out multi-buy or multi-delete moves. The proof is similar to the proof of Lemma 1 in [17].

Proposition 1. *If an agent cannot decrease its expected cost by buying (deleting) one edge in the Adv-NCG, then buying (deleting) $k > 1$ edges cannot decrease the agent's expected cost.*

Lemma 2. *If $1 - \frac{1}{|E|+1} < \alpha < 1 + \frac{1}{|E|(|E|-1)}$ and if agent u is not an endpoint of any double-edge in the Adv-NCG, then buying the minimum number of edges such that u's expected distance to all nodes in $V \setminus N_u$ is 2 and to nodes in N_u is $1 + \frac{1}{|E|}$ is u's best response.*

Now we show that computing the best possible strategy-change is intractable.

Theorem 3. *1. It is NP-hard to compute the best response of agent u in the Adv-NCG.*
2. It is $W[2]$-hard to compute the best multi-swap of agent u in the Adv-NCG.

3.2   Game Dynamics

We investigate the dynamic properties of the Adv-NCG. That is, we turn the model into a sequential version which starts with some initial network (G, α) and then agents move sequentially in some order and perform improving moves, if possible. One natural question is, if this process is guaranteed to converge to a Nash equilibrium of the game.

For the game dynamics of the Adv-NCG we prove the strongest possible negative result, which essentially shows that there is no hope for convergence if agents stick to performing improving moves only. In particular, we prove that the order of the agents moves or any tie-breaking between different improving moves does not help for achieving convergence. This result is even stronger than the best known non-convergence results for the NCG [14].

Theorem 4. *The $Adv - NCG$ is not weakly acyclic.*

4   Analysis of Networks in Nash Equilibrium

In this section we establish the existence of networks in Nash Equilibrium for almost the whole parameter space and we compare NE networks in the Adv-NCG with NE networks from the NCG [10] and Kliemann's adversarial model [15]. Moreover, we investigate structural properties which allow us to provide bounds on the Price of Stability and the Price of Anarchy.

We start with the existence result:

Theorem 5. *The networks DG_n and DS_n are in pure Nash Equilibrium if $\alpha \leq \frac{1}{n(n-1)-1}$ and $\alpha \geq 1 - \frac{1}{2n-1}$, respectively.*

Next, we show that NE in the Adv-NCG are not comparable with NE from the NCG or Kliemann's model.

Theorem 6. *There is a NE in the Adv-NCG which is not an NE in the NCG and vice versa. The analogous statement also holds for Kliemann's model.*

4.1 Relation Between the Diameter and the Social Cost

We prove a property which relates the diameter of a network with its social cost. With this, we prove that one of the most useful tools for analyzing NE in the NCG [10] can be carried over to the Adv-NCG.

Before we start, we analyze the diameter increase induced by removing a single edge in a 2-edge-connected network.

Lemma 3. *Let $G = (V, E)$ be any 2-edge-connected network having diameter D and let $G - e$ be the network G where some edge $e \in E$ is removed. Then the diameter of $G - e$ is at most $2D$.*

Next, we will focus on edges which are part of cuts of the network of size two. Remember that a bridge is an edge whose removal from a network increases the number of connected components of that network. Let $G = (V, E)$ be any 2-edge-connected network. We say that an edge $e \in E$ is a *2-cut-edge* if there exists a cut of G of size 2 which contains edge e. Equivalently, e is a 2-cut-edge of G if its removal from G creates at least one bridge in $G - e$. We now bound the number of 2-cut-edges in any 2-edge-connected network G. This is an important structural result, since this proves that the amount of edge-overbuilding due to the adversary is sharply limited.

Lemma 4. *Any 2-edge-connected network G with n nodes can have at most $2(n - 1)$ edges which are 2-cut-edges.*

Proof. Let e be any 2-cut-edge in network G. By definition, the removal of e creates one or more bridges in $G - e$. Let $b_1, \ldots, b_l$ denote those bridges. Note, that $b_1, \ldots, b_l$ also must be 2-cut-edges in G. Moreover, it follows that there must be a shortest cycle C in G which contains all the edges $e, b_1, \ldots, b_l$. If there are more than one such cycles, then fix one of them. We call the fixed cycle C a cut-cycle.

Notice that any 2-cut-edge corresponds to exactly one cut-cycle in the network and that every cut-cycle contains at least two 2-cut-edges. We show in the following that if any cut-cycle in the network contains at least three 2-cut-edges, then we can modify the network to obtain strictly more 2-cut-edges and strictly more cut-cycles. This implies that the number of 2-cut-edges is maximized if the number of cut-cycles is maximized and every cut-cycle contains exactly two 2-cut-edges.

Now we describe the procedure which converts any network with at least one cut-cycle containing at least three 2-cut-edges into a modified network with a strictly increased number of cut-cycles and 2-cut-edges (see Fig. 2). Let G be

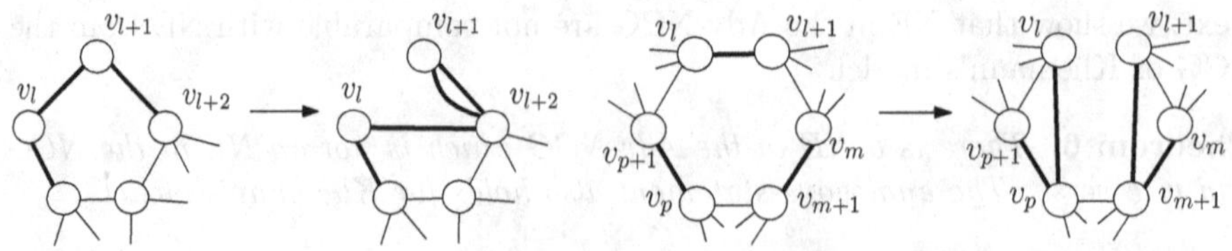

a) Conversion of adjacent 2-cut-edges   b) Conversion of non-adjacent 2-cut-edges

Fig. 2. Increasing the number of 2-cut-edges by splitting up a cut-cycle.

any network with at least one cut-cycle $C = v_1, \ldots, v_k, v_1$ containing at least three 2-cut-edges. If there are two adjacent 2-cut-edges $\{v_l, v_{l+1}\}, \{v_{l+1}, v_{l+2}\}$ in cycle C, then delete the 2-cut-edge $\{v_l, v_{l+1}\}$ and insert two new edges $\{v_l, v_{l+2}\}$ and $\{v_{l+1}, v_{l+2}\}$. First of all, note that these new edges have not been present in network G before the insertion since otherwise $\{\{v_l, v_{l+1}\}, \{v_{l+1}, v_{l+2}\}\}$ cannot be a cut of G. We claim that both new edges are 2-cut-edges and that the cycle C is divided into two new cut-cycles $v_1, \ldots, v_l, v_{l+2}, \ldots, v_k, v_1$ and $v_{l+1}, v_{l+2}, v_{l+1}$. Indeed, there are at least two bridges $\{v_{l+1}, v_{l+2}\}$ and $\{v_k, v_{k+1}\}$ in the cut-cycle C after deleting $\{v_l, v_{l+1}\}$, and both of them end up in different new cut-cycles. Hence, deleting any of the newly inserted edges $\{v_l, v_{l+2}\}$ or $\{v_{l+1}, v_{l+2}\}$ implies that $\{v_k, v_{k+1}\}$ or $\{v_{l+1}, v_{l+2}\}$ becomes a bridge. Thus, both new edges are 2-cut-edges and both of new cycles are cut-cycles.

If there are three pairwise non-adjacent 2-cut-edges $\{v_l, v_{l+1}\}, \{v_m, v_{m+1}\}, \{v_p, v_{p+1}\}$ in cycle C, then delete one 2-cut-edge $\{v_l, v_{l+1}\}$ and insert two new edges $\{v_l, v_{p+2}\}$ and $\{v_{l+1}, v_{m+1}\}$. Analogous to above, both new edges cannot be already present in G and both are 2-cut-edges because deleting any of them renders edge $\{v_m, v_{m+1}\}$ or $\{v_p, v_{p+1}\}$ a bridge. Moreover, cut-cycle C is divided into two new cut-cycles.

Finally, we claim that the maximum number of cut-cycles in any n-vertex network G is at most $n - 1$. Since we know that every such cut-cycle contains exactly two 2-cut-edges this then implies that there can be at most $2(n - 1)$ 2-cut-edges in any network G.

Now we prove the above claim. Note that applying our transformation does not disconnect the network. Thus, we know that network G after all transformations is connected. Now we iteratively choose any cut-cycle C in G and we delete the two 2-cut-edges contained in C. This deletion increases the number of connected components of the current network by exactly 1. We repeat this process until we have destroyed all cut-cycles in G. Note that deleting edges from G may create new cut-cycles, but we never destroy more than one of them at a time. Thus, since each iteration increases the number of connected components of the network by 1, it follows that there can be at most $n - 1$ iterations since network G with n vertices cannot have more than n connected components. □

Remark 1. Lemma 4 is tight, since a path of length $n - 1$, where all neighboring nodes are connected via double edges, has exactly $2(n - 1)$ 2-cut-edges.

Now we relate the diameter with the social cost.

Theorem 7. *Let* (G, α) *be any NE network on* n *nodes having diameter* D *and let* $OPT(n, \alpha)$ *be the optimum network on* n *nodes for the same edge-cost* α. *Then we have that*

$$\frac{cost(G, \alpha)}{cost(OPT(n, \alpha))} \in \mathcal{O}(\mathcal{D}).$$

Proof. Since $OPT(n, \alpha)$ must be 2-edge-connected, it must have at least n edges. Moreover, the minimum expected distance between each pair of vertices in $OPT(n, \alpha)$ is at least 1. Thus, we have that $cost(OPT(n, \alpha)) \in \Omega(\alpha \cdot n + n^2)$.

Now we analyze the social cost of the NE network (G, α), where $G = (V, E)$. We have $cost(G, \alpha) = edge(G, \alpha) + dist(G)$ and we will analyze both terms separately. We start with an upper bound on $dist(G)$.

Since (G, α) has diameter D and since (G, α) is 2-edge-connected, Lemma 3 implies that the expected distance between each pair of vertices in (G, α) is at most $2D$. Thus, we have that $dist(G) \in \mathcal{O}(n^2 \cdot D)$.

Now we analyze $edge(G, \alpha)$. By Lemma 4 we have at most $2n$ many 2-cut-edges in G. Buying all those edges yields cost of at most $2n \cdot \alpha$.

We proceed with bounding the number of non-2-cut-edges in G. We consider any agent v and analyze how many non-2-cut-edges agent v can have bought. We claim that this number is in $\mathcal{O}\left(\frac{nD}{\alpha}\right)$, which yields total edge-cost of $\mathcal{O}(nD)$ for agent v. Summing up over all n agents, this yields total edge-cost of $\mathcal{O}(n^2 D)$ for all non-2-cut-edges of G. This then implies an upper bound of $\mathcal{O}(\alpha \cdot n + n^2 D)$ on the social cost of G which finishes the proof.

Now we prove our claim. Fix any non-2-cut-edge $e = \{v, w\}$ of G which is owned by agent v. Let $V_e \subset V$ be the set of nodes of G to which all shortest paths from v traverse the edge e.

We first show that removing the edge e increases agent v's expected distance to any node in V_e to at most $4D$. By Lemma 3, removing edge e increases the diameter of G from D to at most $2D$. Since e is a non-2-cut-edge, we have that $G - e$ is still 2-edge-connected. Thus, again by Lemma 3, it follows that agent v's expected distance to any other node in $G - e$ is at most $4D$.

However, removing edge e not only increases v's expected distance towards all nodes in V_e, instead, since $G - e$ has a less many edges than G, agent v's expected distance to *all* other nodes in $V \setminus (V_e \cup \{v\})$ increases as well. We now proceed to bound this increase in expected distance cost.

We compare agent v's expected distance cost in network G and in network $G - e$. Let m denote the number of edges in G. Thus, $G - e$ has $m - 1$ many edges. For network G agent v's expected distance cost is

$$dist_G(v) = \frac{1}{m} \sum_{f \in E} \delta_{G-f}(v) = \frac{1}{m} \sum_{f \in E \setminus \{e\}} \delta_{G-f}(v) + \frac{\delta_{G-e}(v)}{m}.$$

In network $G - e$, we have $dist_{G-e}(v) = \frac{1}{m-1} \sum_{f \in E \setminus \{e\}} \delta_{G-e-f}(v)$. Now we upper bound the increase in expected distance cost for agent v due to removal of edge e from G. $dist_{G-e}(v) - dist_G(v)$ is

$$\frac{1}{m-1} \sum_{f \in E \setminus \{e\}} \delta_{G-e-f}(v) - \left(\frac{1}{m} \sum_{f \in E \setminus \{e\}} \delta_{G-f}(v) + \frac{\delta_{G-e}(v)}{m} \right)$$

$$= \sum_{f \in E \setminus \{e\}} \left(\frac{\delta_{G-e-f}(v)}{m-1} - \frac{\delta_{G-f}(v)}{m} \right) - \frac{\delta_{G-e}(v)}{m}.$$

We have that $\delta_{G-e-f}(v) \leq \delta_{G-f}(v) + |V_e| \cdot 4D$, since in $G - e - f$ only the distances to nodes in V_e increase, compared to the network $G - f$ and since e is a non-2-cut-edge in G. Moreover, by Lemma 3, the distances to nodes in V_e in $G - e - f$ increase to at most $4D$ for each node in V_e. Thus, we have that $dist_{G-e}(v) - dist_G(v)$ is

$$\sum_{f \in E \setminus \{e\}} \left(\frac{\delta_{G-e-f}(v)}{m-1} - \frac{\delta_{G-f}(v)}{m} \right) - \frac{\delta_{G-e}(v)}{m}$$

$$\leq \sum_{f \in E \setminus \{e\}} \left(\frac{\delta_{G-f}(v) + |V_e|4D}{m-1} - \frac{\delta_{G-f}(v)}{m} \right)$$

$$= |V_e|4D + \sum_{f \in E \setminus \{e\}} \left(\frac{\delta_{G-f}(v)}{m(m-1)} \right) \leq |V_e|4D + \sum_{f \in E \setminus \{e\}} \left(\frac{2D \cdot n}{n(m-1)} \right)$$

$$\leq |V_e|4D + 4D = (|V_e|+1)4D.$$

Since G is in Nash Equilibrium, we know that removing edge e is not an improving move for agent v. Thus, we have that

$$\alpha \leq (|V_e|+1)4D \iff |V_e| \geq \frac{\alpha}{4D} - 1.$$

Thus, for all non-2-cut-edges e which are bought by agent v, we have that $|V_e| \in \Omega(\frac{\alpha}{D})$. Since all these sets V_e are disjoint, it follows that v can have bought at most $\frac{n}{\Omega(\frac{\alpha}{D})} \in \mathcal{O}(\frac{nD}{\alpha})$ many non-2-cut-edges. $\qquad\square$

4.2 Price of Stability and Price of Anarchy

Theorem 8. *If $\alpha \leq \frac{1}{n(n-1)-1}$, then the PoS is 1. If $\frac{1}{n(n-1)-1} < \alpha < \frac{2}{n(n-1)-1}$, then PoS is strictly larger than 1, if $\alpha > 1 - \frac{1}{2n-1}$, then the PoS is at most 2.*

We now show how to adapt two techniques from the NCG for bounding the diameter of equilibrium networks to our adversarial version. This can be understood as a proof of concept showing that the Adv-NCG can be analyzed as rigorously as the NCG. However, carrying over the currently strongest general diameter bound of $2^{\mathcal{O}(\sqrt{\log n})}$ due to Demaine et al. [8], which is based on interleaved region-growing arguments seems challenging due to the fact that we can only work with expected distances.

We start with a simple diameter upper bound based on [10].

Theorem 9. *The diameter of any NE network* (G, α) *is in* $\mathcal{O}(\sqrt{\alpha})$.

Proof. We prove the statement by contradiction. Assume that there are agents u and v in network G with $d_G(u, v) \geq 4\ell$, for some ℓ. Since expected distances cannot be shorter than distances in G, it follows that u's expected distance to v is at least 4ℓ. If u buys an edge to v for the price of α then u's decrease in expected distance cost is at least $\frac{|E|}{|E|+1}(4\ell - 1 + 4\ell - 3 + \cdots + 1) = \frac{|E|}{|E|+1} 2\ell^2$.

Thus, if $d_G(u, v) > 4\sqrt{\alpha}$, then $u's$ decrease in expected distance cost by buying the edge uv is at least $\frac{|E|}{|E|+1} 2\alpha > \alpha$. Thus, if the diameter of G is at least $4\sqrt{\alpha}$ then there is some agent who has an improving move. □

Together with Theorem 7 this yields the following statement:

Corollary 1. *The Price of Anarchy of the Adv-NCG is in* $\mathcal{O}(\sqrt{\alpha})$.

Adapting a technique by Albers et al. [1] yields a stronger statement, which implies constant PoA for $\alpha \in \mathcal{O}(\sqrt{n})$.

Theorem 10. *The Price of Anarchy of the Adv-NCG is in* $\mathcal{O}\left(1 + \frac{\alpha}{\sqrt{n}}\right)$.

We conclude with a weak lower bound on the PoA, which is tight for high α.

Theorem 11. *The Price of Anarchy of the Adv-NCG is at least* 2 *and for very large* α *this bound is tight.*

5   Conclusion

Our work is the first step towards incorporating both centrality and robustness aspects in a simple and accessible model for selfish network creation. In essence we proved that many properties and techniques can be carried over from the non-adversarial NCG and we indicated that the landscape of optimum and equilibrium networks in the Adv-NCG is much more diverse than without adversary. As for the NCG, proving strong upper or lower bounds on the PoA is very challenging. Especially surprising is the hardness of constructing higher lower bounds than in the NCG since by introducing suitable gadgets it is always possible to enforce that no agent wants to swap or delete edges. A non-constant lower bound on the PoA seems possible if α is linear in n.

It would also be interesting to consider different adversaries. An obvious candidate for this is node-removal at random. Another promising choice is a local adversary, where every agent considers that some of its incident edges may fail. This local perspective combined with a centrality aspect could explain why many selfishly built networks have a high clustering coefficient.

Another direction is to consider the swap version [9,20] of the Adv-NCG, especially in the case where all agents own at least 2 edges. We note in passing, that the swap-version of the Adv-NCG is not a potential game (an improving move cycle can be found in the full version [6]) and that creating equilibrium networks having diameter 4 is already very challenging.

References

1. Albers, S., Eilts, S., Even-Dar, E., Mansour, Y., Roditty, L.: On nash equilibria for a network creation game. ACM TEAC **2**(1), 2 (2014)
2. Alon, N., Demaine, E.D., Hajiaghayi, M.T., Leighton, T.: Basic network creation games. SIAM J. Discrete Math. **27**(2), 656–668 (2013)
3. Anshelevich, E., Bhardwaj, O., Usher, M.: Friend of my friend: Network formation with two-hop benefit. Theor. Comput. Syst. **57**(3), 711–752 (2015)
4. Bala, V., Goyal, S.: A noncooperative model of network formation. Econometrica **68**(5), 1181–1229 (2000)
5. Bilò, D., Gualà, L., Proietti, G.: Bounded-distance network creation games. ACM Trans. Econ. Comput. **3**(3), 16 (2015)
6. Chauhan, A., Lenzner, P., Melnichenko, A., Münn, M.: On selfish creation of robust networks. arXiv preprint (2016). arXiv:1607.02071
7. Corbo, J., Parkes, D.: The price of selfish behavior in bilateral network formation. In: Proceedings of the twenty-fourth annual ACM symposium on Principles of distributed computing, PODC 2005, pp. 99–107. ACM, New York, NY, USA (2005)
8. Demaine, E.D., Hajiaghayi, M.T., Mahini, H., Zadimoghaddam, M.: The price of anarchy in network creation games. ACM Trans. Algorithms **8**(2), 13 (2012)
9. Ehsani, S., Fadaee, S.S., Fazli, M., Mehrabian, A., Sadeghabad, S.S., Safari, M.A., Saghafian, M.: A bounded budget network creation game. ACM Trans. Algorithms **11**(4), 34 (2015)
10. Fabrikant, A., Luthra, A., Maneva, E., Papadimitriou, C.H., Shenker, S.: On a network creation game. In: PODC 2003, pp. 347–351. ACM, New York, USA (2003)
11. Flammini, M., Gallotti, V., Melideo, G., Monaco, G., Moscardelli, L.: Mobile network creation games. In: Even, G., Halldórsson, M.M. (eds.) SIROCCO 2012. LNCS, vol. 7355, pp. 159–170. Springer, Heidelberg (2012)
12. Goyal, S., Jabbari, S., Kearns, M., Khanna, S., Morgenstern, J.: Strategic network formation with attack and immunization. arXiv preprint 2015. arXiv:1511.05196
13. Jackson, M.O., Wolinsky, A.: A strategic model of social and economic networks. J. Econ. Theor. **71**(1), 44–74 (1996)
14. Kawald, B., Lenzner, P.: On dynamics in selfish network creation. In: SPAA 2013, pp. 83–92. ACM, New York, NY, USA (2013)
15. Kliemann, L.: The price of anarchy for network formation in an adversary model. Games **2**(3), 302–332 (2011)
16. Kliemann, L.: The price of anarchy in bilateral network formation in an adversary model. arXiv preprint (2013). arXiv:1308.1832
17. Lenzner, P.: Greedy selfish network creation. In: Goldberg, P.W. (ed.) WINE 2012. LNCS, vol. 7695, pp. 142–155. Springer, Heidelberg (2012)
18. Meirom, E.A., Mannor, S., Orda, A.: Network formation games with heterogeneous players and the internet structure. In: Proceedings of the 15th ACM Conference on Economics and Computation, pp. 735–752. ACM (2014)
19. Meirom, E.A., Mannor, S., Orda, A.: Formation games of reliable networks. In: INFOCOM 2015, pp. 1760–1768. IEEE (2015)
20. Mihalák, M., Schlegel, J.C.: The price of anarchy in network creation games Is (Mostly) constant. In: Kontogiannis, S., Koutsoupias, E., Spirakis, P.G. (eds.) SAGT 2010. LNCS, vol. 6386, pp. 276–287. Springer, Heidelberg (2010)

Dynamic Resource Allocation Games

Guy Avni[1]([✉]), Thomas A. Henzinger[1], and Orna Kupferman[2]

[1] IST Austria, Klosterneuburg, Austria
guy.avni@ist.ac.at
[2] The Hebrew University, Jerusalem, Israel

Abstract. In *resource allocation games*, selfish players share resources that are needed in order to fulfill their objectives. The cost of using a resource depends on the load on it. In the traditional setting, the players make their choices concurrently and in one-shot. That is, a strategy for a player is a subset of the resources. We introduce and study *dynamic resource allocation games*. In this setting, the game proceeds in phases. In each phase each player chooses one resource. A scheduler dictates the order in which the players proceed in a phase, possibly scheduling several players to proceed concurrently. The game ends when each player has collected a set of resources that fulfills his objective. The cost for each player then depends on this set as well as on the load on the resources in it – we consider both congestion and cost-sharing games. We argue that the dynamic setting is the suitable setting for many applications in practice. We study the stability of dynamic resource allocation games, where the appropriate notion of stability is that of subgame perfect equilibrium, study the inefficiency incurred due to selfish behavior, and also study problems that are particular to the dynamic setting, like constraints on the order in which resources can be chosen or the problem of finding a scheduler that achieves stability.

1 Introduction

Resource allocation games (RAGs, for short) [21] model settings in which selfish agents share resources that are needed in order to fulfill their objectives. The cost of using a resource depends on the load on it. Formally, a k-player RAG G is given by a set E of resources and a set of possible strategies for each player. Each strategy is a subset of resources, fulfilling some objective of the player. Each resource $e \in E$ is associated with a latency function $\ell_e : \mathbb{N} \to \mathbb{R}$, where $\ell_e(\gamma)$ is the cost of a single use of e when it has load γ. For example, in *network formation games* (NFGs, for short) [2], a network is modeled by a directed graph, and each player has a source and a target vertex. In the corresponding RAG, the resources are the edges of the graph and the objective of each player is to connect his source and target. Thus, a strategy for a player is a set of edges that

---

This research was supported in part by the European Research Council (ERC) under grants 267989 (QUAREM) and 278410 (QUALITY), and by the Austrian Science Fund (FWF) under grants S11402-N23 (RiSE) and Z211-N23 (Wittgenstein Award).

© Springer-Verlag Berlin Heidelberg 2016
M. Gairing and R. Savani (Eds.): SAGT 2016, LNCS 9928, pp. 153–166, 2016.
DOI: 10.1007/978-3-662-53354-3_13

form a simple path from the source to the target. When an edge e is used by m players, each of them pays $\ell_e(m)$ for his use.

A key feature of RAGs is that the players choose how to fulfill their objectives *in one shot* and *concurrently*. Indeed, a strategy for a player is a subset of the resources – chosen as a whole, and the players choose their strategies simultaneously. In many settings, however, resource sharing proceeds in a different way. First, in many settings, the choices of the players are made resource by resource as the game evolves. For example, when the network in an NFG models a map of roads and players are drivers choosing routes, it makes sense to allow each driver not to commit to a full route in the beginning of the game but rather to choose one road (edge) at each junction (vertex), gradually composing the full route according to the congestion observed. Second, players may not reach the junctions together. Rather, in each "turn" of the game, only a subset of the players (say, these that have a green light) proceed and chose their next road.

As another example to a rich composition and scheduling of strategies, consider the setting of *synthesis from component libraries* [16], where a designer synthesizes a system from existing components rather than from scratch as in the traditional problem [20]. It is shown in [4,6] that when multiple designers use the same library, a RAG arises. Here too, the choice of components may be made during the design process and may evolve according to choices of other designers.

In this work we introduce and study *dynamic resource allocation games*, which allow the players to choose resources in an iterative and non-concurrent manner. A dynamic RAG is given by a pair $\mathcal{G} = \langle G, \nu \rangle$, where G is a k-player RAG and $\nu : \{1, \dots, k\} \to \{1, \dots, k\}$ is a *scheduler*. A dynamic RAG proceeds in *phases*. In each phase, each player chooses one resource. A phase is partitioned into at most k *turns*, and the scheduler dictates which players proceed in each turn: Player i moves at turn $\nu(i)$. Note that the scheduler may assign the same turn to several players, in which case they choose a resource concurrently in a phase. Once all turns have been taken, a phase is concluded and a new phase begins. A *strategy* for a player in a dynamic RAG is a function that takes the history of choices made by the players so far (in the current phase as well as previous ones), and returns the next choice the player makes. A player finishes playing once the resources he has chosen forms a strategy in the underlying RAG. In an outcome of the game, each player selects a set of resources. His cost depends on their load and latency functions as in usual RAGs.

Example 1. Consider the 4-player network formation game that is depicted in Fig. 1. The interesting edges have names, e.g., $a, b, c \dots$, and their latency function is depicted below the edge. For example, we have $\ell_a(x) = x$ and $\ell_{c_1}(x) = 10x$. The other edges have latency function 0. The source and target of a node of Player i are depicted with a node called s and t, respectively, and with a subscript i. For example, Player 2's source is $s_{1,2}$ and he has two targets t_2^L and t_2^R. The players' strategies are paths from one of their sources to one of their targets.

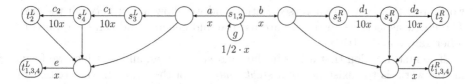

Fig. 1. A network formation game in which it is beneficial to select a path that is not simple.

Consider a dynamic version of the game in which Player i chooses an edge at turn i. At first look, it seems that edge g will never be chosen. However, we show that Player 1's optimal strategy uses it. Player 1 has three options in the first turn, either choose g, a, or b^1. Assume he chooses a (and dually b). Then, we claim that Player 2 will choose b. Note that Players 3 and 4 move oposite of Player 2 no matter how Player 1 moves, as they prefer avoiding a load of 2 on c_1 and c_2, which costs 20 each, even at the cost of a load of 3 on f, which costs only 3. Knowing this, Player 2 prefers using b alone over sharing a with Player 1. Since the loads on a and e are 1 and 3, respectively, Player 1's cost is $1 + 3 = 4$.

On the other hand, if Player 1 chooses g in the first phase, he postpones revealing his choice between left and right. If Player 2 proceeds left, then Players 3 and 4 proceed right, and Player 1 proceeds left in the second phase. Now, the load on a and e is 2 and 1, respectively, thus Player 1's cost is $\frac{1}{2} + 2 + 1 = 3\frac{1}{2}$. □

The concept of what we refer to as a dynamic game is old and dates back to Von Neumann's work on *extensive form games* [18]. Most work on RAGs considers the simultaneous setting. However, there have been different takes on adding dynamicity to RAGs. In [17], the authors refine the notion of NE by considering *lookahead* equilibria; a player predicts the reactions of the other players to his deviations, and he deviates only if the outcome is beneficial. The depth of lookahead is bounded and is a parameter to the equilibria. A similar setting was applied to RAGs in [7], where the players are restricted to choose a *best-response* move rather than a deviation that might not be immediately beneficial. Concurrent ongoing games are commonly used in formal methods to model the interaction between different components of a system (c.f., [1]). In such a game, multiple players move a token on a graph. At each node, each player selects a move, and the transition function determines the next position of token, given the vector of moves the players selected. The objectives of the players refer to the generated path and no costs are involved. Closest to our model is the model of [15], and its subsequent works [8,10]. They study RAGs in which players arrive and select strategies one by one, yet in one shot.

Our dynamic games differ from all of these games in two aspects. We allow the players to reveal their choices of resources in parts, thus we allow "breaking" the strategies. Moreover, the choices the players make in all the games in earlier

---

^{1} In this example we require the players to choose their paths incrementally, which is not the general definition we use in the paper.

work are either concurrent or sequential, and we allow a mix between the two. These new aspects we introduce are natural and general, and can be applied to other games and settings.

The first question that arises in the context of games, and on which we focus in this work, is the existence of a *stable outcome* of the game. In the context of RAGs, the most prominent stability concept is that of a *Nash equilibrium* (NE, for short) – a profile such that no player can decrease his cost by unilaterally deviating from his current strategy. It is well known that every RAG has an NE [21]. The definition of an NE applies to all games, and can also be applied to our dynamic RAGs. As we demonstrate in Example 2, the dynamic setting calls for a different stability concept, and the prominent one is *subgame perfect equilibrium* (SPE, for short) [24], which we define formally in Sect. 2.

Classifying RAGs, we refer to the type of their latency functions as well as the type of the objectives of the players. *Congestion games* [22] are RAGs in which the latency functions are increasing, whereas in *cost-sharing games* [2], each resource has a cost that is split between the players that use it (in particular, the latency functions are decreasing). In terms of objectives, we consider *singleton* RAGs, in which the objectives of the players are singletons of resources, and *symmetric* RAGs, in which all players have the same objective.

Our most interesting results are in terms of equilibrium existence. It is easy to show, and similar results are well known, that every dynamic RAG with a *sequential scheduler* has an SPE. The proof uses backwards induction on the tree of all possible outcomes of the game. One could hope to achieve a similar proof also for schedulers that are not sequential, especially given the fact that every RAG has an NE. Quite surprisingly, however, we show that this is not the case. For congestion games, we show examples of a singleton congestion game and a symmetric congestion game with no SPE. Moreover, the latency function in both cases is linear. On the positive side, we show that singleton and symmetric congestion games are guaranteed to have an SPE for every scheduler. For cost-sharing games, we also show an example with no SPE. In the cost-sharing setting, however, we show that singleton objectives are sufficient to guarantee the existence of an SPE in all schedules. It follows that singleton dynamic congestion games are less stable than singleton dynamic cost-sharing games. This is interesting, as in the one-shot concurrent setting, congestion games are known to be more stable than cost-sharing games in various parameters. One would expect that this "order of stability" would carry over to the dynamic setting, as is the case in other extensions of the traditional setting. For example, an NE is not guaranteed for *weighted* cost-sharing games [9] as well as very restrictive classes of *multiset* cost-sharing games [5], whereas every linear weighted congestion game [12] and even linear multiset congestion game is guaranteed to have an NE [6].

It is well known that decentralized decision-making may lead to solutions that are sub-optimal from the point of view of society as a whole. In simultaneous games, the standard measures to quantify the inefficiency incurred due to selfish behavior is the *price of anarchy* (PoA) [14] and *price of stability* (PoS) [2]. In both measures we compare against the *social optimum* (SO, for short), namely the cheapest profile. The PoA is the worst-case inefficiency of an NE (that is, the ratio between the cost of a worst NE and the SO). The PoS is the best-case inefficiency of an Nash equilibrium (that is, the ratio between the cost of a best NE and the social optimum). For the dynamic setting, we adjust these two measures to consider SPEs rather than NEs, and we refer to them as DPoA and DPoS. We study the equilibrium inefficiency in the classes of games that have SPEs. We show that the DPoA and DPoS in dynamic singleton cost-sharing games as well as dynamic singleton congestion games coincide with the PoA and PoS in the corresponding simultaneous class. As mentioned above, [8,10,15] study games in which players arrive one after the other. Since their games are sequential, they always have an SPE. They study the *sequential PoA*, and show that it can either be equal, below, or above the PoA of the corresponding class of RAGs.

We then turn to study the computational complexity of deciding whether a given dynamic RAG has an SPE. We show that the problem is PSPACE-complete for both congestion and cost-sharing games. Our lower bound for cost-sharing games implies that finding an SPE in sequential games is PSPACE-hard. To the best of our knowledge, while this problem was solved in [15] for congestion games, we are the first to solve it for cost-sharing games.

Due to lack of space, some proofs and examples are given in the full version, which can be found in the authors' homepages.

2   Preliminaries

Resource Allocation Games. For $k \geqslant 1$, let $[k] = \{1,\ldots,k\}$. A *resource-allocation game* (RAG, for short) is a tuple $G = \{[k], E, \{\Sigma_i\}_{i \in [k]}, \{\ell_e\}_{e \in E}\}$, where $[k]$ is a set of k players; E is a set of resources; for $i \in [k]$, the set $\Sigma_i \subseteq 2^E$ is a set of objectives[2] for Player i; and, for $e \in E$, we have that $\ell_e : \mathbb{N} \to \mathbb{R}$ is a latency function. The game proceeds in one-round in which the players select simultaneously one of their objectives. A *profile* $P = \langle \sigma_1,\ldots,\sigma_k \rangle \in \Sigma_1 \times \ldots \times \Sigma_k$ is a choice of an objective for each player. For $e \in E$, we denote by $nused(P,e)$ the number of times e is used in P, thus $nused(P,e) = |\{i \in [k] : e \in \sigma_i\}|$. For $i \in [k]$, the *cost* of Player i in P, denoted $cost_i(P)$, is $\sum_{e \in \sigma_i} \ell_e(nused(P,e))$.

Classes of RAGs are characterized by the type of latency functions and objectives. In *congestion* games (CGs, for short), the latency functions are increasing. An exceptionally stable class of CGs are ones in which the latency functions are affine (c.f., [6,12]); every resource $e \in E$ has two constants a_e and b_e, and the latency function is $\ell_e(x) = a_e \cdot x + b_e$. In *cost-sharing* games (SG, for short), each resource $e \in E$ has a *cost* c_e and the players that use the resource share its cost, thus the latency function for e is $\ell_e(x) = \frac{c_e}{x}$, and in particular is decreasing. We use DCGs and DSGs to refer to dynamic CGs and dynamic SGs, respectively. In terms of objectives, we study *symmetric* games, where the players' sets of

---

[2] We use "objectives" rather than "strategies" as the second will later be used for dynamic games.

objectives are equal, thus $\Sigma_i = \Sigma_j$ for all $i, j \in [k]$, and *singleton* games, where each $\sigma \in \Sigma_i$ is a singleton, for every $i \in [k]$.

Dynamic Resource Allocation Games. A *dynamic* RAG is pair $\mathcal{G} = \langle G, \nu \rangle$, where G is a RAG and $\nu : [k] \to [k]$ is a *scheduler*. Intuitively, in a dynamic game, rather than revealing their objectives at once, the game proceeds in *phases*: in each phase, each player reveals one resource in his objective. Each phase is partitioned into at most k *turns*. The scheduler dictates the order in which the players proceed in a phase by assigning to each player his turn in the phases. If the scheduler assigns the same turn to several players, they select a resource concurrently. Once all players take their turn, a phase is concluded and a new phase begins. There are two "extreme" schedulers: (1) players get different turns, i.e., ν is a permutation, (2) all players move in one turn, i.e., $\nu \equiv 1$. We refer to games with these schedulers as *sequential* and *concurrent*, respectively. Note that ν might not be an onto function. For simplicity, we assume that, for $j > 1$, if turn j is assigned a player, then so is turn $j - 1$. We use t_ν to denote the last turn according to ν, thus $t_\nu = \max_i \nu(i)$.

Let $E_\perp = E \cup \{\perp\}$, where $\perp$ is a special symbol that represents the fact that a player finished playing. Consider a turn $j \in [k]$. We denote by before(j) the set of players that play before turn j; thus before$(j) = \{i \in [k] : \nu(i) < j\}$. A player has full knowledge of the resources that have been chosen in previous phases and the resources chosen in previous turns in the current phase. A strategy for Player i in $\mathcal{G}$ is a function $f_i : (E_\perp^{[k]})^* \cdot (E_\perp^{\text{before}(\nu(i))}) \to E_\perp$. A *profile* $P = \langle f_1, \dots, f_k \rangle$ is a choice of a strategy for each player. The *outcome* of the game given a profile P, denoted $out(P)$, is an infinite sequence of functions $\pi^1, \pi^2, \dots$, where for $i \geqslant 1$, we have $\pi^i : [k] \to E_\perp$. We define the sequence inductively as follows. Let $m \geqslant 1$ and $j \in [k]$. Assume $m - 1$ phases have been played as well as $j - 1$ turns in the m-th phase, thus $\pi^1, \pi^2, \dots, \pi^{m-1}$ are defined as well as $\pi_{j-1}^m : \text{before}(j) \to E_\perp$. We define π_j^m as follows. Consider a player i with $\nu(i) = j$. The resource Player i chooses in the m-th phase is $f_i(\pi^1, \dots, \pi^{m-1}, \pi_{j-1}^m)$. Finally, we define $\pi^m = \pi_{t_\nu}^m$.

We restrict attention to *legal* strategies for the players, namely ones in which the collection of resources chosen by Player i in all phases is an objective in Σ_i.[3] Also, once Player i chooses $\perp$, then he has finished playing and all his choices in future phases must also be $\perp$. Formally, for a profile $P = \langle f_1, \dots, f_k \rangle$ with $out(P) = \pi^1, \pi^2, \dots$ and $i \in [k]$, let $out_i(P)$ be $\pi^1(i), \pi^2(i), \dots$. For $j \geqslant 1$, let $e_j = \pi^j(i)$ be the resource Player i selects in the j-th phase. Thus, $out_i(P)$ is an infinite sequence over $E_\perp$. We say that f_i is legal if (1) there is an index m such that $e_j \in E$ for all $j < m$ and $e_j = \perp$ for all $j \geqslant m$, and (2) the set $\{e_1, \dots, e_{m-1}\}$ is an objective in Σ_i. (In particular, a player cannot select a resource multiple times nor a resource that is not a member in his chosen objective). We refer to

---

[3] It is interesting to allow players to use "redundant resources"; a player's choice of resources should contain one of his objectives. While in the traditional setting, using a redundant resource cannot be beneficial, in the dynamic setting, it is, as a variant of Example 1 demonstrates.

an outcome in which the players use legal strategies as a *legal outcome* and a prefix of a legal outcome as a *legal history*.

In $out(P)$, every player selects a set of resources. The cost of a player is calculated similarly to RAGs. That is, his cost for a resource e, assuming the load on it is γ, is $\ell_e(\gamma)$, and his total cost is the sum of costs of the resources he uses. When the outcome of a profile P in a dynamic RAG coincides with the outcome of a profile Q in a RAG G, we say that P and Q are *matching* profiles.

Equilibrium Concepts. A *Nash equilibrium*[4] (NE, for short) in a game is a profile in which no player has an incentive to unilaterally deviate from his strategy. Formally, for a profile P, let $P[i \leftarrow f_i']$ be the profile in which Player i switches to the strategy f_i' and all other players use their strategies in P. Then, a profile P is an NE if for every $i \in [k]$ and every legal strategy f_i' for Player i, we have $cost_i(P) \leqslant cost_i(P[i \leftarrow f_i'])$. It is well known that every RAG is guaranteed to have an NE [21].

The definition of NE applies to all games, in particular to dynamic ones. Every NE Q in a RAG G matches an NE in a dynamic game $\langle G, \nu \rangle$, for some scheduler ν, in which the players ignore the history of the play and follow their objectives in Q. However, such a strategy is not rational. Thus, one could argue that an NE is not necessarily achievable in a dynamic setting. We illustrate this in the following example.

Example 2. Consider a two-player DCG with resources $\{a, b\}$, latency functions $\ell_a(x) = x$ and $\ell_b(x) = 1.5x$, and objectives $\Sigma_1 = \Sigma_2 = \{\{a\}, \{b\}\}$. Consider the sequential scheduling in which Player 1 moves first followed by Player 2. Since the players' objectives are singletons, the dynamic game consists of one phase. Consider the Player 2 strategy f_2 that "promises" to select the resource a no matter what Player 1 selects, thus $f_2(a) = f_2(b) = a$. Let f_1^a and f_1^b be the Player 1 strategies in which he selects a and b, respectively, thus $f_1^a(\epsilon) = a$ and $f_1^b(\epsilon) = b$, where ϵ denotes the empty history. Note that these are all of Player 1's possible strategies. The profile $P = \langle f_1^b, f_2 \rangle$ is an NE. Indeed, Player 2 pays 1, which is the least possible payment, so he has no incentive to deviate. Also, by deviating to f_1^a, Player 1's payoff increases from 1.5 to 2, so he has no incentive to deviate either. Note, however, that this strategy of Player 2 is not rational. Indeed, when it is Player 2's turn, he is aware of Player 1's choice. If Player 1 plays f_1^a, then a rational Player 2 is not going to choose a, as this results in a cost of 2, whereas by b, his cost will be 1.5. Thus, an NE profile with f_2 may not be achievable.   □

To overcome this issue, the notion of *subgame perfect equilibrium* (SPE, for short) was introduced. In order to define SPE, we need to define a subgame of a dynamic game. Let $\mathcal{G} = \langle G, \nu \rangle$. It is helpful to consider the *outcome tree* $\mathcal{T}_{\mathcal{G}}$ of $\mathcal{G}$, which is a finite rooted tree that contains all the legal histories of $\mathcal{G}$. Each internal node in $\mathcal{T}_{\mathcal{G}}$ corresponds to a legal history, its successors correspond to

---

[4] Throughout this paper, we consider *pure* strategies as is the case in the vast literature on RAGs.

possible extensions of the history, and each leaf corresponds to a legal outcome. Consider a legal history h. We define a dynamic RAG $\mathcal{G}_h$, which, intuitively, is the same as $\mathcal{G}$ after the history h has been played. More formally, the outcome tree of $\mathcal{G}_h$ is the subtree $T_{\mathcal{G}}^h$ whose root is the node h. We define the costs in $\mathcal{G}_h$ so that the costs of the players in the leaves of $T_{\mathcal{G}}^h$ are the same as the corresponding leaves in $T_{\mathcal{G}}$. Assume that h ends at the m-th turn. A profile P in $\mathcal{G}$ corresponds to a trimming of $T_{\mathcal{G}}$ in which the internal node h has exactly one child $h \cdot \overline{\sigma}$, where $\overline{\sigma}$ is the set of choices of the players in $\nu^{-1}(m)$ when they play according to their strategies in P. The profile P induces a profile P^h in $\mathcal{G}_h$, where the trimming of $T_{\mathcal{G}}^h$ according to P^h coincides with the trimming of $\mathcal{G}$ according to P. We formally define the outcome tree and a subgame in the full version.

Definition 1. *A profile P is an SPE if for every legal history h, the profile P^h is an NE in $\mathcal{G}_h$.*

Note that the profile $P = \langle f_1^b, f_2 \rangle$ in the example above is an NE but not an SPE. Indeed, for the history $h = a$, the profile P^h is not an NE in $\mathcal{G}_h$ as Player 2 can benefit from unilaterally deviating as described above.

3   Existence of SPE in Dynamic Congestion Games

It is easy to show that every sequential dynamic game has an SPE by unwinding the outcome tree, and similar results have been shown before (c.f., [15]). The proof can be found in the full version.

Theorem 1. *Every sequential dynamic game has an SPE.*

One could hope to prove that a general dynamic game $\mathcal{G}$ also has an SPE using a similar unwinding of $T_{\mathcal{G}}$, possibly using the well-known fact that every CG is guaranteed to have an NE [21]. Unfortunately, and somewhat surprisingly, we show that this is not possible. We show that (very restrictive) DCGs might not have an SPE. For the good news, we identify a maximal fragment that is guaranteed to have an SPE.

Recall that a CG is singleton when the players' objectives consist of singletons of resources, and a CG is symmetric if all the players agree on their objectives. We start with the bad news and show that symmetric DCGs and singleton DCGs need not have an SPE, even with linear latency functions. We then show that the combination of these two restrictions is sufficient for existence of an SPE in a DCG.

Theorem 2. *There are symmetric and singleton linear DCGs with no SPE.*

Proof. We first describe a linear DCG with no SPE, and then alter it to make it symmetric. The proof for singleton linear DCG is given in the full version. Consider the following three-player linear CG G with resources $E = \{a, a', b, b', c\}$ and linear latency functions $\ell_a(x) = \ell_b(x) = x$, $\ell_{a'}(x) = \frac{3}{4}x$, $\ell_{b'}(x) = 1\frac{1}{4}$, and

$\ell_c(x) = x + \frac{2}{3}$. Let $\Sigma_1 = \Sigma_2 = \{\{a, a'\}, \{b, b'\}, \{c\}\}$ and $\Sigma_3 = \{\{c\}, \{a', b\}\}$. Consider the dynamic game $\mathcal{G}$ in which Players 1 and 2 move concurrently followed by Player 3. Formally, $\mathcal{G} = \langle G, \nu \rangle$, where $\nu(1) = \nu(2) = 1$ and $\nu(3) = 2$.

We claim that there is no SPE in $\mathcal{G}$. Note that since the players' objectives are disjoint, then once a player reveals the first choice of resource, he reveals the whole objective he chooses, thus we analyze the game as if it takes place in one phase in which the players' reveal their whole objective. The profiles in which Players 1 and 2 choose the same objective are clearly not a SPE as they are not an NE in the game $\mathcal{G}_\epsilon$. As for the other profiles, in Fig. 2, we go over half of them, and show that none of them is an SPE. The other half is analogous. The root of each tree is labeled by the objectives of Players 1 and 2, and its branches according to Player 3's objectives. In the leaves we state Player 3's payoff. In an SPE, Player 3 performs a best-response according to the objectives he observes as otherwise the subgame is not in an NE. We depict his choice with a bold edge. Beneath each tree we note the payoffs of all the players in the profile, and the directed edges represent the player that can benefit from unilaterally deviating. In the full version, we construct a symmetric DCG $\mathcal{G}'$ by altering the game $\mathcal{G}$ above. We do this by adding a fourth player and three new resources so that $\mathcal{G}'$ simulates $\mathcal{G}$.   □

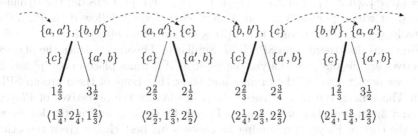

Fig. 2. Profiles in the game $\mathcal{G}$ with no SPE.

We now prove that combining the two restrictions does guarantee the existence of SPE. We note that while our negative results hold for linear DCGs, which tend to be stabler than other DCGs, our positive result holds for every increasing latency function.

Theorem 3. *Every symmetric singleton DCG has an SPE.*

Proof. Consider a symmetric singleton DCG $\mathcal{G} = \langle G, \nu \rangle$. Recall that since G is a singleton game, every outcome of $\mathcal{G}$ consists of one phase. Let P be an NE in G (recall that according to [21] an NE exists in every CG). Since G is symmetric, we can assume that, for $1 \leqslant j < k$, the players that move in the j-th turn do not pay more than the players that move after them. Formally, for $i, i' \in [k]$, if $\nu(i) < \nu(i')$, then $cost_i(P) \leqslant cost_{i'}(P)$. In particular, the players who move in the first turn pay the least, and the players that move in the last turn pay the most. We construct a profile Q in $\mathcal{G}$ and show that it is an SPE. Intuitively, in Q, the players follow their objectives in P assuming the previous players also follow it. Since the costs are increasing with turns, if Player i deviates, a following Player j will prefer switching resources with Player i and also switching the costs. Thus, the deviation is not beneficial for Player i. In the full version, we construct Q formally and prove that it is an SPE. □

4   Existence of SPE in Dynamic Cost-Sharing Games

Cost sharing games tend to be less stable than congestion games in the concurrent setting; for example, very simple fragments of multiset cost-sharing games do not have an NE [5] while linear multiset congestion games are guaranteed to have an NE [6]. In this section we are going to show that, surprisingly, there are classes of games in which an SPE exists only in the cost-sharing setting. Still, SPE is not guaranteed to exist in general DSGs. We start with the bad news.

Theorem 4. *There is a DSG with no SPE.*

Proof. Consider the following four-player SG G with resources $E = \{a, a', a'', b, b', b'', c, c', c''\}$ and costs $c_a = c_b = c_c = 6$, $c_{a'} = c_{b'} = c_{c'} = 4$, and $c_{a''} = c_{b''} = c_{c''} = 3$. Let $\Sigma_1 = \{\{a, a'\}, \{b, b''\}\}$, $\Sigma_2 = \{\{b, b'\}, \{c, c''\}\}$, $\Sigma_3 = \{\{c, c'\}, \{a, a''\}\}$, and $\Sigma_4 = \{\{a, a'\}, \{b, b'\}, \{c, c'\}\}$. Consider the dynamic game $\mathcal{G}$ in which players 1,2, and 3 move concurrently followed by Player 4. Formally, $\mathcal{G} = \langle G, \nu \rangle$, where $\nu(1) = \nu(2) = \nu(3) = 1$ and $\nu(4) = 2$.

We claim that there is no SPE in $\mathcal{G}$. Similar to Theorem 2, since the players' objectives are disjoint, we analyze the game as if it takes place in one phase. In Fig. 3, we depict some of the profiles and show that none of them are an SPE. As in Theorem 2, the root of each tree is labeled by the objectives of Players 1, 2, and 3, its branches according to Player 4's choices, and in the leaves we state the cost of Player 4 assuming he chooses his best choice given the other players' choices. Finally, it is not hard to show that every profile not on the cycle of profiles cannot be an SPE. □

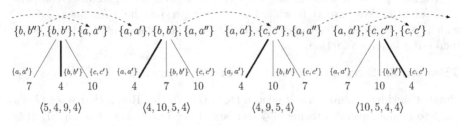

Fig. 3. Profiles in the game with no SPE. Bold edges depict Player 4's best choice given the other players choices. Directed edges represent the player that can benefit from unilaterally deviating.

Recall that singleton DCGs are not guaranteed to have an SPE (Theorem 2). On the other hand, we show below that singleton DSGs are guaranteed to have an SPE. In order to find an SPE in such a game, we use a firmer notion of an equilibria in SGs.

A *strong equilibrium* (SE, for short) [3] is a profile that is stable against deviations of *coalitions* of players rather than deviations of a single player as in NEs (see the full version for a formal definition). We show a connection between strong equilibria and SPEs in singleton SGs. It is shown in [13] that every singleton SG has an SE.

Theorem 5. *Consider a singleton DSG $\mathcal{G} = \langle G, \nu \rangle$. Then, every strong equilibrium in G matches an SPE of $\mathcal{G}$. In particular, every singleton DSG has an SPE.*

Proof. We describe the intuition of the proof and the details can be found in the full version. Consider a singleton DSG $\mathcal{G} = \langle G, \nu \rangle$, and let Q be an SE in G. We describe a profile P in $\mathcal{G}$ that matches Q, and we claim that it is an SPE. Consider a history h that ends in the i-th turn. Assume the players that play in h follow their objective in Q. Then, the players who play next, namely these in $\nu^{-1}(i+1)$, also follow Q. Thus, P matches Q. The definition of the strategies in P for histories that do not follow Q is inductive: assume only the players in $\nu^{-1}(i)$ choose differently than in Q, then the subgame $\mathcal{G}_h$ is a singleton DSG. We find a strong equilibrium in $\mathcal{G}_h$ and let the players in $\nu^{-1}(i+1)$ choose according to it. In order to show that no Player i can unilaterally benefit from deviating to a resource e from P, we observe that it is not possible that all players that deviate into e decrease their costs (as Q is an SE). So, there must be a Player j_1 that deviates from some resource e' to e and increases his cost. This can only happen if there is a Player j_2 that also uses e' in Q and deviates to e'' while decreasing his cost. The same reasoning holds for players deviating to e''. Thus, we find a sequence of resources, which must contain a loop as there are finitely many resources. Using it we can reach a contradiction to the fact that Player i benefits.   ☐

5   Equilibrium Inefficiency

It is well known that decentralized decision-making may lead to sub-optimal solutions from the point of view of society as a whole. We define the cost of a profile P, denoted $cost(P)$, to be $\sum_{i \in [k]} cost_i(P)$. We denote by OPT the cost of a social-optimal solution; i.e., $OPT = \min_P cost(P)$. Two standard measures that quantify the inefficiency incurred due to self-interested behavior are the *price of anarchy* (PoA) [14,19] and *price of stability* (PoS) [2,23]. The PoA is the worst-case inefficiency of an NE; The PoA of a game G is the ratio between the cost of the most expensive NE and the cost of the social optimum. The PoS measures the best-case inefficiency of an NE, and is defined similarly with the cheapest NE. The PoA of a family of games $\mathcal{F}$ is $\sup_{G \in \mathcal{F}} PoA(G)$, and the definition is similar for PoS.

In dynamic games we consider SPE rather than NE. We adapt the definitions above accordingly, and we refer to the new measures as *dynamic PoA* and *dynamic PoS* (DPoA and DPoS, for short). We study the equilibrium inefficiency in the classes of games that are guaranteed to have an SPE, namely singleton DSGs and symmetric singleton DCGs.

The lower bounds for the PoA and PoS for singleton SG and singleton symmetric CGs follow to the dynamic setting as we can consider the scheduler in which all players choose simultaneously in the first turn. For the upper bound we start with the DPoS. In the congestion setting, we show that every NE in the underlying RAG matches an SPE. In the cost-sharing setting, recall that an SE in the traditional game matches an SPE in the dynamic game, and by [26], a singleton SG has an SE whose cost is at most $\log(k) \cdot OPT$. This matches the $\log(k)$ lower bound. We continue to study DPoA. In the cost-sharing setting, the upper bound follows from the same argument as traditional games. For congestion games, it follows by applying a recent result by [10] to our setting. The details can be found in the full version.

Theorem 6. *The DPoA and DPoS in singleton DSGs and singleton symmetric DCGs coincide with the PoA and PoS in singleton SGs and singleton symmetric CGs, respectively.*

Proof. Thus, for singleton DSGs we have DPoA $= k$ and DPoS $= \log k$ [2], and for singleton symmetric DCGs we have DPoA $= 4/3$ [11] and we are not aware of bounds for the PoS in the corresponding CGs.

6   Deciding the Existence of SPE

In the previous sections we showed that dynamic RAGs are not guaranteed to have an SPE. A natural decision problem arises, which we refer to as ∃SPE: given a dynamic RAG, decide whether it has an SPE. We show that the problem is PSPACE-complete in DSGs and DCGs. We start with the lower bound. The crux of the proof is given in the following lemma. For DCGs, such a construction is described in [15], which uses a construction by [25] in order to simulate the logic of a NAND gate by means of a CG. For SGs we are not aware of a similar known result. We describe the construction in the full version, which is inspired by the construction in [15].

Lemma 1. *Given a QBF instance ψ, there is a fully sequential game $\mathcal{G}_\psi$ that is either a DCG or a DSG, and two constants $\gamma, \delta > 0$, such that in every SPE P in $\mathcal{G}_\psi$, (1) if ψ is true, then $cost_1(P) < \gamma$, and (2) if ψ is false, then $cost_1(P) > \delta$.*

To conclude the lower-bound proof, we combine the game that is constructed in Lemma 1 and a game that has no SPE as in the examples we show in the previous sections. For the upper bound, consider a dynamic RAG $\mathcal{G}$, and let $\mathcal{T}_\mathcal{G}$ be the outcome tree of $\mathcal{G}$. Recall that there is a one-to-one correspondence between leaves in $\mathcal{T}_\mathcal{G}$ and legal outcomes of $\mathcal{G}$. In order to decide in PSPACE whether $\mathcal{G}$ has an SPE, we guess a leaf l in $\mathcal{T}_\mathcal{G}$ and verify that it is an outcome of an SPE. Thus, we ask if there is an SPE P in $\mathcal{G}$ whose outcome corresponds to l.

Theorem 7. *The ∃SPE problem is PSPACE-complete for dynamic RAGs.*

7 Extensions

In the full version we study two extensions of the dynamic setting. In the first, we consider the problem of finding a schedule that admits an SPE under given constraints on the order the players move, and show that this problem is also PSPACE-complete. Then, we consider dynamic RAGs in which there is an order on the resources that the players choose. So, if for two resources e_1 and e_2, we have $e_1 \prec e_2$, then a player cannot choose e_1 in a later phase than e_2. The motivation for an order on resources is natural. For example, returning to network formation games, a driver can only extend the path he chooses as the choices are made during driving. We show that all our results carry over to the ordered case.

References

1. Alur, R., Henzinger, T.A., Kupferman, O.: Alternating-time temporal logic. J. ACM **49**(5), 672–713 (2002)
2. Anshelevich, E., Dasgupta, A., Kleinberg, J., Tardos, E., Wexler, T., Roughgarden, T.: The price of stability for network design with fair cost allocation. SIAM J. Comput. **38**(4), 1602–1623 (2008)
3. Aumann, R.: Acceptable points in games of perfect information. Contrib. Theory Games **4**, 287–324 (1959)
4. Avni, G., Kupferman, O.: Synthesis from component libraries with costs. In: Baldan, P., Gorla, D. (eds.) CONCUR 2014. LNCS, vol. 8704, pp. 156–172. Springer, Heidelberg (2014)
5. Avni, G., Kupferman, O., Tamir, T.: Network-formation games with regular objectives. In: Muscholl, A. (ed.) FOSSACS 2014 (ETAPS). LNCS, vol. 8412, pp. 119–133. Springer, Heidelberg (2014)
6. Avni, G., Kupferman, O., Tamir, T.: Congestion games with multisets of resources and applications in synthesis. In: Proceeding of 35th FSTTCS, pp. 365–379 (2015)
7. Bilò, V., Fanelli, A., Moscardelli, L.: On lookahead equilibria in congestion games. In: Chen, Y., Immorlica, N. (eds.) WINE 2013. LNCS, vol. 8289, pp. 54–67. Springer, Heidelberg (2013)
8. Correa, J., de Jong, J., de Keijzer, B., Uetz, M.: The curse of sequentiality in routing games. In: Markakis, E., Schäfer, G. (eds.) WINE 2015. LNCS, vol. 9470, pp. 258–271. Springer, Heidelberg (2015). doi:10.1007/978-3-662-48995-6_19
9. Chen, H., Roughgarden, T.: Network design with weighted players. Theory Comput. Syst. **45**(2), 302–324 (2009)
10. de Jong, J., Uetz, M.: The sequential price of anarchy for atomic congestion games. In: Liu, T.-Y., Qi, Q., Ye, Y. (eds.) WINE 2014. LNCS, vol. 8877, pp. 429–434. Springer, Heidelberg (2014)
11. Fotakis, D.A.: Stackelberg strategies for atomic congestion games. In: Arge, L., Hoffmann, M., Welzl, E. (eds.) ESA 2007. LNCS, vol. 4698, pp. 299–310. Springer, Heidelberg (2007)

12. Harks, T., Klimm, M.: On the existence of pure Nash equilibria in weighted congestion games. Math. Oper. Res. **37**(3), 419–436 (2012)
13. Holzman, R., Law-Yone, N.: Strong equilibrium in congestion games. Games Econ. Behav. **21**(1–2), 85–101 (1997)
14. Koutsoupias, E., Papadimitriou, C.: Worst-case equilibria. CS Rev. **3**(2), 65–69 (2009)
15. Leme, R.P., Syrgkanis, V., Tardos, E.: The curse of simultaneity. In: Proceedings of 3rd ITCS (2012)
16. Lustig, Y., Vardi, M.Y.: Synthesis from component libraries. STTT **15**, 603–618 (2013)
17. Mirrokni, V., Thain, N., Vetta, A.: A theoretical examination of practical game playing: lookahead search. In: Serna, M. (ed.) SAGT 2012. LNCS, vol. 7615, pp. 251–262. Springer, Heidelberg (2012)
18. Neumann, J.: Mathematische Annalen. Zur Theorie der Gesellschaftsspiele **100**(1), 295–320 (1928)
19. Papadimitriou, C.: Algorithms, games, and the internet (extended abstract). In: Orejas, F., Spirakis, P.G., Leeuwen, J. (eds.) ICALP 2001. LNCS, vol. 2076, p. 1. Springer, Heidelberg (2001)
20. Pnueli, A., Rosner, R.: On the synthesis of a reactive module. In: Proceedings of 16th POPL, pp. 179–190 (1989)
21. Rosenthal, R.W.: A class of games possessing pure-strategy Nash equilibria. Int. J. Game Theory **2**, 65–67 (1973)
22. Roughgarden, T., Tardos, E.: How bad is selfish routing? JACM **49**(2), 236–259 (2002)
23. Schulz, A.S., Stier Moses, N.E.: On the performance of user equilibria in traffic networks. In: Proceedings of 14th SODA, pp. 86–87 (2003)
24. Selten, R.: Spieltheoretische Behandlung eines Oligopolmodells mit Nachfrageträgheit. Zeitschrift für die gesamte Staatswissenschaft **121** (1965)
25. Skopalik, A., Vöcking, B.: Inapproximability of pure Nash equilibria. In: Proceedings of 40th STOC, pp. 355–364 (2008)
26. Syrgkanis, V.: The complexity of equilibria in cost sharing games. In: Saberi, A. (ed.) WINE 2010. LNCS, vol. 6484, pp. 366–377. Springer, Heidelberg (2010)

Matching and Voting

Analyzing Power in Weighted Voting Games with Super-Increasing Weights

Yoram Bachrach[1]([⊠]), Yuval Filmus[2], Joel Oren[3], and Yair Zick[4]

[1] Microsoft Research, Cambridge, UK
yobach@microsoft.com
[2] Technion - Israel Institute of Technology, Haifa, Israel
yuvalfi@cs.technion.ac.il
[3] University of Toronto, Toronto, Canada
oren@cs.toronto.edu
[4] Carnegie Mellon University, Pittsburgh, USA
yairzick@cs.cmu.edu

Abstract. Weighted voting games (WVGs) are a class of cooperative games that capture settings of group decision making in various domains, such as parliaments or committees. Earlier work has revealed that the effective decision making power, or influence of agents in WVGs is not necessarily proportional to their weight. This gave rise to measures of influence for WVGs. However, recent work in the algorithmic game theory community have shown that computing agent voting power is computationally intractable. In an effort to characterize WVG instances for which polynomial-time computation of voting power is possible, several classes of WVGs have been proposed and analyzed in the literature. One of the most prominent of these are *super increasing weight sequences*. Recent papers show that when agent weights are super-increasing, it is possible to compute the agents' voting power (as measured by the Shapley value) in polynomial-time. We provide the first set of explicit closed-form formulas for the Shapley value for super-increasing sequences. We bound the effects of changes to the quota, and relate the behavior of voting power to a novel function. This set of results constitutes a complete characterization of the Shapley value in weighted voting games, and answers a number of open questions presented in previous work.

1 Introduction

Weighted voting games (WVGs) are a class of cooperative games, commonly used to model large group decision making systems, such as parliaments. Alternatively, one can think of each player as controlling some resource, with winning coalitions being ones that have sufficient resources in order to complete a task. One of the main challenges in the WVG setting is the measurement of *player influence*, or *power*. It is a well known fact that one's ability to affect decisions may not necessarily be proportional to one's weight. As an intuitive example, consider a parliament with three parties, A, B and C: A and B both have 50 seats, while C has 20 (a government must control a majority of the house, i.e.,

© Springer-Verlag Berlin Heidelberg 2016
M. Gairing and R. Savani (Eds.): SAGT 2016, LNCS 9928, pp. 169–181, 2016.
DOI: 10.1007/978-3-662-53354-3_14

have at least 60 votes). If one equates voting power with weight, then A and B are significantly more powerful than C; a government can be formed by any two coalitions, and no single party can form a government on its own. Based on this observation, it can be reasonably argued that all parties have equal electoral power. Formal measures of voting influence, such as the *Shapley value*, aim to capture exactly these effects, providing a formal measure of player influence in WVGs. The Shapley value is considered by many to be a particularly appealing method of measuring voting power, as it satisfies several desired properties. However, it is well-known that computing the Shapley value in WVGs is computationally intractable [1]. This has naturally led to works identifying classes of WVGs for which computing voting influence is computationally tractable. In particular, an interesting sufficient condition on weights has been identified, which, if satisfied, guarantees the polynomial-time computability of the Shapley value. More formally, polynomial-time computability of the Shapley value is guaranteed if player weights are known to be *super-increasing*: a sequence of weights $w_1, \ldots, w_n$ is said to be super-increasing if $w_i > \sum_{j=i+1}^{n} w_j$ for all $i \in \{1, \ldots, n-1\}$.

1.1 Our Contributions

We provide a complete characterization of the Shapley values in a game where weights form a super-increasing sequence (Sect. 3). We provide a closed-form formula for the Shapley value when weights are super-increasing (extending techniques and observations on such games discussed in earlier work [2–4]). This formula is derived by exploiting an interesting relation between general super-increasing sequences, and the WVG obtained when weights are exponents of 2. We show several implications of our analysis to the results by [2,4], as well as a relation to a curious fractal function (Fig. 1). We significantly improve our understanding of this function, showing its various analytical properties, and its relation to Shapley values in WVGs with super-increasing weights. On a technical level, we employ several non-trivial combinatorial techniques, as well as surprising insights on the bit representation of fractions.

1.2 Related Work

Our work generalizes several results appearing in [2–4] with respect to WVGs with weights that are powers of 2. We use the Shapley value [5,6] to measure voting power; this follows the extensive literature in mathematical economics and, more recently, the AI community (see [7, Chap. 4] and [8] for a literature review), on measuring influence in cooperative games. The computational complexity of computing the Shapley value is a well-studied problem, with several works on either establishing its intractability [1,9,10], approximating it [11–13], or computing it exactly for some class of cooperative game [14–20], using it to measure importance and assign gains or costs [21–26] or analyzing its behavior in the face of various types of uncertainty [27–31] (this list is by no means exhaustive, for a comprehensive review see [7,8]).

2   Preliminaries

We generally refer to vectors as lowercase, boldface letters and sets as uppercase letters. Given a positive integer m we denote $[m] = \{1, \ldots, m\}$. A *weighted voting game* (WVG) is given by a set of agents $N = \{1, \ldots, n\}$, a non-negative weight vector $\mathbf{w} = (w_1, \ldots, w_n)$, where w_i is the weight of player $i \in N$ (and we let $\mathbf{w}$ denote the length-n weight vector), and a *quota* (or *threshold*) q. Thus, we refer to a WVG over N as the tuple $\langle \mathbf{w}; q \rangle$. Unless otherwise specified, we assume that $w_1 \geq \cdots \geq w_n$. For a subset of agents $S \subseteq N$ (also referred to as a *coalition*), we define $w(S) = \sum_{i \in S} w_i$.

A coalition $S \subseteq N$ is called *winning* (has a value $v(S) = 1$) if $w(S) \geq q$ and is called *losing* (has a value $v(S) = 0$) otherwise. To define the Shapley value, we require the following notation. Given a coalition $S \subseteq N$ and some $i \in N \backslash S$, we let the *marginal contribution* of i to S be $m_i(S) = v(S \cup \{i\}) - v(S)$; for WVGs, $m_i(S) \in \{0, 1\}$, and $m_i(S) = 1$ iff $w(S) < q$ but $w(S) + w_i \geq q$. If $m_i(S) = 1$ we say that i is *pivotal* for S. Given a permutation $\sigma \colon N \to N$, we let $P_i(\sigma) = \{j \in N \mid \sigma(i) > \sigma(j)\}$ be the set of i's *predecessors* in σ. By letting $m_i(\sigma) = m_i(P_i(\sigma))$, we have that $m_i(\sigma) = 1$ iff i is pivotal for its predecessors in σ, in which case we simply say that i is pivotal for σ. Let Sym_n be the set of all permutations of N. The *Shapley value* of player i is the probability that i is pivotal for a randomly selected permutation $\sigma \in \mathrm{Sym}_n$: $\varphi_i(\mathbf{w}; q) = \frac{1}{n!} \sum_{\sigma \in \mathrm{Sym}_n} m_i(\sigma)$. For $i \in N$, we write $\varphi_i(q)$ whenever $\mathbf{w}$ is clear from the context, and assume that $q \in (0, w(N)]$ (as otherwise $\varphi_i(\mathbf{w}; q) = 0$).

3   A Formula for the Shapley Value Under Super-Increasing Sequences

Given a vector of weights $\mathbf{w} = (w_1, \ldots, w_n)$, we say that $\mathbf{w}$ is *super-increasing* (SI) if $w_i > \sum_{j=i+1}^{n} w_j$ for all $i \in \{1, \ldots, n-1\}$. We henceforth assume that $\mathbf{w}$ is a super-increasing sequence.[1]

In Lemma 2, we show that computing the Shapley value for SI weight sequences is essentially equivalent to doing so for the sequence $\beta = (2^{n-1}, 2^{n-2}, \ldots, 1)$ (for a subset $S \subseteq N$, recall that $\beta(S) = \sum_{i \in S} 2^{n-i}$). Given an integer value $q \in (0, 2^n - 1 = \beta(N)]$, we note that there exists a unique subset $S_q \subseteq N$ such that $\beta(S_q) = q$. Given an SI vector $\mathbf{w}$, not every number q in the range $(0, w(N)]$ can be written as a sum of members of $\{w_1, \ldots, w_n\}$; however, there are certain naturally defined intervals that partition $(0, w(N)]$.

We begin by proving the following two simple lemmas (the proof of Lemma 1 is omitted due to space constraints).

Lemma 1. *Let $\mathbf{w}$ be an SI weight vector. For every $S, T \subseteq N$, $\beta(S) < \beta(T)$ if and only if $w(S) < w(T)$.*

---

[1] Our definition actually results in *super-decreasing* weight sequences; for consistent notation with [2,4] and others, we refer to our sequences as super-increasing.

For a non-empty set of agents $S \subseteq N$, we let $S^- \subseteq N$ be the unique subset of agents satisfying $\beta(S^-) = \beta(S) - 1$. For example, assuming $n = 4$, if $S = \{1, 3, 4\}$, then $\beta(S) = 2^{4-1} + 2^{4-3} + 2^{4-4} = 2^3 + 2^1 + 2^0 = 11$; thus $S^- = \{1, 3\}$ since $\beta(\{1, 3\}) = 2^{4-1} + 2^{4-3} = 2^3 + 2^1 = 10$. Lemma 1 shows that for every quota $q \in (0, w(N)]$ there exists a *unique* set $A(q) \subseteq N$ such that q is in $(w(A(q)^-), w(A(q))]$. Whenever we write $A(q) = \{a_0, \ldots, a_r\}$, we will always assume that $a_0 < \cdots < a_r$.

Lemma 2. *Given an SI vector* w, *then for every* $i \in N$ *and* $q \in (0, w(N)]$, $\varphi_i(\mathrm{w}; q) = \varphi_i(\boldsymbol{\beta}; \beta(A(q)))$.

Proof. Recall that $P_i(\sigma)$ is the set of agents appearing before agent i in a given permutation $\sigma \in \mathrm{Sym}_n$. The Shapley value $\varphi_i(\mathrm{w}; q)$ is the probability that $w(P_i(\sigma)) \in [q - w_i, q)$, or equivalently, that $q \in (w(P_i(\sigma)), w(P_i(\sigma)) + w_i]$. The intervals $(w(C^-), w(C)]$ partition $(0, w(N)]$; thus q is in $(w(P_i(\sigma)), w(P_i(\sigma)) + w_i]$ if and only if $w(P_i(\sigma)) \leq w(A(q)^-)$ and $w(A(q)) \leq w(P_i(\sigma) \cup \{i\})$. Lemma 1 shows that this is equivalent to checking whether $\beta(P_i(\sigma)) \leq \beta(A(q)^-)$ and $\beta(A(q)) \leq \beta(P_i(\sigma) \cup \{i\})$. Now, note that $\beta(A(q)^-) = \beta(A(q)) - 1$, so the above condition simply states that i is pivotal for σ under β when the quota is $\beta(A(q))$.

Lemma 2 implies that for any SI w, computing $\varphi_i(\mathrm{w}; q)$ only requires finding $A(q)$; this can be done using Algorithm 1. This is a straightforward greedy algorithm, whose proof of correctness is omitted due to space constraints.

Algorithm 1. Algorithm Find-Set for finding $A(q)$

Input: w, q
$A \leftarrow \emptyset$
for $i \leftarrow 1$ **to** n **do**
   if $q > w(A \cup \{i + 1, \ldots, n\})$ **then**
      $A \leftarrow A \cup \{i\}$
   end
end
return A

We now present our main result, a closed form formula for the Shapley values in the super-increasing case. The resulting Shapley values are illustrated in Fig. 1.

Theorem 1. *Given an SI vector* w *and a threshold* q, *let* $A(q) = \{a_0, \ldots, a_r\}$. *If* $i \notin A(q)$ *then:*

$$\varphi_i(\mathrm{w}; q) = \sum_{\substack{t \in \{0, \ldots, r\}: \\ a_t > i}} \frac{1}{a_t \binom{a_t - 1}{t}}.$$

If $i \in A(q)$, *say* $i = a_s$, *then:*

$$\varphi_i(\mathrm{w}; q) = \frac{1}{a_s \binom{a_s - 1}{s}} - \sum_{t > s} \frac{1}{a_t \binom{a_t - 1}{t - 1}}.$$

Proof. We present here the case where $i \notin A(q)$; the case where $i \in A(q)$ is similar, and its proof is omitted due to space constraints. Lemma 2 shows that $\varphi_i(\mathrm{w}; q) = \varphi_i(\beta; \beta(A(q)))$, where $\beta = 2^{n-1}, \ldots, 1$. Therefore we can assume w.l.o.g. that $\mathrm{w} = \beta$ and that the threshold is $q^* = \sum_{j \in A(q)} 2^{n-j}$.

Recall that $\varphi_i(\mathrm{w}; q)$ is the probability that $w(P_i(\sigma)) \in [q - w_i, q)$, where σ is chosen randomly from Sym_n, and $P_i(\sigma)$ is the set of predecessors of i in σ. The idea of the proof is to consider the maximal $\tau \in \{1, \ldots, r+1\}$ such that $a_t \in P_i(\sigma)$ for all $t < \tau$. We will show that when $i \notin A(q)$, each possible value of τ corresponds to one summand in the expression for $\varphi_i(\mathrm{w}; q)$.

Suppose that i is pivotal for σ. We start by showing that $\tau \le r$, ruling out the case $\tau = r+1$. If $\tau = r+1$ then by definition $\beta(P_i(\sigma)) \ge \sum_{j \in A(q)} 2^{n-j} = q^*$, contradicting the assumption $\beta(P_i(\sigma)) < q^*$. Thus $\tau \le r$, and so a_τ is well-defined. We claim that if $k \in P_i(\sigma)$ for some agent $k < a_\tau$ then $k \in A(q)$. Indeed, otherwise:

$$\beta(P_i(\sigma)) \ge \sum_{t=0}^{\tau-1} 2^{n-a_t} + 2^{n-k} \ge \sum_{t=0}^{\tau-1} 2^{n-a_t} + 2^{n-a_\tau+1}$$

$$\ge \sum_{t=0}^{\tau-1} 2^{n-a_t} + \sum_{j=a_\tau}^{n} 2^{n-j} \ge \beta(A(q)) = q^*,$$

again contradicting $\beta(P_i(\sigma)) < q^*$; thus, if $k \in P_i(\sigma) \setminus A(q)$, then $k > a_\tau$.

Furthermore, we claim that $a_\tau \ge i$. Otherwise:

$$\beta(P_i(\sigma)) \le \sum_{t=0}^{\tau-1} 2^{n-a_t} + \sum_{j=a_\tau+1}^{n} 2^{n-j} - 2^{n-i}$$

$$< \sum_{t=0}^{\tau} 2^{n-a_t} - 2^{n-i} \le q^* - w_i,$$

contradicting the assumption $w(P_i(\sigma)) \ge q^* - w_i$.

Summarizing, we have that if i is pivotal for σ, then $\tau \le r$, $a_\tau \ge i$ and

$$P_i(\sigma) \cap \{1, \ldots, a_\tau\} = \{a_0, \ldots, a_{\tau-1}\}. \tag{1}$$

Denote this event E_τ, and call a $\tau \le r$ satisfying $a_\tau \ge i$ *legal*.

Recall that $i \notin A(q)$; we have shown above that if i is pivotal for σ then E_τ occurs for some legal τ. We claim that the converse is also true; that is, if there exists some legal τ such that (1) holds with respect to σ, then i is pivotal for σ. Indeed, given E_τ defined with respect to a permutation σ, and for some legal τ, the weight of $P_i(\sigma)$ can be bounded as follows.

$$\sum_{t=0}^{\tau-1} 2^{n-a_t} \le \beta(P_i(\sigma)) \le \sum_{t=0}^{\tau-1} 2^{n-a_t} + \sum_{j=a_\tau+1}^{n} 2^{n-j} < \sum_{t=0}^{\tau} 2^{n-a_t},$$

where the last expression is at most q^*. The second inequality follows from the definition of τ. As $i < a_\tau$, the lower bound satisfies:

$$\sum_{t=0}^{\tau-1} 2^{n-a_t} \ge q^* - \sum_{j=a_\tau}^{n} 2^{n-j} > q^* - 2^{n-a_\tau+1} \ge q^* - 2^{n-i},$$

It remains to calculate $\Pr[E_\tau]$. The event E_τ states that the restriction of σ to $\{1, \ldots, a_\tau\}$ consists of the elements $\{a_0, \ldots, a_{\tau-1}\}$ in some order, followed by i (recall that $i \leq a_\tau$). For each of the $\tau!$ possible orders, the probability of this is $1/a_\tau \cdots (a_\tau - \tau) = (a_\tau - \tau - 1)!/a_\tau!$, and so

$$\Pr[E_\tau] = \frac{\tau!(a_\tau - \tau - 1)!}{a_\tau!} = \frac{1}{a_\tau \binom{a_\tau - 1}{\tau}}. \tag{2}$$

Summing over all legal τ, we obtain the formula in the statement of the theorem. This completes the proof in the case $i \notin A(q)$.

Example 1. Consider a 10 agent game where $w_i = 2^{n-i}$. Let us compute the Shapley value of agent 7 when the quota is $q = 27$. We can write $q = 16 + 8 + 2 + 1 = w_6 + w_7 + w_9 + w_{10}$, hence $A(q) = \{a_0 = 6, a_1 = 7, a_2 = 9, a_3 = 10\}$. Since agent 7 is in $A(q)$, it must be the case that: $\varphi_7(27) = \frac{1}{7\binom{6}{1}} - \frac{1}{9\binom{8}{1}} - \frac{1}{10\binom{9}{2}} \approx 0.007143$.

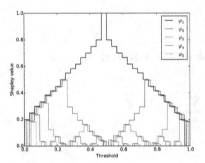

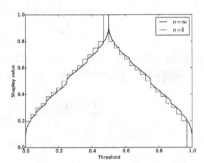

(a) Shapley values for $n = 5$, $w_i = 2^{-i}$. Values $\varphi_i(q)$ for different i are slightly nudged to show the effects of Lemma 4.

(b) Shapley values $\varphi_1(q)$ for $n = 5$, $w_i = 2^{-i}$ compared to the limiting case $n = \infty$.

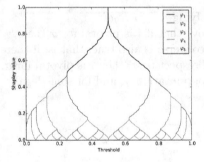

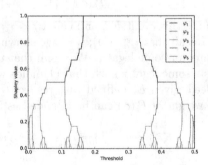

(c) Shapley values in the case $w_i = 2^{-i}$.

(d) Shapley values in the case $w_i = 3^{-i}$.

Fig. 1. Examples Shapley values corresponding to super-increasing sequences.

4   Shapley Values Under Super-Increasing Weights

Zuckerman et al. [4] prove a nice property of super-increasing sets (Lemma 19):

Theorem 2 (*given in*[4]). Suppose that $n \geq 3$; if the weights w are SI, then for every quota $q \in (0, w(N)]$, either $\varphi_n(q) = \varphi_{n-1}(q)$ or $\varphi_{n-1}(q) = \varphi_{n-2}(q)$.

We generalize this result using Theorem 1, showing how to determine in which cases $\varphi_i(q) = \varphi_{i+1}(q)$. We prove Lemma 4 using a combinatorial identity.

Lemma 3. *Let p, t be integers satisfying $p > t \geq 1$. Then*

$$\frac{1}{p\binom{p-1}{t}} + \frac{1}{p\binom{p-1}{t-1}} = \frac{1}{(p-1)\binom{p-2}{t-1}}.$$

Lemma 4. *Given a quota $q \in (0, w(N)]$, let $A(q) = \{a_0, \ldots, a_r\}$. Given some $i \in N \backslash \{n\}$, (a) if $i, i+1 \in A(q)$ or $i, i+1 \notin A(q)$ then $\varphi_i(q) = \varphi_{i+1}(q)$; (b) if $i \notin A(q)$ and $i+1 \in A(q)$ then $\varphi_i(q) \geq \varphi_{i+1}(q)$, with equality if and only if $i+1 = a_r$; (c) if $i \in A(q)$ and $i+1 \notin A(q)$ then $\varphi_i(q) > \varphi_{i+1}(q)$.*

Proof. We write $A(q) = \{a_1, \ldots, a_r\}$. Let us assume that neither i nor $i+1$ are in $A(q)$; the other cases are similar and their proof is omitted due to space constraints. For every $t \in \{0, \ldots, r\}$, $a_t > i$ if and only if $a_t > i+1$. Employing the formula used in Theorem 1, we have that

$$\varphi_i(q) = \sum_{\substack{t \in \{0,\ldots,r\}: \\ a_t > i}} \frac{1}{a_t \binom{a_t - 1}{t}}$$

$$= \sum_{\substack{t \in \{0,\ldots,r\}: \\ a_t > i+1}} \frac{1}{a_t \binom{a_t - 1}{t}} = \varphi_{i+1}(q).$$

Next, if $i, i+1 \in A(q)$ then there is some s such that $i = a_s$ and $i+1 = a_{s+1}$, so:

$$\varphi_i(q) = \frac{1}{a_s \binom{a_s - 1}{s}} - \sum_{\substack{t \in \{0,\ldots,r\}: \\ a_t > i}} \frac{1}{a_t \binom{a_t - 1}{t-1}}$$

$$= \frac{1}{a_s \binom{a_s - 1}{s}} - \frac{1}{a_{s+1} \binom{a_{s+1} - 1}{s}} - \sum_{\substack{t \in \{0,\ldots,r\}: \\ a_t > i+1}} \frac{1}{a_t \binom{a_t - 1}{t-1}}$$

$$= \frac{1}{a_s \binom{a_s - 1}{s}} - \frac{1}{(a_s + 1) \binom{a_s}{s}} - \sum_{\substack{t \in \{0,\ldots,r\}: \\ a_t > i+1}} \frac{1}{a_t \binom{a_t - 1}{t-1}}$$

According to Lemma 3 this equals:

$$\frac{1}{(a_s + 1) \binom{a_s}{s+1}} - \sum_{\substack{t \in \{0,\ldots,r\}: \\ a_t > i+1}} \frac{1}{a_t \binom{a_t - 1}{t-1}} = \frac{1}{(a_s+1) \binom{a_{s+1} - 1}{s+1}} - \sum_{\substack{t \in \{0,\ldots,r\}: \\ a_t > i+1}} \frac{1}{a_t \binom{a_t - 1}{t-1}} = \varphi_{i+1}(q),$$

where the last equality uses Theorem 1.

Next, we show that Lemma 4 generalizes Theorem 2. We can, in fact, show the following stronger corollary (proof omitted).

Corollary 1. *Let w be a vector of super-increasing weights. Let $A(q) = \{a_0, \ldots, a_r\}$. Then for all $i \geq a_r$, either $\varphi_i(q) = \varphi_{i-1}(q)$, or $\varphi_{i-1}(q) = \varphi_{i-2}(q)$.*

Invoking Corollary 1 with $i = n$ gives Theorem 2.

Another interesting implication of Corollary 1 is the following. Suppose that $A(q) = \{a_0, \ldots, a_r\}$, then for all $i, j > a_r$, $\varphi_i(q) = \varphi_j(q)$.

It is often desirable that WVGs exhibit *separability*: if two players have different weights, then they should have different voting power. [4] show that separability is not attainable under SI weights; Corollary 1 implies that some quotas offer more separability than others: if $A(q)$ does not consist of low-weight agents, then low-weight agents are not separable under q. For example, if weights are exponents of 2 and $q = \ell 2^{n-m}$, where ℓ is an odd number, then $\varphi_{n-m+1} = \cdots = \varphi_n(q)$. Our results allow us to bound the difference in voting power that one may achieve by changing the quota under SI weights. Recall that given a set $S \subseteq N$, S^- is the set for which $\beta(S) = \beta(S^-) + 1$. As the Shapley values are constant in the interval $(w(S^-), w(S)]$, in order to analyze the behavior of $\varphi_i(q)$, one needs only determine the rate of increase or decrease at quotas of the form $w(S)$ for $S \subseteq N$. These are given by the following lemma.

Lemma 5. *For every $S \subseteq N$, and any $i \in N$, if $i \notin S^-$ then $\varphi_i(w(S^-)) < \varphi_i(w(S))$. If $i \in S^-$ then $\varphi_i(w(S^-)) > \varphi_i(w(S))$.*

Moreover, $|\varphi_i(w(S)) - \varphi_i(w(S^-))| = \frac{1}{n}$ if one of the following holds: (a) $S = \{n\}$; (b) $i < n$ and $S = \{1, \ldots, i\}$ or $S = \{i, n\}$; or (c) $i = n$ and $S = \{n-1\}$. Otherwise, $|\varphi_i(w(S)) - \varphi_i(w(S^-))| \leq \frac{1}{n(n-1)}$.

Proof. Given a non-empty set $S \subseteq N$, we define $\varphi_+ = \varphi_i(w(S))$ and $\varphi_- = \varphi_i(w(S^-))$. Let $S = \{a_0, \ldots, a_r\}$. We have $S^- = \{a_0, \ldots, a_{r-1}, a_r + 1, \ldots, n\}$.

Suppose first that $i > a_r$, and let s be the index of i in the sequence S^-. According to Theorem 1, $\varphi_+ = 0$ and

$$\varphi_- = \frac{1}{i\binom{i-1}{s}} - \sum_{\ell=1}^{n-i} \frac{1}{(i+\ell)\binom{i+\ell-1}{s+\ell-1}} = \frac{1}{n\binom{n-1}{s+n-i}};$$

thus $\varphi_- > \varphi_+$. Furthermore, $|\varphi_+ - \varphi_-| \leq \frac{1}{n(n-1)}$, unless $s+n-i \in \{0, n-1\}$. If $s+n-i = 0$ then $s = 0$ and $i = n$, implying $S^- = \{n\}$ and so $S = \{n-1\}$. If $s+n-i = n-1$ then $s = i-1$ and so $S^- = \{1, \ldots, n\}$, which is impossible. The cases where $i = a_r$ and $i < a_r$ are similarly analyzed, and provide the complete case analysis; we omit the details due to space constraints.

5   The Limiting Behavior of the Shapley Value Under Super-Increasing Weights

Given a super-increasing sequence $w_1, \ldots, w_n$ (where again, $w_1 > w_2 > \cdots > w_n$) and some $m \in N$, let us write $w|_m$ for $(w_1, \ldots, w_m)$ and $[m]$ for $\{1, \ldots, m\}$.

We write $\varphi_i(\mathsf{w}|_m; q)$ for the Shapley value of agent $i \in [m]$ in the weighted voting game in which the set of agents is $[m]$, the weights are $\mathsf{w}|_m$, and the quota is q. We also write $A|_m(q)$ for the set $S \subseteq [m]$ such that $q \in (w|_m(S^-), w|_m(S)]$.

The following lemma (proof omitted) relates $\varphi_i(\mathsf{w}; q)$ and $\varphi_i(\mathsf{w}|_m; q)$.

Lemma 6. *Let $m \in N$ and $i \in [m]$, and let $q \in (0, w([m])]$. Then*

$$\varphi_i(\mathsf{w}|_m; q) = \varphi_i(\mathsf{w}; w(A|_m(q))).$$

Therefore the plot of $\varphi_i(\mathsf{w}|_m; q)$ (as a function of q) can be readily obtained from that of $\varphi_i(\mathsf{w}; q)$. This suggests looking at the limiting case of an *infinite* super-increasing sequence $(w_i)_{i=1}^{\infty}$, which is a sequence satisfying $w_i > 0$ and $w_i \geq \sum_{j=i+1}^{\infty} w_j$ for all $i \geq 1$. In this section we make some normalizing assumptions that will be useful. Just like in the preceding subsections, we assume that weights are arranged in decreasing order; furthermore, we assume that $w_1 = \frac{1}{2}$. This is no loss of generality: it is an easy exercise to see that given a weight vector w and some positive constant α, $\varphi_i(\mathsf{w}; q) = \varphi_i(\alpha\mathsf{w}; \alpha q)$. Thus, instead of the weight vector $(2^{n-1}, 2^{n-2}, \ldots, 1)$, we now have $(\frac{1}{2}, \frac{1}{4}, \ldots, \frac{1}{2^{n-1}})$. The super-increasing condition implies that the infinite sequence sums to some value $w(\infty) \leq 1$. Lemma 6 suggests how to define $\varphi_i(q)$ in this case. For $q \in (0, w(\infty))$ and $i \geq 1$, define: $\varphi_i^{(\infty)}(q) = \lim_{n \to \infty} \varphi_i(\mathsf{w}|_n; q)$.

We show that the limit exists by providing an explicit formula for it, as given in the main result of this section, Theorem 3. Under this definition, Lemma 6 easily extends to the case $n = \infty$ (proof omitted):

Lemma 7. *Let $m \geq 1$ be an integer, let $i \in [m]$, and let $q \in (0, w([m])]$. Then $\varphi_i(\mathsf{w}|_m; q) = \varphi_i^{(\infty)}(w(A|_m(q)))$.*

Below, we consider possibly infinite subsets $S = \{a_0, \ldots, a_r\}$ of the positive integers, ordered in increasing order; when $r = \infty$, the subset is infinite. Also, the notation $\{a, \ldots, \infty\}$ (or $\{a, \ldots, r\}$ when $r = \infty$) means all integers larger than or equal to a.

Given a finite sequence of integers $S = \{a_0, \ldots, a_r\}$, such that $a_0 < a_1 < \cdots < a_r$, we define S^- to be $\{a_0, \ldots, a_{r-1}\} \cup \{a_{r+1}, \ldots, \infty\}$; note the analogy to the finite case: when we had a finite sequence of agents N, S^- was the maximal weight set such that $w(S^-) < w(S)$. This is also the case for S^- as defined above. For a (possibly infinite) subset S of the positive integers, define $\beta_{\infty}(S) = \sum_{i \in S} 2^{-i}$. First, we show an analog of Lemma 1 (proof omitted).

Lemma 8. *Suppose $S, T \subseteq N$ are two subsets of the positive integers. Then $\beta_{\infty}(S) \leq \beta_{\infty}(T)$ if and only if $w(S) \leq w(T)$. Further, if $\beta_{\infty}(S) < \beta_{\infty}(T)$ then $w(S) < w(T)$.*

There is a subtlety involved here: unlike the finite case explored in Lemma 1, we can have $\beta_{\infty}(S) = \beta_{\infty}(T)$ for $S \neq T$. This is because dyadic rationals (numbers of the form $\frac{a}{2^b}$ for some positive integer a) have two different binary expansions. For example, $\frac{1}{2} = (0.1000\ldots)_2 = (0.0111\ldots)_2$. The lemma states (in this case) that $w(\{1\}) \geq w(\{2, 3, 4, \ldots\})$, but there need not be equality.

Next, we use the fact that any real $r \in (0,1)$ has a binary expansion with infinitely many 0s (alternatively, a set S_r such that $\beta_\infty(S_r) = \sum_{n \in S_r} 2^{-n} = r$ and there are infinitely many $n \notin S_r$), and a binary expansion with infinitely many 1s (alternatively, a set T_r such that $\beta_\infty(T_r) = \sum_{n \in T} 2^{-n} = r$ and there are infinitely many $n \in T_r$). If r is not dyadic, then it has a unique binary expansion which has infinitely many 0s and 1s. If r is dyadic, say $r = \frac{1}{2}$, then it has one expansion $(0.1000\ldots)_2$ with infinitely many 0s and another expansion $(0.0111\ldots)_2$ with infinitely many 1s. The following lemma describes the analog of the intervals $(w(S^-), w(S)]$ in the infinite case.

Lemma 9. *Let $q \in (0, w(\infty))$. There exists a non-empty subset S of the positive integers such that either $q = w(S)$ or $S = \{a_0, \ldots, a_r\}$ is finite and $q \in (w(S^-), w(S)]$.*

Proof. Since $q < w(\infty)$, there exists some finite m such that $q \le w([m])$. For any $n \ge m$, let $A|_n = A|_n(q)$. Let $Q|_n$ be the subset of $[n]$ preceding $A|_n$, and let $R|_n$ be the subset of $[n+1]$ preceding $A|_n$; here "preceding" is in the sense of $X \mapsto X^-$. The interval $(w(Q|_n), w(A|_n)]$ splits into $(w(Q|_n), w(R|_n)] \cup (w(R|_n), w(A|_n)]$, and so $A|_{n+1} \in \{R|_n, A|_n\}$. Also $\beta_\infty(A|_{n+1}) \le \beta_\infty(A|_n)$, with equality only if $A|_{n+1} = A|_n$. We consider two cases. The first case is when for some integer M, for all $n \ge M$ we have $A|_n = A = \{a_0, \ldots, a_r\}$. In that case for all $n \ge M$ $\sum_{t=0}^{r-1} w_{a_t} + \sum_{t=a_r+1}^{n} w_t < q \le \sum_{t=0}^{r} w_{a_t}$, and taking the limit $n \to \infty$ we obtain $q \in (w(A^-), w(A)]$. The other case is when $A|_n$ never stabilizes. The sequence $\beta_\infty(A|_n)$ is monotonically decreasing, and reaches a limit b satisfying $b < \beta_\infty(A|_n)$ for all n. Since $w(A|_m) \in (w(Q|_n), w(A|_n)]$ for all integers $m \ge n \ge 1$, Lemma 8 implies that $b \in [\beta_\infty(Q|_n), \beta_\infty(A|_n))$.

Let L be a subset such that $b = \beta_\infty(L)$ and there are infinitely many $i \notin L$, and define $L|_n = L \cap [n]$. We have $b \in [\beta_\infty(L|_n), \beta_\infty(L|_n) + 2^{-n})$. Thus $Q|_n = L|_n$, and so $q > w(Q|_n) = w(L|_n)$. Taking the limit $n \to \infty$, we deduce that $q \ge w(L)$. If $n \notin L$ then $A|_n = Q|_n \cup \{n\}$, and so $q \le w(A|_n) = w(L|_n) + w_n$. There are infinitely many such n, so taking the limit $n \to \infty$ we conclude that $q \le w(L)$ and so $q = w(L)$.

We can now give an explicit formula for $\varphi_i^{(\infty)}$. We extend our notation to accommodate the notions given in Lemma 9. The proof of Theorem 3 is similar in spirit to the proof of Theorem 1, with one important subtlety: given some $q \in (0, w(\infty))$, we write $A(q) \subseteq \mathbb{N}$ to be an infinite set S such that $q = w(S)$, or the finite set S for which $q \in (w(S^-), w(S)]$. In the first case there may be *more than one set* S such that $q = w(S)$; Theorem 3 holds for any of the possible representations of q using w.

Theorem 3. *Let $q \in (0, w(\infty))$ and let i be a positive integer. Let $A(q) = \{a_0, \ldots, a_r\}$ be the set defined in Lemma 9. Then:*

(a) the limit $\varphi_i^{(\infty)}(q) = \lim_{n \to \infty} \varphi_i(\mathrm{w}|_n; q)$ exists.
(b) if $i \notin A(q)$ then

$$\varphi_i^{(\infty)}(q) = \sum_{\substack{t \in \{0, \ldots, r\}: \\ a_t > i}} \frac{1}{a_t \binom{a_t - 1}{t}}.$$

If $i \in A(q)$, say $i = a_s$, then

$$\varphi_i^{(\infty)}(q) = \frac{1}{a_s\binom{a_s-1}{s}} - \sum_{\substack{t\in\{0,\ldots,r\}:\\ a_t>i}} \frac{1}{a_t\binom{a_t-1}{t-1}}.$$

We conclude by stating that the limiting functions $\varphi_i^{(\infty)}$ are continuous; the proof is omitted due to space constraints.

Theorem 4. *Let i be a positive integer. The function $\varphi_i^{(\infty)}$ is continuous on $(0, w(\infty))$, and $\lim_{q\to 0} \varphi_i^{(\infty)}(q) = \lim_{q\to w(\infty)} \varphi_i^{(\infty)}(q) = 0$.*

Summarizing, we can extend the functions $\varphi_i(\mathbf{w}|_n; q)$ to a continuous function $\varphi_i^{(\infty)}$ which agrees with $\varphi_i(\mathbf{w}|_n; q)$ on the points $w(S)$ for $S \subseteq \{1, \ldots, n\}$.

When $w_i = 2^{-i}$ the plot of $\varphi^{(\infty)}$ has no flat areas, but when $w_i = d^{-i}$ for $d > 2$, the limiting function is constant on intervals $(w(S^-), w(S)]$. This is reflected in Fig. 1. These flat areas highlight a curious phenomenon. When $w_1 > \sum_{j=2}^{\infty} w_j$, we have $w(\{2, 3, \ldots, \infty\}) < w(\{1\})$, which corresponds to the strict inequality $0.0111\ldots < 0.1$ in binary, or $0.4999\ldots < 0.5$ in decimal. The infinitesimal difference is expanded to an interval $(w(\{1\}^-), w(\{1\}))$ of non-zero width $w_1 - \sum_{j=2}^{\infty} w_j$. When $w_i > \sum_{j=i+1}^{\infty} w_j$ for all i, this phenomenon happens around every dyadic number.

6 Conclusions and Future Work

In this paper we present a series of novel results characterizing the behavior of the Shapley value in WVGs when weights are super-increasing. We derive an explicit formula for the Shapley value in this case, and use it to gain several insights, bounding the gain in value as the quota changes, and explaining our results via the behavior of an interesting fractal function. While our technical results are interesting on their own, they offer some instructive insights on the study of WVGs in the AI lens. For example, our combinatorial techniques can inform the study of annexation and merging in WVGs [32–35], as well as other AI domains such as combinatorial auctions and boolean threshold logic.

References

1. Matsui, Y., Matsui, T.: *NP*-completeness for calculating power indices of weighted majority games. Theor. Comput. Sci. **263**(1–2), 305–310 (2001)
2. Aziz, H., Paterson, M.: Computing voting power in easy weighted voting games. CoRR abs/0811.2497 (2008)
3. Chakravarty, N., Goel, A., Sastry, T.: Easy weighted majority games. Math. Soc. Sci. **40**(2), 227–235 (2000)
4. Zuckerman, M., Faliszewski, P., Bachrach, Y., Elkind, E.: Manipulating the quota in weighted voting games. Artif. Intell. **180–181**, 1–19 (2012)

5. Shapley, L.: A value for n-person games. In: Contributions to the Theory of Games, vol. 2. Annals of Mathematics Studies, vol. 28, pp. 307–317. Princeton University Press, Princeton (1953)
6. Shapley, L., Shubik, M.: A method for evaluating the distribution of power in a committee system. Am. Polit. Sci. Rev. **48**(3), 787–792 (1954)
7. Chalkiadakis, G., Elkind, E., Wooldridge, M.: Computational Aspects of Cooperative Game Theory. Morgan and Claypool (2011)
8. Chalkiadakis, G., Wooldridge, M.: Weighted voting games. In: Brandt, F., Conitzer, V., Endriss, U., Lang, J., Procaccia, A. (eds.) Handbook of Computational Social Choice. Cambridge University Press (2016)
9. Elkind, E., Goldberg, L., Goldberg, P., Wooldridge, M.: Computational complexity of weighted threshold games. In: Proceedings of the 22nd AAAI Conference on Artificial Intelligence (AAAI 2007), pp. 718–723 (2007)
10. See, A., Bachrach, Y., Kohli, P.: The cost of principles: analyzing power in compatibility weighted voting games. In: AAMAS (2014)
11. Bachrach, Y., Markakis, E., Resnick, E., Procaccia, A., Rosenschein, J., Saberi, A.: Approximating power indices: theoretical and empirical analysis. Auton. Agent. Multi-Agent Syst. **20**(2), 105–122 (2010)
12. Fatima, S., Wooldridge, M., Jennings, N.: An approximation method for power indices for voting games. In: Proceedings of the 2nd International Workshop on Agent-Based Complex Automated Negotiations (ACAN 2009), pp. 72–86 (2009)
13. Maleki, S., Tran-Thanh, L., Hines, G., Rahwan, T., Rogers, A.: Bounding the estimation error of sampling-based shapley value approximation with/without stratifying. CoRR abs/1306.4265 (2013)
14. Deng, X., Papadimitriou, C.: On the complexity of cooperative solution concepts. Math. Oper. Res. **19**(2), 257–266 (1994)
15. Littlechild, S.C., Owen, G.: A simple expression for the shapely value in a special case. Manage. Sci. **20**(3), 370–372 (1973)
16. Szczepański, P., Michalak, T., Rahwan, T.: Efficient algorithms for game-theoretic betweenness centrality. Artif. Intell. **231**, 39–63 (2016)
17. Bachrach, Y., Parkes, D.C., Rosenschein, J.S.: Computing cooperative solution concepts in coalitional skill games. Artif. Intell. **204**, 1–21 (2013)
18. Blocq, G., Bachrach, Y., Key, P.: The shared assignment game and applications to pricing in cloud computing. In: AAMAS (2014)
19. Bachrach, Y.: Honor among thieves: collusion in multi-unit auctions. In: AAMAS (2010)
20. Bachrach, Y., Lev, O., Lovett, S., Rosenschein, J.S., Zadimoghaddam, M.: Cooperative weakest link games. In: AAMAS (2014)
21. Bachrach, Y., Graepel, T., Kasneci, G., Kosinski, M., Van Gael, J.: Crowd IQ: aggregating opinions to boost performance. In: AAMAS (2012)
22. Elkind, E., Pasechnik, D., Zick, Y.: Dynamic weighted voting games. In: Proceedings of the 2013 international conference on Autonomous agents and multi-agent systems, International Foundation for Autonomous Agents and Multiagent Systems, pp. 515–522 (2013)
23. Bachrach, Y., Kohli, P., Graepel, T.: Rip-off: playing the cooperative negotiation game. In: AAMAS, pp. 1179–1180 (2011)
24. Bachrach, Y., Elkind, E., Faliszewski, P.: Coalitional voting manipulation: a game-theoretic perspective (2011)
25. Bachrach, Y., Zuckerman, M., Wooldridge, M., Rosenschein, J.S.: Proof systems and transformation games. Ann. Math. Artif. Intell. **67**(1), 1–30 (2013)

26. Bachrach, Y., Porat, E.P., Rosenschein, J.S.: Sharing rewards in cooperative connectivity games. J. Artif. Intell. Res. **47**, 281–311 (2013)
27. Bachrach, Y., Meir, R., Feldman, M., Tennenholtz, M.: Solving cooperative reliability games. In: UAI (2012)
28. Bachrach, Y., Shah, N.: Reliability weighted voting games. In: Vöcking, B. (ed.) SAGT 2013. LNCS, vol. 8146, pp. 38–49. Springer, Heidelberg (2013)
29. Zick, Y.: On random quotas and proportional representation in weighted voting Games. IJCAI **13**, 432–438 (2013)
30. Bachrach, Y., Kash, I., Shah, N.: Agent failures in totally balanced games and convex games. In: Goldberg, P.W. (ed.) WINE 2012. LNCS, vol. 7695, pp. 15–29. Springer, Heidelberg (2012)
31. Bachrach, Y., Savani, R., Shah, N.: Cooperative max games and agent failures. In: AAMAS (2014)
32. Aziz, H., Bachrach, Y., Elkind, E., Paterson, M.: False-name manipulations in weighted voting games. J. Artif. Intell. Res. **40**, 57–93 (2011)
33. Lasisi, R.O., Allan, V.H.: Manipulation of weighted voting games via annexation and merging. In: Filipe, J., Fred, A. (eds.) ICAART 2012. CCIS, vol. 358, pp. 364–378. Springer, Heidelberg (2013)
34. Lasisi, R., Allan, V.: New bounds on false-name manipulation in weighted voting games. In: Proceedings of the 27th International Florida Artificial Intelligence Research Society Conference (FLAIRS 2014), pp. 57–62 (2014)
35. Rey, A., Rothe, J.: False-name manipulation in weighted voting games is hard for probabilistic polynomial time. J. Artif. Intell. Res. **50**, 573–601 (2014)

Strong and Weak Acyclicity in Iterative Voting

Reshef Meir[⊠]

Technion—Israel Institute of Technology, Haifa, Israel
`reshefm@ie.technion.ac.il`

Abstract. We cast the various different models used for the analysis of iterative voting schemes into a general framework, consistent with the literature on acyclicity in games. More specifically, we classify convergence results based on the underlying assumptions on the agent scheduler (the order of players) and the action scheduler (the response played by the agent).

Our main technical result is proving that Plurality with randomized tie-breaking (which is not guaranteed to converge under arbitrary agent schedulers) is weakly-acyclic. I.e., from any initial state there is *some* path of better-replies to a Nash equilibrium. We thus show a separation between restricted-acyclicity and weak-acyclicity of game forms, thereby settling an open question from [17]. In addition, we refute another conjecture by showing the existence of strongly-acyclic voting rules that are not separable.

1 Introduction

Voting protocols are commonly used to aggregate agents' preferences over outcomes. In iterative voting, the voting rules and the voters' preferences are fixed, but voters are strategic and are allowed to change their vote one at a time after observing the interim outcome [25]. The main questions in the field are regarding which voting rules guarantee convergence of the iterative process to a Nash equilibrium, and under what conditions (e.g.,[19, 28, 31]).

Long before that, researchers in economics and game theory since Cournot [7] had been developing a formal framework to study questions about acyclicity and convergence of local improvement dynamics in games [1, 2, 9, 17, 26, 27]. However, these two lines of work remained largely detached from one another. Bridging this gap is the main conceptual contribution of this work.

Intuitively put, strong-acyclicity means that the game will converge regardless of the order of players/voters and how they select their action (as long as they are improving their utility), i.e. that there are no cycles of better-replies whatsoever; Weak-acyclicity means that while cycles may occur, from any initial state (voting profile) there is at least one path of better-replies that leads to a Nash equilibrium; Restricted-acyclicity is a middle ground, requiring convergence for any order of players (agent scheduler), but allowing the action scheduler to restrict the way they choose among several available replies (e.g., only allowing

© Springer-Verlag Berlin Heidelberg 2016
M. Gairing and R. Savani (Eds.): SAGT 2016, LNCS 9928, pp. 182–194, 2016.
DOI: 10.1007/978-3-662-53354-3_15

best-replies). Most relevant to us is the work of Kukushkin [15–17], who studied general characterizations of game forms that guarantee various notions of acyclicity.

Papers on iterative voting typically focus on a specific voting rule (game form), and study its convergence properties. Most results, both positive and negative, are about restricted acyclicity (under various notions of restriction) and include the work of Meir et al. [25], Reyhani and Wilson [31], and Lev and Rosenschein [19].

More recent work on iterative voting deals with voters who are uncertain, truth-biased, lazy-biased, bounded-rational, non-myopic, or apply some other restrictions and/or heuristics that diverge from the standard notion of better-reply in games [11,12,22,24,28–30]. Although the framework is suitable for studying such iterative dynamics as well, this paper deals exclusively with myopic better-reply dynamics.

Building on the formalism of Kukushkin [17] for strong/restricted/weak-acyclicity of game forms, we re-interpret most of the known results on convergence of better- and best-reply in voting games.

1.1   Contribution and Structure

The paper unfolds as follows. In Sect. 2, we define the iterative voting model within the more general framework of acyclic and weakly acyclic game forms. In Sect. 3 we consider strong acyclicity, and settle an open question regarding the existence of acyclic non-separable game forms by explicitly constructing such a game form. Section 2.4 briefly discusses restricted acyclicity. Our main technical contribution is in Sect. 4, where we use variations of Plurality to show a strict separation between restricted acyclicity and weak acyclicity, thereby settling another open question. We conclude in Sect. 5.

2   Preliminaries

We build upon the basic notations and definitions of Meir et al. [25] and Kukushkin [17]. We usually denote sets by uppercase letters and vectors by bold letters, e.g., $\mathbf{a} = (a_1, \ldots, a_n)$.

2.1   Voting Rules and Game Forms

There is a set C of m alternatives (or *candidates*), and a set N of n strategic agents, or *voters*. A game form (also called a *voting rule*) f allows each agent $i \in N$ to select an action from a set A_i. Thus the input to f is a vector $\mathbf{a} = (a_1, \ldots, a_n)$ called an *action profile*. Mixed strategies are not allowed. We also refer to a_i as the *vote* of agent i in profile $\mathbf{a}$. Then, f chooses a winning alternative—i.e., it is a function $f : \mathcal{A} \to C$, where $\mathcal{A} = \times_{i \in N} A_i$. See Fig. 1 for examples.

The definitions in this section apply to all voting rules unless stated otherwise. For a permutation $P \in \pi(C)$, We denote by $top(P)$ the first element in P.

Plurality. In the Plurality voting rule we have that $A = C$, and the winner is the candidate with the most votes. We allow for a broader set of "Plurality game forms" by considering both weighted and fixed voters, and varying the tie-breaking method. Each of the strategic voters $i \in N$ has an integer weight $w_i \in \mathbb{N}$. In addition, there are $\hat{n}$ "fixed voters" who do not play strategically or change their vote. The vector $\hat{s} \in \mathbb{N}^m$ (called "initial score vector") specifies the number of fixed votes for each candidate. Weights and initial scores are part of the game form.[1] These extensions also apply to other positional scoring rules.

f_1	a	b	c		f_2	a	b	c		f_3	x	y		f_4	x	y	z	w
a	a	a	a		a	a	a	a		a	a	b		a	ax	ay	az	aw
b	b	b	b		b	a	b	b		b	b	c		b	bx	by	bz	bw
c	c	c	c		c	a	b	c		c	c	a		c	cx	cy	cz	cw

Fig. 1. Four examples of game forms with two agents. f_1 is a dictatorial game form with 3 candidates (the row agent is the dictator). f_2 is the Plurality voting rule with 3 candidates and lexicographic tie-breaking. In f_3, $A_1 = C = \{a, b, c\}$, $A_2 = \{x, y\}$. Note that f_4 is completely general (there are 3×4 possible outcomes in C, one for each voting profile) and can represent any 3-by-4 game.

The *final score* of c for a given profile $\mathbf{a} \in A^n$ in the Plurality game form $f_{\mathbf{w}, \hat{s}}$ is the total weight of voters that vote c. We denote the final score vector by $\mathbf{s}_{\hat{s}, \mathbf{w}, \mathbf{a}}$ (often just $\mathbf{s_a}$ or $\mathbf{s}$ when the other parameters are clear from the context), where $s(c) = \hat{s}(c) + \sum_{i \in N : a_i = c} w_i$. Thus the Plurality rule selects some candidate from $W = \text{argmax}_{c \in C}\, s_{\hat{s}, \mathbf{w}, \mathbf{a}}(c)$, breaking ties according to some specified method. The two variations we consider are $f_{\hat{s}, \mathbf{w}}^{PL}$ which breaks ties lexicographically, and $f_{\hat{s}, \mathbf{w}}^{PR}$ which selects a winner from W uniformly at random. As with $\mathbf{s}$, we omit the scripts $\mathbf{w}$ and $\hat{s}$ when they are clear from the context.

2.2   Incentives

Games are attained by adding either cardinal or ordinal utility to a game form. The linear order relation $Q_i \in \pi(C)$ reflects the preferences of agent i. That is, i prefers c over c' (denoted $c \succ_i c'$) if $(c, c') \in Q_i$. The vector containing the preferences of all n agents is called a *preference profile*, and is denoted by $\mathbf{Q} = (Q_1, \ldots, Q_n)$. The game form f, coupled with a preference profile $\mathbf{Q}$, defines an ordinal utility normal form game $G = \langle f, \mathbf{Q} \rangle$ with n agents, where agent i prefers outcome $f(\mathbf{a})$ over outcome $f(\mathbf{a}')$ if $f(\mathbf{a}) \succ_i f(\mathbf{a}')$.

---

[1] All of our results still hold if there are no fixed voters, but allowing fixed voters enables the introduction of simpler examples. For further discussion on fixed voters see [8].

Manipulation and Stability. Having defined a normal form game, we can now apply standard solution concepts. Let $G = \langle f, \mathbf{Q} \rangle$ be a game, and let $\mathbf{a} = (\mathbf{a}_{-i}, a_i)$ be a joint action in G. We denote by $\mathbf{a} \overset{i}{\to} \mathbf{a}'$ an *individual improvement step*, if (1) $\mathbf{a}, \mathbf{a}'$ differ only by the action of player i; and (2) $f(a_{-i}, a'_i) \succ_i f(a_{-i}, a_i)$. We sometimes omit the actions of the other voters $\mathbf{a}_{-i}$ when they are clear from the context, only writing $a_i \overset{i}{\to} a'_i$. We denote by $I_i(\mathbf{a}) \subseteq A$ the set of actions a'_i s.t. $a_i \overset{i}{\to} a'_i$ is an improvement step of agent i in $\mathbf{a}$, and $I(\mathbf{a}) = \bigcup_{i \in N} \bigcup_{a'_i \in I_i(\mathbf{a})} (\mathbf{a}_{-i}, a'_i)$. $\mathbf{a} \overset{i}{\to} a'_i$ is called a *best reply* if a'_i is i's most preferred candidate in $I_i(\mathbf{a})$.

A joint action $\mathbf{a}$ is a (pure) *Nash equilibrium* (NE) in G if $I(\mathbf{a}) = \emptyset$. That is, no agent can gain by changing his vote, provided that others keep their strategies unchanged. A priori, a game with pure strategies does not have to admit any NE.

Now, observe that in most common voting rules the preference profile $\mathbf{Q}$ induces a special joint action $\mathbf{a}^* = \mathbf{a}^*(\mathbf{Q})$, termed the *truthful state*, where a_i^* is implied by Q_i. E.g. in Plurality $a_i^* = top(Q_i)$. We refer to $f(\mathbf{a}^*)$ as the *truthful outcome* of the $\langle f, \mathbf{Q} \rangle$.

2.3   Iterative Games

We consider natural *dynamics* in iterative games. Assume that agents start by announcing some initial profile $\mathbf{a}^0$, and then proceed as follows: at each step t a single agent i may change his vote to $a'_i \in I_i(\mathbf{a}^{t-1})$, resulting in a new state (joint action) $\mathbf{a}^t = (\mathbf{a}_{-i}^{t-1}, a'_i)$. The process ends when no agent has objections, and the outcome is set by the last state. Such a restriction makes sense in many computerized environments, where voters can log-in and change their vote at any time.

Local Improvement Graphs and Schedulers. Any game G induces a directed graph whose vertices are all action profiles (states) A^n, and edges are all local improvement steps [1,35]. The pure Nash equilibria of G are all states with no outgoing edges. Since a state may have multiple outgoing edges ($|I(\mathbf{a})| > 1$), we need to specify which one is selected in a given play. A *scheduler* ϕ selects which edge is followed at state $\mathbf{a}$ at any step of the game [2]. The scheduler can be decomposed into two parts, namely selecting an agent i to play (agent scheduler ϕ^N), and selecting an action in $I_i(\mathbf{a})$ (action scheduler ϕ^A), where $\phi = (\phi^N, \phi^A)$.

Convergence and Acyclicity. Given a game G, an initial action profile $\mathbf{a}^0$ and a scheduler ϕ, we get a unique (possibly infinite) path of steps. Also, it is immediate to see that the path is finite if and only if it reaches a Nash equilibrium (which is the last state in the path). We say that the triple $\langle G, \mathbf{a}^0, \phi \rangle$ *converges* if the induced path is finite.

Following [26,27], a game G has the *finite individual improvement property* (we say that G *is* FIP), if $\langle G, \mathbf{a}^0, \phi \rangle$ converges for *any* $\mathbf{a}^0$ and scheduler ϕ. Games that are FIP are also known as *acyclic games* and as *generalized ordinal potential games* [27].

It is quite easy to see that not all Plurality games are FIP [25]. However, there are alternative, weaker notions of acyclicity and convergence.

- A game G is *weakly-FIP* if there is *some* scheduler ϕ such that $\langle G, \mathbf{a}^0, \phi \rangle$ converges for any $\mathbf{a}^0$. Such games are known as *weakly acyclic*, or as ϕ-potential games [2].
- A game G is *restricted-FIP* if there is *some action scheduler* ϕ^A such that $\langle G, \mathbf{a}^0, (\phi^N, \phi^A) \rangle$ converges for any $\mathbf{a}^0$ and ϕ^N [17]. We term such games as *order-free acyclic*.

Intuitively, restricted FIP means that there is some restriction players can adopt s.t. convergence is guaranteed regardless of the order in which they play. Kukushkin identifies a particular restriction of interest, namely restriction to best-reply improvements, and defines the *finite best-reply property* (FBRP) and its weak and restricted analogs. We emphasize that an action scheduler *must* select an action in $I_i(\mathbf{a})$, if one exists. Thus restricted dynamics that may disallow all available actions (as in [11,12]) do not fall under the definition of restricted-FIP (but can be considered as separate dynamics).

For the Plurality rule we identify a different restriction, namely *direct reply*. Formally, a step $\mathbf{a} \overset{i}{\to} \mathbf{a}'$ is a direct reply if $f(\mathbf{a}') = a_i'$, i.e., if i votes for the new winner. ϕ^A is direct if it always selects a direct reply. We get the following definitions for a Plurality game G, where FDRP stands for *finite direct reply property*:

- G is *FDRP* if $\langle G, \mathbf{a}^0, (\phi^N, \phi^A) \rangle$ converges for any $\mathbf{a}^0$, any ϕ^N, and any direct ϕ^A.
- G is *weakly-FDRP* if there is a direct ϕ such that $\langle G, \mathbf{a}^0, \phi \rangle$ converges for any $\mathbf{a}^0$.
- G is *restricted-FDRP* if there is a direct ϕ^A such that $\langle G, \mathbf{a}^0, (\phi^N, \phi^A) \rangle$ converges for any $\mathbf{a}^0$ and ϕ^N.
- FDBRP means that replies are both best and direct. Note that it is unique and thus cannot be further restricted.

Finally, a game form f has the X property (where X is any of the above versions of finite improvement) if $\langle f, \mathbf{Q} \rangle$ is X for any preference profile $\mathbf{Q}$. Since some convergence properties entail others, we describe these entailments in Fig. 2.

Kukushkin notes that there are no known examples of game forms that are weak-FIP, but not restricted-FIP. We settle this question later in Sect. 4.2.

Convergence from the Truth. We say that a game G is FIP *from state* $\mathbf{a}$ if $\langle G, \mathbf{a}, \phi \rangle$ converges for any ϕ. Clearly a game is FIP iff it is FIP from $\mathbf{a}$ for any $\mathbf{a} \in A^n$. The definitions for other all other notions of finite improvement properties are analogous.

We are particularly interested in convergence from the truthful state $\mathbf{a}^*$. This is since: a. it is rather plausible to assume that agents will start by voting truthfully, especially when not sure about others' preferences; and b. even with complete information, they may be inclined to start truthfully, as they can always later change their vote.

$$
\begin{array}{c|c|c}
\text{FBRP} & \text{restricted-FBRP} \Rightarrow \text{weak-FBRP} \\
\Uparrow & \Downarrow \quad\quad\quad \Downarrow \\
\text{FIP} \;\Rightarrow\; \text{FDBRP} \;\Rightarrow\; & \text{restricted-FIP} \;\Rightarrow\; \text{weak-FIP} \;\Rightarrow\; \text{pure Nash} \\
\Downarrow & \Uparrow \quad\quad\quad \Uparrow \quad\quad \text{exists} \\
\text{FDRP} & \text{restricted-FDRP} \Rightarrow \text{weak-FDRP}
\end{array}
$$

Fig. 2. A double arrow $X \Rightarrow Y$ means that any game or game form with the X property also has the Y property. A triple arrow means any property on the premise side entails all properties on the conclusion side. The third row is only relevant for Plurality/Veto, where direct-reply is well defined.

2.4 Known Results for Plurality

Restricted better-replies and order-free acyclicity have been heavily studied in [14,19,25,28,31], mainly under the restrictions of direct-reply and best-reply. We summarize most known results in Table 1. We only list here explicitly two results on the convergence of Plurality with random tie-breaking (see Sect. 4.1 for a formal definition), that we will use later.

Theorem 1 (Meir et al. [23,25]). $f_{\hat{s}}^{PR}$ is FBRP from the truth.

Theorem 2 (Meir et al. [23,25]). f^{PR} is not restricted-FIP.

3 Strong Acyclicity

A game form f is called "separable" [17] if there are mappings $g_i : A_i \to C$ for $i \in N$ s.t. for all $\mathbf{a} \in \mathcal{A}$, $f(\mathbf{a}) \in \{g_1(a_1), g_2(a_2), \ldots, g_n(a_n)\}$. That is, the vote of each voter is mapped to a single candidate via some function g_i, and the outcome is always one of the candidates in the range. Examples of separable rules include Plurality and dictatorial rules, in both of which g_i are the identity functions.

Conjecture 1 (Kukushkin [17]). Any FIP game form is separable.

Some weaker variations of this conjecture have been proved. In particular, for game forms with finite *coalitional improvement* property [17], and for FIP game forms with $n = 2$ voters [3] (separable game forms are called "assignable" there). We next show that for sufficiently large n, there are non-separable FIP game forms, thereby refuting the conjecture.

Theorem 3. *For any $n \geq 20$, there is a non-separable game form f_n s.t. f_n is FIP.*

Proof sketch. Let $C = \{a^1, \ldots, a^{2n}\} \cup \{z\}$. Let $A_i = \{x, y\}$ for each voter. Thus f_n is a function from the n dimensional binary cube $\mathcal{B} = \{x, y\}^n$ to C. Our proof uses the probabilistic method: we define a game form f_n by sampling $2n$ specific profiles in which the outcome is distinct, and in all other $2^n - 2n$ profiles the outcome is z. We then show that with positive probability, f_n must be non-separable and FIP. The first property follows from a counting argument, and the latter since any cycle would have to go through two consequent profiles where the outcome remains z. $\qquad\square$

4   Weak Acyclicity

Except for Plurality and Veto, convergence is not guaranteed even under restrictions on the action scheduler and the initial state. In contrast, simulations [12,14,24] show that iterative voting almost always converges even when this is not guaranteed by theory. We believe that weak acyclicity is an important part of the explanation to this gap.

4.1   Plurality with Random Tie-Breaking

Formally, the game form $f^{PR}_{\hat{s},\mathbf{w}}$ maps any state $\mathbf{a} \in A^n$ to the set $\operatorname{argmax}_{c \in C} s_{\hat{s},\mathbf{w},\mathbf{a}}(c)$. Let $W^t = f^{PR}(\mathbf{a}^t) \subseteq C$ denote the set of winners at time t. We define a direct reply $a^{t-1}_i \xrightarrow{i} a^t_i$ as one where $a^t_i \in W^t$. Note that Q_i does *not* induce a complete order over set outcomes. For instance, the order $a \succ_i b \succ_i c$ does not determine if i will prefer $\{b\}$ over $\{a,c\}$. However, we can naturally extend Q_i to a *partial preference order* over subsets. There are several standard extensions, using the following axioms:[2]

K (Kelly [13]): $(\forall a \in X, b \in Y, a \succ_i b) \Rightarrow X \succ_i Y$;
G (Gärdenfors [10]): $(\forall b \in Y, a \succ_i b) \Rightarrow \{a\} \succ_i (\{a\} \cup Y) \succ_i Y$;
R (Responsiveness [32]): $a \succ_i b \iff \forall X \subseteq C \setminus \{a,b\}, (\{a\} \cup X) \succ_i (\{b\} \cup X)$.

Note that G entails K. The axioms reflect various beliefs a rational voter may have on the tie-breaking procedure: the K axiom reflects no assumptions whatsoever; The G axiom is consistent with tie-breaking according to a fixed but unknown order; and K+G+R axioms are consistent with random tie-breaking with equal probabilities [21,25]. For the next result we assume all K+G+R axioms hold (as in [25]), however our results do not depend on the above interpretations, and we do not specify the voter's preferences in cases not covered by the above axioms.

By Theorems 1+2, f^{PR} is FBRP from the truthful initial state, but order-free convergence is not guaranteed under any action scheduler. Our main theorem shows that under a certain scheduler (of agents+actions), convergence is guaranteed from any state. Further, this still holds if actions are restricted to direct-replies.

Lemma 4. *Consider any game* $G = \langle f^{PR}_{\hat{s}}, \mathbf{Q} \rangle$. *Consider some candidate* a^*, *and suppose that in* $\mathbf{a}^0$, *there are* x,y *s.t.* $s^0(x) \geq s^0(y) \geq s^0(a^*) + 2$. *Then for any sequence of direct replies,* $a^* \notin f(\mathbf{a}^t)$.

Proof sketch. We show by induction that a^* is not considered a possible winner by any voter (and thus does not get any additional votes), and that the gap remains.   □

Theorem 5. $f^{PR}_{\hat{s}}$ *is weak-FDRP.*

---

[2] We thank an anonymous reviewer for the references.

Proof. Consider a game $G = \langle f_{\hat{s}}^{PR}, \mathbf{Q} \rangle$, and an initial state $\mathbf{a}^0$. For a state $\mathbf{a}$, denote by $B(\mathbf{a}) \subseteq A^n$ all states reachable from $\mathbf{a}$ via paths of direct replies. Let $B = B(\mathbf{a}^0)$, and assume towards a contradiction that B does not contain a Nash equilibrium. For every $\mathbf{b} \in B$, let $C(\mathbf{b}) = \{c \in C : \exists \mathbf{a} \in B(\mathbf{b}) \wedge c \in f(\mathbf{a})\}$, i.e. all candidates that are winners in some state reachable from $\mathbf{b}$.

For any $\mathbf{b} \in B(\mathbf{a}^0)$, define a game $G_\mathbf{b}$ by taking G and eliminating all candidates *not in* $C(\mathbf{b})$. Since we only consider direct replies, for any $\mathbf{a} \in B(\mathbf{b})$, the set of outgoing edges $I(\mathbf{a})$ is the same in G and in $G_\mathbf{b}$ (as any direct reply must be to candidate in $C(\mathbf{b})$). Thus by our assumption, the set $B(\mathbf{b})$ in game $G_\mathbf{b}$ does not contain an NE.

For any $\mathbf{b} \in B(\mathbf{a}^0)$, let $\mathbf{b}^*$ be the truthful state of game $G_\mathbf{b}$, and let $T(\mathbf{b}) \subseteq N$ be the set of agents who are truthful in $\mathbf{b}$. That is, $i \in T(\mathbf{b})$ if $b_i = b_i^*$.

Let $\mathbf{b}^0$ be some state $\mathbf{b} \in B(\mathbf{a}^0)$ s.t. $|T(\mathbf{b})|$ is maximal, and let $T^0 = T(\mathbf{b}^0)$. If $|T^0| = n$ then $\mathbf{b}^0$ is the truthful state of $G_{\mathbf{b}^0}$, and thus by Theorem 1 all best-reply paths from $\mathbf{b}^0$ in $G_{\mathbf{b}^0}$ lead to an NE, in contradiction to $B(\mathbf{b}^0)$ not containing any NE. Thus $T^0 < n$. We will prove that there is a path from $\mathbf{b}^0$ to a state $\mathbf{b}'$ s.t. $|T(\mathbf{b}')| > |T^0|$.

Let $i \notin T(\mathbf{b}^0)$ (must exist by the previous paragraph). Consider the score of candidate b_i^* at state $\mathbf{b}^0$. We divide into 5 cases. All scores specified below are in the game $G_{\mathbf{b}^0}$.

Case 1. $|f(\mathbf{b}^0)| > 1$ and $b_i^* \in f(\mathbf{b}^0)$ (i.e. b_i^* is one of several winners). Then consider the step $\mathbf{b}^0 \xrightarrow{i} b_i^*$. This make b_i^* the unique winner, and thus it is a direct best-reply for i. In the new state $\mathbf{b}' = (\mathbf{b}_{-i}^0, b_i^*)$ we have $T(\mathbf{b}') = T(\mathbf{b}^0) \cup \{i\}$.

Case 2. $s^0(b_i^*) = sw^0 - 1$ (i.e., b_i^* needs one more vote to become a winner). By Axioms G+R, i prefers $f(\mathbf{b}_{-i}^0, b_i^*)$ over $f(\mathbf{b}^0)$. Then similarly to case 1, i has a direct step $\mathbf{b}^0 \xrightarrow{i} b_i^*$, which results in a "more truthful" state $\mathbf{b}'$.

Case 3. $b_i^* = f(\mathbf{b}^0)$ (i.e. b_i^* is the unique winner). Then the next step $\mathbf{b}^0 \xrightarrow{j} \mathbf{b}^1$ will bring us to one of the two previous cases. Moreover, it must hold that $j \notin T(\mathbf{b}^0)$ since otherwise $b_j^0 = b_j^* = f(\mathbf{b}^0)$ which means $I_j(\mathbf{b}^0) = \emptyset$. Thus $|T(\mathbf{b}')| = |T(\mathbf{b}^1)| + 1 \geq |T(\mathbf{b}^0)| + 1$.

Case 4. $f(\mathbf{b}^0) = x \neq b_i^*$, and $s^0(x) = s^0(b_i^*) + 2$. We further divide into:

Case 4.1. $s^0(b_i^*) \geq s^0(y)$ for all $y \neq x$. Then the next step by j must be from x, which brings us to one of the two first cases (as in Case 3).

Case 4.2. There is $y \neq x$ s.t. $s^0(x) = s^0(y) + 1 = s^0(b_i^*) + 2$. Then we continue the sequence of steps until the winner's score decreases. Since all steps that maintain sw^t select a more preferred candidate, this most occur at some time t, and $T(\mathbf{b}^0) \subseteq T(\mathbf{b}^t)$. Then at $\mathbf{b}^t$ we are again in Case 1 or 2.

Case 4.3. There is $y \neq x$ s.t. $s^0(x) = s^0(y) = s^0(b_i^*) + 2$. Then by Lemma 4 b_i^* can never be selected, in contradiction to $b_i^* \in C(\mathbf{b}^0)$.

Case 5. $f(\mathbf{b}^0) = x \neq b_i^*$, and $s^0(x) \geq s^0(b_i^*) + 3$. We further divide into:

Case 5.1. For all $y \neq x$, $s^0(y) \leq s^0(x) - 3$. In this case no reply is possible.

Case 5.2. There is some $y \neq x$ s.t. $s^0(y) \geq s^0(b_i^*) + 2$. Then by Lemma 4 b_i^* can never be selected, in contradiction to $b_i^* \in C(\mathbf{b}^0)$.

Case 5.3. There is some $y \neq x$ s.t. $s^0(y) \geq s^0(b_i^*) + 1$ Then the next step must be from x to such y. Which means $s^1(x) = s^1(y) = sw^0 - 1 \geq s^0(b_i^*) + 2 = s^1(b_i^*) + 2$. Thus again by Lemma 4 we reach a contradiction.

Therefore we either construct a path of direct replies to $\mathbf{b}' \in B(\mathbf{b}^0)$ with $|T(\mathbf{b}')| > |T(\mathbf{b}^0)|$ in contradiction to our maximality assumption, or we reach another contradiction. Thus $B(\mathbf{b}^0)$ must contain some NE (both in $G_{\mathbf{b}^0}$ and in G), which means by construction that G is weakly-FDRP from $\mathbf{b}^0$. However since $\mathbf{b}^0 \in B(\mathbf{a}^0)$, we get that G is weakly-FDRP from $\mathbf{a}^0$ as well.   □

4.2   Weighted Plurality

When voters are weighted, cycles of direct responses can emerge [23,25]. We conjecture that such cycles must depend on the order of agents, and that certain orders will break such cycles and reach an equilibrium, at least from the truthful state.

Conjecture 2. $f_{\hat{\mathbf{s}},\mathbf{w}}^{PL}$ is weak-FDRP (in particular weak-FIP).

We leave the proof of the general conjecture for future work. Yet, we want to demonstrate the power of weak acyclicity over restricted acyclicity, even when there are no randomness or restrictions on the utility space. That is, to provide a definite (negative) answer to Kukushkin's question of whether weak acyclicity entails restricted acyclicity. To do so, we will use a slight variation of Plurality with weighted voters and lexicographic tie-breaking.

Theorem 6. *There exist a game form f^* s.t. f^* is weak-FIP but not restricted-FIP.*

Proof. Consider the following game G: The initial fixed score of candidates $\{a, b, c, d\}$ is $\hat{\mathbf{s}} = (0, 1, 2, 3)$. The weight of each voter $i \in \{1, 2, 3\}$ is i. The preference profile is as follows: $c \succ_1 d \succ_1 b \succ_1 a$, $b \succ_2 c \succ_2 a \succ_2 d$, and $a \succ_3 b \succ_3 c \succ_3 d$. This game was used in [25] to demonstrate that Plurality with weighted voters is not FDRP, however it can be verified that G is restricted-FIP so it is not good enough for our use.

If we ignore agents' preferences, we get a particular game form $f_{\hat{\mathbf{s}},\mathbf{w}}^{PL}$ where $N = \{1, 2, 3\}$, $M = \{a, b, c, d\}$, $\hat{\mathbf{s}} = (0, 1, 2, 3)$ and $\mathbf{w} = (1, 2, 3)$. We define f^* by modifying $f_{\hat{\mathbf{s}},\mathbf{w}}^{PL}$ with the following restrictions on agents' actions: $A_1 = \{c, d\}, A_2 = \{b, c\}, A_3 = \{a, b, d\}$. Thus f^* is a $2 \times 2 \times 3$ game form, presented in Fig. 3(a).

Q_3		state	action	new state
1	$b \succ d$	(d, b, a)	b	(d, b, b)
2	$d \succ b$ & $d \succ a$	(c, b, b)	d	(c, b, d)
3	$a \succ d \succ b \succ c$	(d, c, b)	d	(d, c, d)

In either case, agent 3 moves from a state on the cycle to a Nash equilibrium. □

We first show that f^* is not restricted-FIP. Indeed, consider the game G^* accepted from f^* with the same preferences from game G (Fig. 3(b)). We can see that there is a cycle of length 6 (in bold). An agent scheduler that always selects the agent with the bold reply guarantees that convergence does not occur, since in all 6 relevant states the selected agent has no alternative replies.

Next, we show that f^* is weak-FIP. That is, for any preference profile there is some scheduler that guarantees convergence. We thus divide into cases according to the preferences of agent 3. In each case, we specify a state where the scheduler selects agent 3, the action of the agent, and the new state.

We note that since all thick edges must be oriented in the same direction, $a \succ_3 b$ if and only if $b \succ_3 c$. Thus the following three cases are exhaustive.

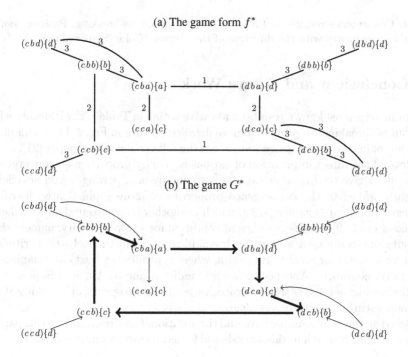

(a) The game form f^*

(b) The game G^*

Fig. 3. In each state we specify the actions of all 3 agents, and the outcome in curly brackets. Agent 1 controls the horizontal axis, agent 2 the vertical axis, and agent 3 the in/out axis. We omit edges between states with identical outcomes, since such moves are impossible for any preferences. A directed edge in (b) is a better-reply in G^*.

Table 1. Positive results carry to the right side, negative to the left side. We assume lexicographic tie breaking in all rules except Plurality. FDBRP is only well-defined for Plurality and Veto. See [21] for details.

Voting rule	FIP	FBRP	FDBRP	restricted-FIP	Weak-FIP
Dictator	V	V	-	V	V
Plurality (lex.)	X	X [25]	V [25]	V	V
Plurality (rand.)	X	X	X	X [25]	V (Thm. 5)
Weighted Plurality (lex.)	X	X	X [25]	?	?
Veto	X	X [23]	V [31, 19]	V	V
k-approval ($k \geq 2$)	X	X [19, 18]	-	X	X [23]
Borda	X	X [19, 31]	-	X	X [31]
Approval	X	X [23]	-	V [23]	V
many other rules	X	X [19, 18, 14]	-	?	?

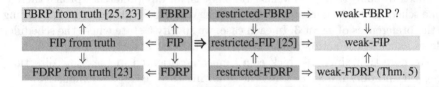

Fig. 4. Convergence results for Plurality under random tie-breaking. Positive results (in light green) carry with the direction of the arrows. (Color figure online)

5   Conclusions and Future Work

We summarize most known results on iterative voting in Table 1. For Plurality with random tie-breaking we provide a more detailed picture in Fig. 4. For a complete overview of known convergence results, see the full version of this paper [21].

Beyond the direct implication of various acyclicity properties on convergence in an interactive setting where agents vote one-by-one, [strong/weak] acyclicity is tightly linked to the convergence properties of more sophisticated learning strategies in repeated games [4,20], which is another reason to understand them. Fabrikant et al. [9] provide a sufficient condition for weak-acyclicity, namely that any subgame contains a *unique* Nash equilibrium. Unfortunately, this criterion is not very useful for most voting rules, where typically (at least) all unanimous votes form equilibria. Another sufficient condition due to Apt and Simon [2] is by eliminating never-best-reply strategies, and the prospects of applying it to common voting rules is not yet clear.

Based on the work summarized, and the additional progress made in this paper, we believe that research in this area should focus on three primary directions:

1. Weak-acyclicity seems more indicative than order-free acyclicity to determine convergence in practice. Thus theorists should study which voting rules are weak-FIP, perhaps under reasonable restrictions (as we demonstrated, this property is distinct from restricted-FIP).

2. It is important to experimentally study how people really vote in iterative settings (both in and out of the lab), so that this behavior can be formalized and behavioral models can be improved. The work of [33] is a preliminary step in this direction, but there is much more to learn. Ideally, we would like to identify a few types of voters, such that for each type we can relatively accurately predict the next action in a particular state. It would be even better if these types are not specific to a particular voting rule or contextual details.

3. We would like to know not only if a voting rule converges under a particular dynamics (always or often), but also what are the properties of the attained outcome—in particular, whether the iterative process improves welfare or fairness, avoids "voting paradoxes" [34] and so on. Towards this end, several researchers (e.g., [5,6,14,24,30]) have started to explore these questions via theory and simulations. However, a good understanding of how iterative voting shapes the outcome, whether the population of voters consists of humans or artificial agents, is still missing.

References

1. Andersson, D., Gurvich, V., Hansen, T.D.: On acyclicity of games with cycles. Discrete Appl. Math. **158**(10), 1049–1063 (2010)
2. Apt, K.R., Simon, S.: A classification of weakly acyclic games. In: Serna, M. (ed.) SAGT 2012. LNCS, vol. 7615, pp. 1–12. Springer, Heidelberg (2012)
3. Boros, E., Gurvich, V., Makino, K., Papp, D.: Acyclic, or totally tight, two-person game forms: characterization and main properties. Discrete Math. **310**(6), 1135–1151 (2010)
4. Bowling, M.: Convergence and no-regret in multiagent learning. Adv. Neural Inf. Process. Syst. **17**, 209–216 (2005)
5. Bowman, C., Hodge, J.K., Ada, Y.: The potential of iterative voting to solve the separability problem in referendum elections. Theor. Decis. **77**(1), 111–124 (2014)
6. Brânzei, S., Caragiannis, I., Morgenstern, J., Procaccia, A.D.: How bad is selfish voting? In: Proceeding of 27th AAAI (2013)
7. Cournot, A.-A.: Recherches sur les principes mathématiques de la théorie des richesses par Augustin Cournot. chez L. Hachette (1838)
8. Elkind, E., Grandi, U., Rossi, F., Slinko, A.: Gibbard-satterthwaite games. In: IJCAI 2015 (2015)
9. Fabrikant, A., Jaggard, A.D., Schapira, M.: On the structure of weakly acyclic games. In: Kontogiannis, S., Koutsoupias, E., Spirakis, P.G. (eds.) SAGT 2010. LNCS, vol. 6386, pp. 126–137. Springer, Heidelberg (2010)
10. Gärdenfors, P.: Manipulation of social choice functions. J. Econ. Theory **13**(2), 217–228 (1976)
11. Gohar, N.: Manipulative voting dynamics. PhD thesis, University of Liverpool (2012)
12. Grandi, U., Loreggia, A., Rossi, F., Venable, K.B., Walsh, T.: Restricted manipulation in iterative voting: condorcet efficiency and borda score. In: Perny, P., Pirlot, M., Tsoukiàs, A. (eds.) ADT 2013. LNCS, vol. 8176, pp. 181–192. Springer, Heidelberg (2013)
13. Kelly, J.S.: Strategy-proofness and social choice functions without singlevaluedness. Econometrica: J. Econometric Soc. 439–446 (1977)

14. Koolyk, A., Lev, O., Rosenschein, J.S.: Convergence and quality of iterative voting under non-scoring rules (extended abstract). In: Proceeding of 15th AAMAS (2016)

15. Kukushkin, N.S.: Congestion games: a purely ordinal approach. Econ. Lett. **64**, 279–283 (1999)

16. Kukushkin, N.S.: Perfect information and congestion games. Games Econ. Behav. **38**, 306–317 (2002)

17. Kukushkin, N.S.: Acyclicity of improvements in finite game forms. Int. J. Game Theory **40**(1), 147–177 (2011)

18. Lev, O.: Agent modeling of human interaction: stability, dynamics and cooperation. PhD thesis, The Hebrew University of Jerusalem (2015)

19. Lev, O., Rosenschein, J.S.: Convergence of iterative voting. In: Proceeding of 11th AAMAS, pp. 611–618 (2012)

20. Marden, J.R., Arslan, G., Shamma, J.S.: Regret based dynamics: convergence in weakly acyclic games. In: Proceeding of 6th AAMAS. ACM (2007)

21. Meir, R., Polukarov, M., Rosenschein, J.S., Jennings, N.R.: Acyclic games and iterative voting. ArXiv e-prints (2016)

22. Meir, R.: Plurality voting under uncertainty. In: Proceeding of 29th AAAI, pp. 2103–2109 (2015)

23. Meir, R.: Strong and weak acyclicity in iterative voting. In: COMSOC 2016 (2016)

24. Meir, R., Lev, O., Rosenschein, J.S.: A local-dominance theory of voting equilibria. In: Proceeding of 15th ACM-EC (2014)

25. Meir, R., Polukarov, M., Rosenschein, J.S., Jennings, N.: Convergence to equilibria of plurality voting. In: Proceeding of 24th AAAI, pp. 823–828 (2010)

26. Milchtaich, I.: Congestion games with player-specific payoff functions. Games Econ. Behav. **13**(1), 111–124 (1996)

27. Monderer, D., Shapley, L.S.: Potential games. Games Econ. Behav. **14**(1), 124–143 (1996)

28. Obraztsova, S., Markakis, E., Polukarov, M., Rabinovich, Z., Jennings, N.R.: On the convergence of iterative voting: how restrictive shouldrestricted dynamics be? In: Proceeding of 29th AAAI (2015)

29. Obraztsova, S., Markakis, E., Thompson, D.R.M.: Plurality voting with truth-biased agents. In: Vöcking, B. (ed.) SAGT 2013. LNCS, vol. 8146, pp. 26–37. Springer, Heidelberg (2013)

30. Reijngoud, A., Endriss, U.: Voter response to iterated poll information. In: Proceeding of 11th AAMAS, pp. 635–644 (2012)

31. Reyhani, R., Wilson, M.C.: Best-reply dynamics for scoring rules. In: Proceeding of 20th ECAI. IOS Press (2012)

32. Roth, A.E.: The college admissions problem is not equivalent to the marriage problem. J. Econ. Theory **36**(2), 277–288 (1985)

33. Tal, M., Meir, R., Gal, Y.: A study of human behavior in voting systems. In: Proceeding of 14th AAMAS, pp. 665–673 (2015)

34. Xia, L., Lang, J., Ying, M.: Sequential voting rules and multiple elections paradoxes. In: TARK 2007, pp. 279–288 (2007)

35. Young, H.P.: The evolution of conventions. Econometrica: J. Econometric Soc., 57–84 (1993)

Stable Matching with Uncertain Linear Preferences

Haris Aziz[1,2], Péter Biró[3(✉)], Serge Gaspers[1,2], Ronald de Haan[4],
Nicholas Mattei[1,2], and Baharak Rastegari[5]

[1] Data61/CSIRO, Sydney, Australia
{haris.aziz,nicholas.mattei}@data61.csiro.au
[2] University of New South Wales, Sydney, Australia
sergeg@cse.unsw.edu.au
[3] Hungarian Academy of Sciences, Institute of Economics, Budapest, Hungary
peter.biro@krtk.mta.hu
[4] Technische Universität Wien, Vienna, Austria
dehaan@ac.tuwien.ac.at
[5] School of Computing Science, University of Glasgow, Glasgow, UK
baharak.rastegari@glasgow.ac.uk

Abstract. We consider the two-sided stable matching setting in which there may be uncertainty about the agents' preferences due to limited information or communication. We consider three models of uncertainty: (1) lottery model — in which for each agent, there is a probability distribution over linear preferences, (2) compact indifference model — for each agent, a weak preference order is specified and each linear order compatible with the weak order is equally likely and (3) joint probability model — there is a lottery over preference profiles. For each of the models, we study the computational complexity of computing the stability probability of a given matching as well as finding a matching with the highest probability of being stable. We also examine more restricted problems such as deciding whether a certainly stable matching exists. We find a rich complexity landscape for these problems, indicating that the form uncertainty takes is significant.

1 Introduction

We consider a *Stable Marriage problem (SM)* in which there is a set of men and a set of women. Each man has a linear order over the women, and each woman has a linear order over the men. For the purpose of this paper we assume that the preference lists are complete, i.e., each agent finds each member of the opposite side acceptable.[1] In the stable marriage problem the goal is to compute a *stable matching*; a matching where no two agents prefer to be matched to each other rather than be matched to their current partners. Unlike most of

---

[1] We note that the complexity of all problems that we study are the same for complete and incomplete lists, where non-listed agents are deemed unacceptable—see Proposition 2 in the full version of the paper [1].

© Springer-Verlag Berlin Heidelberg 2016
M. Gairing and R. Savani (Eds.): SAGT 2016, LNCS 9928, pp. 195–206, 2016.
DOI: 10.1007/978-3-662-53354-3_16

the literature on stable matching problems [4,9,11], we assume that men and women may have uncertainty in their preferences which can be captured by various probabilistic uncertainty models. We focus on *linear models* in which each possible deterministic preference profile is a set of linear orders.

Uncertainty in preferences could arise for a number of reasons both practical and epistemological. For example, an agent could express a weak order because the agent did not invest enough time or effort to differentiate between potential matches and therefore one could assume that each linear extension of the weak order is equally likely; this maps to our *compact indifference model*. In many real applications the ties are broken randomly with lotteries, e.g., in the school choice programs in New York and Boston as well as in centralized college admissions in Ireland. However, a central planner may also choose a matching that is optimal in some sense, without breaking the ties in the preference list. For instance, in Scotland they used to compute the maximum size (weakly) stable matching to allocate residents to hospitals [9]. We argue that another natural solution could be the matching which has the highest probability of being stable after conducting a lottery. Alternatively, there may be a cost associated with eliciting preferences from the agents, so a central planner may want to only obtain and provide a recommendation based on a subset of the complete orders [2].

As another example, imagine a group of interns are admitted to a company and allocated to different projects based on their preferences and the preferences of the project leaders. Suppose that after three months the interns can switch projects if the project leaders agree; though the company would prefer not to have swaps if possible. However, both the interns and the project leaders can have better information about each other after the three months, and the assignment should also be stable with regard to the refined preferences. This example motivates our lottery and joint probability models. In the *lottery model*, the agents have *independent* probabilities over possible linear orders (e.g. each project leader has a probability distribution on possible refined rankings over the interns independently from each other). In the *joint probability model*, the probability distribution is over possible preference profiles and can thus accommodate the possibility that the preferences of the agents are refined in a correlated way (e.g. if an intern performs well in the first three months then she is likely to be highly ranked by all project leaders). Uncertainty in preferences has already been studied in voting [6] and for cooperative games [8]. Ehlers and Massó [3] considers many-to-one matching markets under a Bayesian setting. Similarly, in auction theory, it is standard to examine Bayesian settings in which there is a probability distribution over the types of agents.

To illustrate the problem we describe a simple example with four agents. We write $b \succ_a c$ to say that agent a prefers b to c and assume the lottery model.

Example 1. We have two men m_1 and m_2 and two women w_1 and w_2. Each agent assigns a probability to each strict preference ordering as follows.
(i) $p(w_1 \succ_{m_1} w_2) = 0.4$ and $p(w_2 \succ_{m_1} w_1) = 0.6$ (ii) $p(w_1 \succ_{m_2} w_2) = 0.0$ and $p(w_2 \succ_{m_2} w_1) = 1.0$ (iii) $p(m_1 \succ_{w_1} m_2) = 1.0$ and $p(m_2 \succ_{w_1} m_1) = 0.0$ (iv) $p(m_1 \succ_{w_2} m_2) = 0.8$ and $p(m_2 \succ_{w_2} m_1) = 0.2$. This setting admits two

matchings that are stable with positive probability: $\mu_1 = \{(m_1, w_1), (m_2, w_2)\}$ and $\mu_2 = \{(m_1, w_2), (m_2, w_1)\}$. Notice that if each agent submits the preference list that s/he finds most likely to be true, then the setting admits a unique stable matching that is μ_2. The probability of μ_2 being stable, however, is 0.48 whereas the probability of μ_1 being stable is 0.52.

1.1 Uncertainty Models

We consider three different uncertainty models:

- **Lottery Model**: For each agent, we are given a probability distribution over strict preference lists.
- **Compact Indifference Model**: Each agent reports a single weak preference list that allows for ties. Each complete linear order extension of this weak order is assumed to be equally likely.
- **Joint Probability Model**: A probability distribution over preference profiles is specified.

Note that for the Lottery Model and the Joint Probability Model the representation of the input preferences can be exponentially large. However, in settings where similar models of uncertainty are used, including resident matching [2] and voting [6], a limited amount of uncertainty (i.e. small supports) is commonly expected and observed in real world data. Consequently, we consider special cases when the uncertainty is bounded in certain natural ways including the existence of only a small number of uncertain preferences and/or uncertainty on only one side of the market.

   Observe that the compact indifference model can be represented as a lottery model. This is a special case of the lottery model in which each agent expresses a weak order over the candidates (similar to the SMT setting [4,9]). However, the lottery model representation can be exponentially larger than the compact indifference model; for an agent that is indifferent among n agents on the other side of the market, there are $n!$ possible linearly ordered preferences.

1.2 Computational Problems

Given a stable marriage setting where agents have uncertain preferences, various natural computational problems arise. Let *stability probability* denote the probability that a matching is stable. We then consider the following two natural problems for each of our uncertainty models.

- MATCHINGWITHHIGHESTSTABILITYPROBABILITY: Given uncertain preferences of the agents, compute a matching with highest stability probability.
- STABILITYPROBABILITY: Given a matching and uncertain preferences of the agents, what is the stability probability of the matching?

Table 1. Summary of results.

Problems	Lottery Model	Compact Indifference	Joint Probability
STABILITYPROBABILITY	#P-complete	?	in P
	in P for all three models if 1 side is certain		
ISSTABILITYPROBABILITYNON-ZERO	NP-complete	in P	in P
ISSTABILITYPROBABILITYONE	in P	in P	in P
EXISTSPOSSIBLYSTABLEMATCHING	in P	in P	in P
EXISTSCERTAINLYSTABLEMATCHING	in P	in P	NP-complete
MATCHINGWITHHIGHESTSTABILITYPROB	?	NP-hard	NP-hard
	in P for all models if 1 side is certain and there is $O(1)$ number of uncertain agents		

We also consider two specific problems that are simpler than STABILITYPROB-ABILITY: (1) ISSTABILITYPROBABILITYNON-ZERO — For a given matching, is its stability probability non-zero? (2) ISSTABILITYPROBABILITYONE — For a given matching, is its stability probability one?

We additionally consider problems connected to, and more restricted than, MATCHINGWITHHIGHESTSTABILITYPROBABILITY: (1) EXISTSCERTAINLY STABLEMATCHING — Does there exist a matching that has stability probability one? (2) EXISTSPOSSIBLYSTABLEMATCHING — Does there exist a matching that has non-zero stability probability?

Note that EXISTSPOSSIBLYSTABLEMATCHING is straightforward to answer for any of the three uncertainty models we consider here, since there exists a stable matching for each deterministic preference profile that is a possible realization of the uncertain preferences.

1.3 Results

Table 1 summarizes our main findings. Note that the complexity of each problem is considered with respect to the input size, and that under the lottery and joint probability models the input size could be exponential in n, namely $O(n! \cdot 2n)$ for the lottery model and $O((n!)^{2n})$ for the joint probability model, where n is the number of agents on either side of the market. The complete version of the sketched or missing proofs can be found in the full version of the paper [1].

We point out that STABILITYPROBABILITY is #P-complete for the lottery model even when each agent has at most two possible preferences, but in P if one side has certain preferences. Additionally, we show that ISSTABILITYPROBABILITYNON-ZERO is in P for the lottery model if each agent has at most two possible preferences. Note that STABILITYPROBABILITY is open for the compact indifference model when both sides may be uncertain, and we also do not know the complexity of MATCHINGWITHHIGHESTSTABILITYPRO-BILITY in the lottery model, except when only a constant number of agents are uncertain on the same side of the market.

2 Preliminaries

In the Stable Marriage problem, there are two sets of agents. Let M denote a set of n men and W a set of n women. We use the term *agents* when making statements that apply to both men and women, and the term *candidates* to refer to the agents on the opposite side of the market to that of an agent under consideration. Each agent has a linearly ordered preference over the candidates. An agent may be uncertain about his/her linear preference ordering. Let L denote the *uncertain preference profile* for all agents. We denote by $I = (M, W, L)$ an instance of a *Stable Marriage problem with Uncertain Linear Preferences (SMULP)*.

We say that a given uncertainty model is *independent* if any uncertain preference profile L under the model can be written as a product of uncertain preferences L_a for all agents a, where all L_a's are independent. Note that the lottery and the compact indifference models are both independent, but the joint probability model is not.

A *matching* μ is a pairing of men and women such that each man is paired with at most one woman and vice versa; defining a list of (man, woman) pairs (m, w). We use $\mu(m)$ to denote the woman w that is matched to m and $\mu(w)$ to denote the match for w. Given linearly ordered preferences, a matching is *stable* if there is no pair (m, w) not in μ where m prefers w to his current partner in μ, i.e., $w \succ_m \mu(m)$, and vice versa. If such a pair exists, it constitutes a *blocking pair*; as the pair would prefer to defect and match with each other rather than stay with their partner in μ. Given an instance of SMULP, a matching is *certainly stable* if it is stable with probability 1.

The following extensions of SM will come in handy in proving our results. The *Stable Marriage problem with Partially ordered lists (SMP)* is an extension of SM in which agents' preferences are partial orders over the candidates. The *Stable Marriage problem with Ties (SMT)* is a special case of SMP in which incomparability is transitive and is interpreted as indifference. Therefore, in SMT each agent partitions the candidates into different ties (equivalence classes), is indifferent between the candidates in the same tie, and has strict preference ordering over the ties. In some practical settings some agents may find some candidates unacceptable and prefer to remain unmatched than to get matched to the unacceptable ones. *SMP with Incomplete lists (SMPI)* and *SMT with Incomplete lists (SMTI)* capture these scenarios where each agent's partially ordered list contains only his/her acceptable candidates. A matching is *super-stable* in an instance of SMPI if it is stable w.r.t. all linear extensions of the partially ordered lists.

We define the *certainly preferred* relation $\succ_a^{\text{cert}}$ for agent a. We write $b \succ_a^{\text{cert}} c$ if and only if agent a prefers b over c with probability 1. Based on the certainly preferred relation, we can define a dominance relation D: $D_m(w) = \{w\} \cup \{w' : w' \succ_m^{\text{cert}} w\}$; $D_w(m) = \{m\} \cup \{m' : m' \succ_w^{\text{cert}} m\}$. Based on the notion of the dominance relation, we present a useful characterization of certainly stable matchings for independent uncertainty models.

Lemma 1. *A matching μ is certainly stable for an independent uncertainty model if and only if for each pair $\{m, w\}$, $\mu(m) \in D_m(w)$ or $\mu(w) \in D_w(m)$.*

We point out that certainly preferred relation can be computed in polynomial time for all three models studied in this paper.

Certainly stable matchings are closely related to the notion of super-stable matchings [4,7]. In fact we can define a certainly stable matching using a terminology similar to that of super-stability. Given a matching μ and an unmatched pair $\{m, w\}$, we say that $\{m, w\}$ *very weakly blocks (blocks)* μ if $\mu(m) \not\succ_m^{\mathrm{cert}} w$ and $\mu(w) \not\succ_w^{\mathrm{cert}} m$. The next claim then follows from Lemma 1.

Proposition 1. *A matching μ is certainly stable for an independent uncertainty model if and only if it admits no very weakly blocking pair.*

3   General Results

In this section, we present some general results that apply to multiple uncertainty models. First we show that EXISTSCERTAINLYSTABLEMATCHING can be solved in polynomial time for any independent uncertainty model including lottery and compact indifference. Second, when the number of uncertain agents is constant and one side of the market is certain, then we can solve MATCHING-WITHHIGHESTSTABILITYPROBABILITY efficiently for each of the linear models.

3.1   An Algorithm for the Lottery and Compact Indifference Models

Theorem 1. *For any independent uncertainty model in which the certainly preferred relation is transitive and can be computed in polynomial time, EXISTSCERTAINLYSTABLEMATCHING can be solved in polynomial time.*

Proof sketch. We prove this by reducing EXISTSCERTAINLYSTABLEMATCHING to the problem of deciding whether an instance of SMP admits a super-stable matching. The latter problem can be solved in polynomial time using algorithm SUPER-SMP in [10].

Let $I = (M, W, L)$ be an instance of EXISTSCERTAINLYSTABLEMATCHING under an independent uncertainty model, assuming that the certainly preferred relation is transitive and can be computed in polynomial time. We construct an instance $I' = (M, W, p)$ of SMP, in polynomial time, as follows. The set of men and women are unchanged. To create the partial preference ordering p_a for each agent a we do the following. W.l.o.g., assume that a is a man m. For every pair of women w_1 and w_2 (i) if $w_1 \succ_m^{\mathrm{cert}} w_2$ then $(w_1, w_2) \in p_m$, denoting that m (strictly) prefers w_1 to w_2 in I', (ii) if $w_2 \succ_m^{\mathrm{cert}} w_1$ then $(w_2, w_1) \in p_m$, denoting that m (strictly) prefers w_2 to w_1 in I'. We claim, and show, that I' admits a super-stable matching iff I has a certainly stable matching. $\square$

3.2 An Algorithm for a Constant Number of Uncertain Agents

Theorem 2. *When the number of uncertain agents is constant and one side of the market is certain then* MATCHINGWITHHIGHESTSTABILITYPROBABILITY *is polynomial-time solvable for each of the linear models.*

Proof sketch. Let $I = (M, W, L)$ be an instance of MATCHINGWITHHIGHEST-STABILITYPROBABILITY and let $X \subseteq M$ be the set of uncertain agents with $|X| = k$ for a constant k. We consider all the possible matchings between X and W, where their total number is $K = n(n-1)\ldots(n-k)$. Let μ_i be such a matching for $i \in \{1 \ldots K\}$. The main idea of the proof is to show that there exist an extension of μ_i to $M \cup W$ that has stability probability at least as high as any other extension of μ_i. In this way we will need to compute this probability for only a polynomial number of matchings in n, which we can do efficiently for each model when one side has certain preferences—see Theorems 3, 8 and 10, and select the one with the highest probability. □

4 Lottery Model

In this section we focus on the lottery model.

Theorem 3. *For the lottery model, if one side has certain preferences,* STABIL-ITYPROBABILITY *is polynomial-time solvable.*

Proof sketch. W.l.o.g. assume that men are certain. The stability probability of a given matching μ is equal to the probability that none of the possible blocking pairs form. The probability of one blocking pair $\{m, w\}$ forming is equal to the probability that w prefers m to $\mu(w)$ given m also prefers w to $\mu(m)$. □

Theorem 4. *For the lottery model,* ISSTABILITYPROBABILITYONE *can be solved in linear time.*

Theorem 5. *For the lottery model,* ISSTABILITYPROBABILITYNON-ZERO *is polynomial-time solvable when each agent has at most two possible preference orderings.*

Proof sketch. We reduce the problem to 2SAT, that is polynomial-time solvable. For each agent and for both possible preference orderings for that agent, we introduce a variable, and we construct a 2CNF formula that encodes (1) that for each agent exactly one preference ordering is selected, and (2) that the selected preference orderings cause the given matching to be stable. Satisfying assignments then correspond to witnesses for non-zero stability probability. □

Lemma 2. *In polynomial time, we can transform any 2CNF formula φ over the variables $x_1, \ldots, x_n$ to a 2CNF formula φ' over the variables $x_1, \ldots, x_n, y_1, \ldots, y_n$ such that (1) φ and φ' have the same number of satisfying assignments, (2) each clause of φ' contains exactly one variable x_i and one variable y_j, and (3) for any two variables, there is at most one clause in φ' that contains these variables.*

Theorem 6. *For the lottery model,* STABILITYPROBABILITY *is #P-complete, even when each agent has at most two possible preferences.*

Proof. We show how to count the number of satisfying assignments for a 2CNF formula using the problem STABILITYPROBABILITY for the lottery model where each agent has two possible preferences. Since this problem is #P-hard, we get #P-hardness also for STABILITYPROBABILITY.

Let φ be a 2CNF formula over the variables $x_1, \ldots, x_n$. We firstly transform φ to a 2CNF formula φ' over the variables $x_1, \ldots, x_n, y_1, \ldots, y_n$ as specified by Lemma 2. We then construct an instance of STABILITYPROBABILITY. The sets of agents that we consider are $\{x_1, \ldots, x_n, a_1, \ldots, a_n\}$ and $\{y_1, \ldots, y_n, b_1, \ldots, b_n\}$. The matching that we consider matches x_i to b_i and matches y_i to a_i, for each $1 \leq i \leq n$. This is depicted below. Each agent b_i has only a single possible preference, namely one where they prefer x_i over all other agents. Similarly, each agent a_i has a single possible preference where they prefer y_i over all other agents. In other words, the agents a_i and b_i are perfectly happy with the given matching.

The agents x_i and y_i each have two possible preferences, that are each chosen with probability $\frac{1}{2}$. These two possible preferences are associated with setting these variables to true or false, respectively. We describe how these preferences are constructed for the agents x_i. The construction for the preferences of the agents y_i is then entirely analogous.

Take an arbitrary agent x_i. We show how to construct the two possible preferences for agent x_i, which we denote by p_{x_i} and $p_{\neg x_i}$. Both of these possible preferences are based on the following partial ranking: $b_1 > b_2 > \cdots > b_n$, and we add some of the agents $y_1, \ldots, y_n$ to the top of this partial ranking, and the remaining agents to the bottom of this partial ranking.

To the ranking p_{x_i} we add exactly those agents y_j to the top where φ' contains a clause $(\neg x_i \vee y_j)$ or a clause $(\neg x_i \vee \neg y_j)$. All remaining agents we add to the bottom. Similarly, to the ranking $p_{\neg x_i}$ we add exactly those agents y_j to the top where φ' contains a clause $(x_i \vee y_j)$ or a clause $(x_i \vee \neg y_j)$. The rankings p_{y_i} and $p_{\neg y_i}$, for the agents y_i, are constructed entirely similarly.

Now consider a truth assignment $\alpha : \{x_1, \ldots, x_n, y_1, \ldots, y_n\} \rightarrow \{0, 1\}$, and consider the corresponding choice of preferences for the agents $x_1, \ldots, x_n, y_1, \ldots, y_n$, where for each agent x_i the preference p_{x_i} is chosen if and only if $\alpha(x_i) = 1$, and for each agent y_i the preference p_{y_i} is chosen if and only if $\alpha(y_i) = 1$. Then α satisfies φ' if and only if the corresponding choice of preferences leads to the matching being stable. Since each combination of preferences is equally likely to occur, and there are 2^{2n} many combinations of preferences, the probability that the given matching is stable is exactly $q = \frac{s}{2^{2n}}$, where s is the number of satisfying truth assignments for φ. Therefore, given q, s can be obtained by computing $s = q 2^{2n}$. □

If each agent is allowed to have three possible preferences, then even the following problem is NP-complete. The statement can be proved via a reduction from Exact Cover by 3-Sets (X3C).

Theorem 7. *For the lottery model,* ISSTABILITYPROBABILITYNON-ZERO *is NP-complete.*

We obtain the first corollary from Theorem 7 and the second from [12, Proposition 8] and Theorem 7.

Corollary 1. *For the lottery model, unless $P = NP$, there exists no polynomial-time algorithm for approximating* STABILITYPROBABILITY *of a given matching.*

Corollary 2. *For the lottery model, unless $NP = RP$, there is no FPRAS for* STABILITYPROBABILITY.

5 Compact Indifference Model

The compact indifference model is equivalent to assuming that we are given an instance of SMT and each linear order over candidates (each possible preference ordering) is achieved by breaking ties independently at random with uniform probabilities. It is easy to show that ISSTABILITYPROBABLITYNONZERO, ISSTABILITYPROBABLITYONE, and EXISTSCERTAINLYSTABLEMATCHING are all in P. The corresponding claims and the proof can be found in [1].

We do not yet know the complexity of computing the stability probability of a given matching under the compact indifference model, but this problem can be shown to be in P if one side has certain preferences.

Theorem 8. *In the compact indifference model, if one side has certain preferences,* STABILITYPROBABILITY *is polynomial-time solvable.*

Proof. Assume, w.l.o.g., that men have certain preferences. The following procedure gives us the stability probability of any given matching μ. (1) For each uncertain woman w identify those men with whom she can potentially form a blocking pair. That is, those m such that $w \succ_m \mu(m)$ and w is indifferent between m and her partner in μ. Assume there are k of such men. The probability of w not forming a blocking pair with any men is then $\frac{1}{k+1}$. (2) Multiply the probabilities from step 1.   □

We next show that MATCHINGWITHHIGHESTSTABILITYPROBABILITY is NP-hard. For an instance I of SMT and matching μ, let $p(\mu, I)$ denote the probability of μ being stable, and let $p_S(I) = max\{p(\mu, I) | \mu$ is a matching in $I\}$, that is the maximum probability of a matching being stable. A matching μ is said to be weakly stable if there exists a tie-breaking rule where μ is stable. Therefore a matching μ has positive probability of being stable if and only if it is weakly stable. Furthermore, if the number of possible tie-breaking is N then any weakly stable matching has a probability of being stable at least $\frac{1}{N}$.

An extreme case occurs if we have one woman only with n men, where the woman is indifferent between all men. In this case any matching (pair) has a $\frac{1}{n}$ probability of being stable. An even more unfortunate scenario is when we have n men and n women, each women is indifferent between all men, and each man ranks the women in a strict order in the same way, e.g. in the order of their indices. In this case, the probability that the first woman picks her best partner, and thus does not block any matching is $\frac{1}{n}$. Suppose that the first woman picked her best partner, the probability that the second woman also picks her best partner from the remaining $n-1$ men is $\frac{1}{n-1}$, and so on. Therefore, the probability that an arbitrary complete matching is stable is $\frac{1}{n(n-1)...2} = \frac{1}{n!}$.

Theorem 9. *For the compact indifference model* MATCHINGWITHHIGHEST-STABILITYPROBABILITY *is NP-hard, even if only one side of the market has uncertain agents.*

Proof sketch. For an instance I of SMTI, let $opt(I)$ denote the maximum size of a weakly stable matching in I. Halldorsson et al. [5] showed [in the proof of Corollary 3.4] that given an instance I of SMTI of size n, where only one side of the market has agents with indifferences and each of these agents has a single tie of size two, and any arbitrary small positive ϵ, it is NP-hard to distinguish between the following two cases: (1) $opt(I) \geq \frac{21-\epsilon}{27}n$ (2) $opt(I) < \frac{19+\epsilon}{27}n$.

When choosing ϵ so that $0 < \epsilon < \frac{1}{2}$ we can simplify the above cases to (1) $opt(I) > \frac{41}{54}n$, since $opt(I) \geq \frac{21-\epsilon}{27}n > \frac{41}{54}n$ and (2) $opt(I) < \frac{39}{54}n$, since $opt(I) < \frac{19+\epsilon}{27}n < \frac{39}{54}n$.

Therefore, the number of agents left unmatched on either side of the market is less than $\frac{13}{54}n$ in the first case and more than $\frac{15}{54}n$ in the second case. Let us now extend instance I to a larger instance of SMTI I' as follows. Besides the n men $M = \{m_1, \ldots, m_n\}$ and n women $W = \{w_1, \ldots, w_n\}$, we introduce $\frac{13}{54}n$ men $X = \{x_1, \ldots x_k\}$ and another $\frac{n}{27}$ men $Y = \{y_1, \ldots y_l\}$ and $\frac{n}{27}$ women $Z = \{z_1, \ldots z_l\}$. Furthermore, for each $y_j \in Y$, we introduce n men $Y^j = \{y_1^j, \ldots, y_n^j\}$. We create the preferences of I' as follows. The preferences of men M remain the same. For each woman $w \in W$ we append the men X and then Y at the end of her list in the order of their indices. Each man $x_i \in X$ has only all the women W in his list in the order of their indices. Furthermore, each $y_j \in Y$ has all the women W first in his preference list in the order or their indices and then z_j. Let each $z_j \in Z$ has y_j as first choice and then all the men Y^j in one tie of size n. Each man in Y^j has only z_j in his list. We will show that in case one $p_S(I') \geq \frac{1}{2^n}$, whilst in case two $p_S \leq (\frac{1}{n})^{\frac{n}{27}}$. Therefore, for $n > 2^{27}$, it is NP-hard to decide which of the two separate intervals contains the value $p_S(I')$.   □

6   Joint Probability Model

In this section, we examine problems concerning the joint probability model.

Theorem 10. *For the joint probability model,* STABILITYPROBABILITY *can be solved in polynomial time.*

Corollary 3. *For the joint probability model,* IsSTABILITYPROBABILITYNON-
ZERO *and* IsSTABILITYPROBABILITYONE *can be solved in polynomial time.*

For the joint probability model, the problem EXISTSCERTAINLYSTABLEMATCH-
ING is equivalent to checking whether the intersection of the sets of stable matchings of the different preference profiles is empty or not.

Theorem 11. *For the joint probability model,* EXISTSCERTAINLYSTABLE-
MATCHING *is NP-complete.*

Proof sketch. The problem is in NP, since computing STABILITYPROBABILITY
can be done in polynomial time by Theorem 10. The proof is by reduction from
3-Colorability. Let $G = (V, E)$ be a graph specifying an instance of 3-Colorability,
where $V = \{v_1, \ldots, v_n\}$. We construct an instance I of SMULP assuming the
joint probability model.

For each vertex $v_i \in V$, we introduce three men $m_{i,1}, m_{i,2}, m_{i,3}$ and three
women $w_{i,1}, w_{i,2}, w_{i,3}$. Then, we introduce one preference profile P_0 that ensures
that every certainly stable matching matches—for each $i \in [n]$— each $m_{i,j}$ to
some $w_{i,j'}$ and, vice versa, each $w_{i,j}$ to some $m_{i,j'}$, for $j, j' \in [3]$. Moreover,
it ensures that for each $i \in [n]$, exactly one of three matchings between the
men $m_{i,j}$ and the women $w_{i,j}$ must be used:

(1) $m_{i,1}$ is matched to $w_{i,1}, m_{i,2}$ is matched to $w_{i,2}$, and $m_{i,3}$ is matched to $w_{i,3}$;
(2) $m_{i,1}$ is matched to $w_{i,2}, m_{i,2}$ is matched to $w_{i,3}$, and $m_{i,3}$ is matched to $w_{i,1}$; or
(3) $m_{i,1}$ is matched to $w_{i,3}, m_{i,2}$ is matched to $w_{i,1}$, and $m_{i,3}$ is matched to $w_{i,2}$;

Intuitively, choosing one of the matchings (1)–(3) for the agents $m_{i,j}, w_{i,j}$ corresponds to coloring vertex v_i with one of the three colors in $\{1, 2, 3\}$.

Then, for each edge $e = \{v_{i_1}, v_{i_2}\} \in E$, and for each color $c \in \{1, 2, 3\}$,
we introduce a preference profile $P_{e,c}$ that ensures that in any certainly stable matching, the agents $m_{i_1,j}, w_{i_1,j}$ and the agents $m_{i_2,j}, w_{i_2,j}$ cannot both be
matched to each other with matching (c). We let each preference profile appear
with non-zero probability (e.g., we take a uniform lottery). As a result, any
certainly stable matching directly corresponds to a proper 3-coloring of G. A
detailed description of the preference profiles P_0 and $P_{e,c}$ can be found in [1], as
well as a proof of correctness for this reduction.   □
By modifying the proof of Theorem 11, the following can also be proved.

Corollary 4. *For the joint probability model,* EXISTSCERTAINLYSTABLE-
MATCHING *is NP-complete, even when there are only 16 preference profiles in
the lottery.*

7   Future Work

First we note that we left open two outstanding questions, as described in
Table 1. In this paper we focused on the problem of computing a matching with
the highest stability probability. However, a similarly reasonable goal could be
to minimize the expected number of blocking pairs. It would also be interesting

to investigate some further realistic probability models, such as the situation when the candidates are ranked according to some noisy scores (like the SAT scores in the US college admissions). This would be a special case of the joint probability model that may turn out to be easier to solve. Finally, in a follow-up paper we are planning to investigate another probabilistic model that is based on independent pairwise comparisons.

Acknowledgments. Biró is supported by the Hungarian Academy of Sciences under its Momentum Programme (LP2016-3) and the Hungarian Scientific Research Fund, OTKA, Grant No. K108673. Rastegari was supported EPSRC grant EP/K010042/1 at the time of the submission. De Haan is supported by the Austrian Science Fund (FWF), project P26200. The authors gratefully acknowledge the support from European Cooperation in Science and Technology (COST) action IC1205. Serge Gaspers is the recipient of an Australian Research Council (ARC) Future Fellowship (FT140100048) and acknowledges support under the ARC's Discovery Projects funding scheme (DP150101134). Data61/CSIRO (formerly, NICTA) is funded by the Australian Government through the Department of Communications and the ARC through the ICT Centre of Excellence Program.

References

1. Aziz, H., Biró, P., Gaspers, S., de Haan, R., Mattei, N., Rastegari, B.: Stable matching with uncertain linear preferences. Technical Report 1607.02917, Cornell University Library. http://arxiv.org/abs/1607.02917
2. Drummond, J., Boutilier, C.: Preference elicitation and interview minimization in stable matchings. In: Proceedings of the Twenty-Eighth AAAI Conference on Artificial Intelligence, pp. 645–653 (2014)
3. Ehlers, L., Massó, J.: Matching markets under (in)complete information. J. Econ. Theor. **157**, 295–314 (2015)
4. Gusfield, D., Irving, R.: The Stable Marriage Problem: Structure and Algorithms. MIT Press, Cambridge (1989)
5. Halldórsson, M.M., Iwama, K., Miyazaki, S., Yanagisawa, H.: Improved approximation results for the stable marriage problem. ACM Trans. Algorithms **3**(3), 30 (2007)
6. Hazon, N., Aumann, Y., Kraus, S., Wooldridge, M.: On the evaluation of electionoutcomes under uncertainty. Artif. Intell. **189**, 1–18 (2012)
7. Irving, R.: Stable marriage and indifference. Discrete Appl. Math. **48**, 261–272 (1994)
8. Li, Y., Conitzer, V.: Cooperative game solution concepts that maximize stability under noise. In: AAAI, pp. 979–985 (2015)
9. Manlove, D.: Algorithmics of Matching Under Preferences. World Scientific Publishing Company, Hackensack (2013)
10. Rastegari, B., Condon, A., Immorlica, N., Irving, R., Leyton-Brown, K.: Reasoning about optimal stable matching under partial information. In: Proceedings of the ACM Conference on Electronic Commerce (EC), pp. 431–448. ACM (2014)
11. Roth, A.E., Sotomayor, M.A.O.: Two-Sided Matching: A Study in Game Theoretic Modelling and Analysis. Cambridge University Press, Cambridge (1990)
12. Welsh, D.J.A., Merino, C.: The Potts model and the Tutte polynomial. J. Math. Phys. **41**(3), 1127–1152 (2000)

The Stable Roommates Problem
with Short Lists

Ágnes Cseh[1]([⊠]), Robert W. Irving[2], and David F. Manlove[2]

[1] School of Computer Science, Reykjavik University, Reykjavík, Iceland
cseh@ru.is
[2] School of Computing Science, University of Glasgow, Glasgow, Scotland, UK
{Rob.Irving,David.Manlove}@glasgow.ac.uk

Abstract. We consider two variants of the classical Stable Roommates problem with Incomplete (but strictly ordered) preference lists (SRI) that are degree constrained, i.e., preference lists are of bounded length. The first variant, egal d-SRI, involves finding an egalitarian stable matching in solvable instances of SRI with preference lists of length at most d. We show that this problem is NP-hard even if $d = 3$. On the positive side we give a $\frac{2d+3}{7}$-approximation algorithm for $d \in \{3, 4, 5\}$ which improves on the known bound of 2 for the unbounded preference list case. In the second variant of SRI, called d-SRTI, preference lists can include ties and are of length at most d. We show that the problem of deciding whether an instance of d-SRTI admits a stable matching is NP-complete even if $d = 3$. We also consider the "most stable" version of this problem and prove a strong inapproximability bound for the $d = 3$ case. However for $d = 2$ we show that the latter problem can be solved in polynomial time.

1 Introduction

In the *Stable Roommates problem with Incomplete lists* (SRI), a graph $G = (A, E)$ and a set of preference lists $\mathcal{O}$ are given, where the vertices $A = \{a_1, \ldots, a_n\}$ correspond to *agents*, and $\mathcal{O} = \{\prec_1, \ldots, \prec_n\}$, where $\prec_i$ is a linear order on the vertices adjacent to a_i in G ($1 \leq i \leq n$). We refer to $\prec_i$ as a_i's *preference list*. The agents that are adjacent to a_i in G are said to be *acceptable* to a_i. If a_j and a_k are two acceptable agents for a_i where $a_j \prec_i a_k$ then we say that a_i *prefers* a_j to a_k.

Let M be a matching in G. If $a_i a_j \in M$ then we let $M(a_i)$ denote a_j. An edge $a_i a_j \notin M$ *blocks* M, or forms a *blocking edge* of M, if a_i is unmatched or prefers a_j to $M(a_i)$, and similarly a_j is unmatched or prefers a_i to $M(a_j)$. A matching is called *stable* if no edge blocks it. Denote by SR the special case

Á. Cseh—Supported by Icelandic Research Fund grant no. 152679-051, the Hungarian Academy of Sciences under its Momentum Programme (LP2016-3) and COST Action IC1205 on Computational Social Choice. Part of this work was carried out whilst visiting the University of Glasgow.

D.F. Manlove—Supported by EPSRC grant EP/K010042/1.

M. Gairing and R. Savani (Eds.): SAGT 2016, LNCS 9928, pp. 207–219, 2016.
DOI: 10.1007/978-3-662-53354-3_17

of SRI in which $G = K_n$. Gale and Shapley [8] observed that an instance of SR need not admit a stable matching. Irving [13] gave a linear-time algorithm to find a stable matching or report that none exists, given an instance of SR. The straightforward modification of this algorithm to the SRI case is described in [10]. We call an SRI instance *solvable* if it admits a stable matching.

In practice agents may find it difficult to rank a large number of alternatives in strict order of preference. One natural assumption, therefore, is that preference lists are short, which corresponds to the graph being of bounded degree. Given an integer $d \geq 1$, we define d-SRI to be the restriction of SRI in which G is of bounded degree d. This special case of SRI problem has potential applications in organising tournaments. As already pointed out in a paper of Kujansuu et al. [16], SRI can model a pairing process similar to the Swiss system, which is used in large-scale chess competitions. The assumption on short lists is reasonable, because according to the Swiss system, players can be matched only to other players with approximately the same score.

A second variant of SRI, which can be motivated in a similar fashion, arises if we allow ties in the preference lists, i.e., $\prec_i$ is now a strict weak ordering[1] ($1 \leq i \leq n$). We refer to this problem as the *Stable Roommates problem with Ties and Incomplete lists* (SRTI) [15]. As in the SRI case, define d-SRTI to be the restriction of SRTI in which G is of bounded degree d. Denote by SRT the special case of SRTI in which $G = K_n$. In the context of the motivating application of chess tournament construction as mentioned in the previous paragraph, d-SRTI is naturally obtained if a chess player has several potential partners of the same score and match history in the tournament.

In the SRTI context, ties correspond to indifference in the preference lists. In particular, if $a_i a_j \in E$ and $a_i a_k \in E$ where $a_j \not\prec_i a_k$ and $a_k \not\prec_i a_j$ then a_i is said to be *indifferent between* a_j and a_k. Thus preference in the SRI context corresponds to strict preference in the case of SRTI. Relative to the strict weak orders in $\mathcal{O}$, we can define stability in SRTI instances in exactly the same way as for SRI. This means, for example, that if $a_i a_j \in M$ for some matching M, and a_i is indifferent between a_j and some agent a_k, then $a_i a_k$ cannot block M. The term *solvable* can be defined in the SRTI context in an analogous fashion to SRI. Using a highly technical reduction from a restriction of 3-SAT, Ronn [20] proved that the problem of deciding whether a given SRT instance is solvable is NP-complete. A simpler reduction was given by Irving and Manlove [15].

For solvable instances of SRI there can be many stable matchings. Often it is beneficial to work with a stable matching that is fair to all agents in a precise sense [9,14]. One such fairness concept can be defined as follows. Given two agents a_i, a_j in an instance $\mathcal{I}$ of SRI, where $a_i a_j \in E$, let $\mathrm{rank}(a_i, a_j)$ denote the rank of a_j in a_i's preference list (that is, 1 plus the number of agents that a_i prefers to a_j). Let A_M denote the set of agents who are matched in a given stable matching M. (Note that this set depends only on $\mathcal{I}$ and is independent of M by [10, Theorem 4.5.2].) Define $c(M) = \sum_{a_i \in A_M} \mathrm{rank}(a_i, M(a_i))$ to be the *cost* of M. An *egalitarian stable matching* is a stable matching M that minimises $c(M)$

---

[1] That is, $\prec_i$ is a strict partial order in which incomparability is transitive.

over the set of stable matchings in $\mathcal{I}$. Finding an egalitarian stable matching in SR was shown to be NP-hard by Feder [6]. Feder [6,7] also gave a 2-approximation algorithm for this problem in the SRI setting. He also showed that an egalitarian stable matching in SR can be approximated within a factor of α of the optimum if and only if Minimum Vertex Cover can be approximated within the same factor α. Is was proved later that, assuming the Unique Games Conjecture, Minimum Vertex Cover cannot be approximated within $2 - \varepsilon$ for any $\varepsilon > 0$ [17].

Given an unsolvable instance $\mathcal{I}$ of SRI or SRTI, a natural approximation to a stable matching is a *most-stable* matching [1]. Relative to a matching M in $\mathcal{I}$, define $bp(M)$ to be the set of blocking edges of M and let $bp(\mathcal{I})$ denote the minimum value of $|bp(M')|$, taken over all matchings M' in $\mathcal{I}$. Then M is a *most-stable* matching in $\mathcal{I}$ if $|bp(M)| = bp(\mathcal{I})$. The problem of finding a most-stable matching was shown to be NP-hard and not approximable within $n^{k-\varepsilon}$, for any $\varepsilon > 0$, unless P = NP, where $k = \frac{1}{2}$ if $\mathcal{I}$ is an instance of SR and $k = 1$ if $\mathcal{I}$ is an instance of SRT [1].

To the best of our knowledge, there has not been any previous work published on either the problem of finding an egalitarian stable matching in a solvable instance of SRI with bounded-length preference lists or the solvability of SRTI with bounded-length preference lists. This paper provides contributions in both of these directions, focusing on instances of d-SRI and d-SRTI for $d \geq 2$, with the aim of drawing the line between polynomial-time solvability and NP-hardness for the associated problems in terms of d.

Our contribution. In Sect. 2 we study the problem of finding an egalitarian stable matching in an instance of d-SRI. We show that this problem is NP-hard if $d = 3$, whilst there is a straightforward algorithm for the case that $d = 2$. We then consider the approximability of this problem for the case that $d \geq 3$. We give an approximation algorithm with a performance guarantee of $\frac{9}{7}$ for the case that $d = 3$, $\frac{11}{7}$ if $d = 4$ and $\frac{13}{7}$ if $d = 5$. These performance guarantees improve on Feder's 2-approximation algorithm for the general SRI case [6,7]. In Sect. 3 we turn to d-SRTI and prove that the problem of deciding whether an instance of 3-SRTI is solvable is NP-complete. We then show that the problem of finding a most-stable matching in an instance of d-SRTI is solvable in polynomial time if $d = 2$, whilst for $d = 3$ we show that this problem is NP-hard and not approximable within $n^{1-\varepsilon}$, for any $\varepsilon > 0$, unless P = NP. Due to various complications, as explained in Appendix A of the full version of this paper [5], we do not attempt to define and study egalitarian stable matchings in instances of SRTI. A structured overview of previous results and our results (marked by *) for d-SRI and d-SRTI is contained in Table 1. All missing proofs are contained in Appendix B of the full version of this paper [5].

Related work. Degree-bounded graphs, most-stable matchings and egalitarian stable matchings are widely studied concepts in the literature on matching under preferences [18]. As already mentioned, the problem of finding a most-stable matching has been studied previously in the context of SRI [1]. In addition to

Table 1. Summary of results for d-SRI and d-SRTI.

	Finding a stable matching	Finding an egalitarian stable matching
d-SRI	In P [10,13]	in P for $d = 2$ (*)
		NP-hard even for $d = 3$ (*)
		$\frac{2d+3}{7}$-approximation for $d \in \{3,4,5\}$ (*)
		2-approximation for $d \geq 6$ [6,7]
d-SRTI	In P for $d = 2$ (*)	Not well-defined (see [5, Appendix A])
	NP-hard even for $d = 3$ (*)	

the results surveyed already, the authors of [1] gave an $O(m^{k+1})$ algorithm to find a matching M with $|bp(M)| \leq k$ or report that no such matching exists, where $m = |E|$ and $k \geq 1$ is any integer. Most-stable matchings have also been considered in the context of d-SRI [3]. The authors showed that, if $d = 3$, there is some constant $c > 1$ such that the problem of finding a most-stable matching is not approximable within c unless P = NP. On the other hand, they proved that the problem is solvable in polynomial time for $d \leq 2$. The authors also gave a $(2d-3)$-approximation algorithm for the problem for fixed $d \geq 3$. This bound was improved to $2d - 4$ if the given instance satisfies an additional condition (namely the absence of a structure called an *elitist odd party*). Most-stable matchings have also been studied in the bipartite restriction of SRI called the *Stable Marriage problem with Incomplete lists* (SMI) [4,12]. Since every instance of SMI admits a stable matching M (and hence $bp(M) = \emptyset$), the focus in [4,12] was on finding maximum cardinality matchings with the minimum number of blocking edges.

Regarding the problem of finding an egalitarian stable matching in an instance of SRI, as already mentioned Feder [6,7] showed that this problem is NP-hard, though approximable within a factor of 2. A 2-approximation algorithm for this problem was also given independently by Gusfield and Pitt [11], and by Teo and Sethuraman [23]. These approximation algorithms can also be extended to the more general setting where we are given a weight function on the edges, and we seek a stable matching of minimum weight. Feder's 2-approximation algorithm requires monotone, non-negative and integral edge weights, whereas with the help of LP techniques [22,23], the integrality constraint can be dropped, while the monotonicity constraint can be partially relaxed.

2   The Egalitarian Stable Roommates Problem

In this section we consider the complexity and approximability of the problem of computing an egalitarian stable matching in instances of d-SRI. We begin by defining the following problems.

Problem 1. EGAL d-SRI

Input: A solvable instance $\mathcal{I} = \langle G, \mathcal{O} \rangle$ of d-SRI, where G is a graph and $\mathcal{O}$ is a set of preference lists, each of length at most d.
Output: An egalitarian stable matching M in $\mathcal{I}$.

The decision version of EGAL d-SRI is defined as follows:

Problem 2. EGAL d-SRI DEC

Input: $\mathcal{I} = \langle G, \mathcal{O}, K' \rangle$, where $\langle G, \mathcal{O} \rangle$ is a solvable instance $\mathcal{I}'$ of d-SRI and K' is an integer.
Question: Does $\mathcal{I}'$ admit a stable matching M with $c(M) \leq K'$?

In [5, Appendix B] we give a reduction from the NP-complete decision version of Minimum Vertex Cover in cubic graphs to EGAL 3-SRI DEC, deriving the hardness of the latter problem.

Theorem 1. EGAL 3-SRI DEC *is* NP-*complete.*

Theorem 1 immediately implies the following result.

Corollary 2. EGAL 3-SRI *is* NP-*hard.*

We remark that EGAL 2-SRI is trivially solvable in polynomial time: the components of the graph are paths and cycles in this case, and the cost of a stable matching selected in one component is not affected by the matching edges chosen in another component. Therefore we can deal with each path and cycle separately, minimising the cost of a stable matching in each. Paths and odd cycles admit exactly one stable matching (recall that (i) the instance is assumed to be solvable, and (ii) the set of matched agents is the same in all stable matchings [10, Theorem 4.5.2]), whilst even cycles admit at most two stable matchings (to find them, just pick the two perfect matchings and test each for stability) – we can just pick the stable matching with lower cost in such a case. The following result is therefore immediate.

Proposition 3. EGAL 2-SRI *admits a linear-time algorithm.*

Corollary 2 naturally leads to the question of the approximability of EGAL d-SRI. As mentioned in the Introduction, Feder [6,7] provided a 2-approximation algorithm for the problem of finding an egalitarian stable matching in an instance of SRI. As Theorems 4, 6 and 7 show, this bound can be improved for instances with bounded-length preference lists.

Theorem 4. EGAL 3-SRI *is approximable within* 9/7.

Proof. Let $\mathcal{I}$ be an instance of 3-SRI and let M_{egal} denote an egalitarian stable matching in $\mathcal{I}$. First we show that any stable matching in $\mathcal{I}$ is a 4/3-approximation to M_{egal}. We then focus on the worst-case scenario when this ratio 4/3 is in fact realised. Then we design a weight function on the edges of the graph and apply Teo and Sethuraman's 2-approximation algorithm [22,23] to find an approximate solution M' to a minimum weight stable matching M_{opt} for this weight function. This weight function helps M' to avoid the worst case for the 4/3-approximation for a significant amount of the matching edges. We will ultimately show that M' is in fact a 9/7-approximation to M_{egal}.

Claim 5. *In an instance of* EGAL 3-SRI, *any stable matching approximates* $c(M_{egal})$ *within a factor of 4/3.*

Proof. Let M be an arbitrary stable matching in $\mathcal{I}$. Call an edge uv an (i,j)-*pair* $(i \leq j)$ if v is u's ith choice and u is v's jth choice. By Theorem 4.5.2 of [10], the set of agents matched in M_{egal} is identical to the set of agents matched in M. We will now study the worst approximation ratios in all cases of (i,j)-pairs, given that $1 \leq i \leq j \leq 3$ in 3-SRI.

- If $uv \in M_{egal}$ is a $(1,1)$-pair then u and v contribute 2 to $c(M_{egal})$ and also 2 to $c(M)$ since they must be also be matched in M (and in every stable matching).
- If $uv \in M_{egal}$ is a $(1,2)$-pair then u and v contribute 3 to $c(M_{egal})$ and at most 4 to $c(M)$. Since, if $uv \notin M$, then v must be matched to his 1st choice and u to his 2nd or 3rd, because one of u and v must be better off and the other must be worse off in M than in M_{egal}.
- If $uv \in M_{egal}$ is a $(1,3)$-pair then u and v contribute 4 to $c(M_{egal})$ and at most 5 to $c(M)$. Since, if $uv \notin M$, then v must be matched to his 1st or 2nd choice and u to his 2nd or 3rd.
- If $uv \in M_{egal}$ is a $(2,2)$-pair then u and v contribute 4 to $c(M_{egal})$ and at most 4 to $c(M)$. Since, if $uv \notin M$, then one must be matched to his 1st choice and the other to his 3rd.
- If $uv \in M_{egal}$ is a $(2,3)$-pair then u and v contribute 5 to $c(M_{egal})$ and at most 5 to $c(M)$. Since, if $uv \notin M$, then v must be matched to his 1st or 2nd choice and u to his 3rd.
- If $uv \in M_{egal}$ is a $(3,3)$-pair then u and v contribute 6 to $c(M_{egal})$ and also 6 to $c(M)$ since they must be also be matched in M (and in every stable matching – this follows by [10, Lemma 4.3.9]).

It follows that, for every pair $uv \in M_{egal}$,

$$\frac{\text{rank}(u, M(u)) + \text{rank}(v, M(v))}{\text{rank}(u, M_{egal}(u)) + \text{rank}(v, M_{egal}(v))} = \frac{\text{rank}(u, M(u)) + \text{rank}(v, M(v))}{\text{rank}(u, v) + \text{rank}(v, u)} \leq 4/3.$$

Hence $c(M)/c(M_{egal}) \leq 4/3$ and Claim 5 is proved.   □

As shown in Claim 5, the only case when the approximation ratio 4/3 is reached is where M_{egal} consists of $(1,2)$-pairs exclusively, while the stable matching output by the approximation algorithm contains $(1,3)$-pairs only. We will now present an algorithm that either delivers a stable solution M' containing at least a significant amount of the $(1,2)$-pairs in M_{egal} or a certificate that M_{egal} contains only a few $(1,2)$-pairs and thus any stable solution is a good approximation.

To simplify our proof, we execute some basic pre-processing of the input graph. If there are any (1, 1)-pairs in G, then these can be fixed, because they occur in every stable matching and thus can only lower the approximation ratio. Similarly, if an arbitrary stable matching contains a (3, 3)-pair, then this edge appears in all stable matchings and thus we can fix it. Those (3, 3)-pairs that do not belong to the set of stable edges can be deleted from the graph. From this point on, we assume that no edge is ranked first or last by both of its end vertices in G and prove the approximation ratio for such graphs.

Take the following weight function on all $uv \in E$:

$$w(uv) = \begin{cases} 0 & \text{if } uv \text{ is a } (1, 2)\text{-pair,} \\ 1 & \text{otherwise.} \end{cases}$$

We designed $w(uv)$ to fit the necessary U-shaped condition of Teo and Sethuraman's 2-approximation algorithm [22,23]. This condition on the weight function is as follows. We are given a function f_p on the neighbouring edges of a vertex p. Function f_p is *U-shaped* if it is non-negative and there is a neighbour q of p so that f_p is monotone decreasing on neighbours in order of p's preference until q, and f_p is monotone increasing on neighbours in order of p's preference after q. The approximation guarantee of Teo and Sethuraman's algorithm holds for an edge weight function $w(uv)$ if for every edge $uv \in E$, $w(uv)$ can be written as $w(uv) = f_u(uv) + f_v(uv)$, where f_u and f_v are U-shaped functions.

Our $w(uv)$ function is clearly U-shaped, because at each vertex the sequence of edges in order of preference is either monotone increasing or it is $(1, 0, 1)$. Since w itself is U-shaped, it is easy to decompose it into a sum of U-shaped f_v functions, for example by setting $f_v(uv) = f_u(uv) = \frac{w(uv)}{2}$ for every edge uv.

Let M denote an arbitrary stable matching and $M^{(1,2)}$ be the set of (1, 2)-pairs in a matching M and M_{opt} be a minimum weight stable matching with respect to the weight function $w(uv)$. Since M_{opt} is by definition the stable matching with the largest number of (1, 2)-pairs, $|M_{\text{opt}}^{(1,2)}| \geq |M_{\text{egal}}^{(1,2)}|$. We also know that $w(M) = |M| - |M^{(1,2)}|$ for every stable matching M.

Due to Teo and Sethuraman's approximation algorithm [22,23], it is possible to find a stable matching M' whose weight approximates $w(M_{\text{opt}})$ within a factor of 2. Formally,

$$|M| - |M'^{(1,2)}| = w(M') \leq 2w(M_{\text{opt}}) = 2|M| - 2|M_{\text{opt}}^{(1,2)}|.$$

This gives us a lower bound on $|M'^{(1,2)}|$.

$$|M'^{(1,2)}| \geq 2|M_{\text{opt}}^{(1,2)}| - |M| \geq 2|M_{\text{egal}}^{(1,2)}| - |M| \tag{1}$$

We distinguish two cases from here on, depending on the sign of the term on the right. In both cases, we establish a lower bound on $c(M_{\text{egal}})$ and an upper bound on $c(M')$. These will give the desired upper bound of 9/7 on $\frac{c(M')}{c(M_{\text{egal}})}$.

(1) $2|M_{\text{egal}}^{(1,2)}| - |M| \leq 0$

The derived lower bound for $|M'^{(1,2)}|$ is negative or zero in this case. Yet we know that at most half of the edges in M_{egal} are $(1, 2)$-pairs, and $c(e) \geq 4$ for the rest of the edges in M_{egal}. Let us denote $|M| - 2|M_{\text{egal}}^{(1,2)}| \geq 0$ by x. Thus, $|M_{\text{egal}}^{(1,2)}| = \frac{|M|-x}{2}$.

$$c(M_{\text{egal}}) \geq \frac{|M| - x}{2} \cdot 3 + \frac{|M| + x}{2} \cdot 4 = 3.5|M| + 0.5x \qquad (2)$$

We use our arguments in the proof of Claim 5 to derive that an arbitrary stable matching approximates $c(M_{\text{egal}})$ on the $\frac{|M|-x}{2}$ $(1, 2)$-edges within a ratio of $\frac{4}{3}$, while its cost on the remaining $\frac{|M|+x}{2}$ edges is at most 5. These imply the following inequalities for an arbitrary stable matching M.

$$c(M) \leq \frac{|M| - x}{2} \cdot 3 \cdot \frac{4}{3} + \frac{|M| + x}{2} \cdot 5 = 4.5|M| + 0.5x \qquad (3)$$

We now combine (2) and (3). The last inequality holds for all $x \geq 0$.

$$\frac{c(M)}{c(M_{\text{egal}})} \leq \frac{4.5|M| + 0.5x}{3.5|M| + 0.5x} \leq \frac{9}{7}$$

(2) $2|M_{\text{egal}}^{(1,2)}| - |M| > 0$

Let us denote $2|M_{\text{egal}}^{(1,2)}| - |M|$ by $\hat{x}$. Notice that $|M_{\text{egal}}^{(1,2)}| = \frac{\hat{x}+|M|}{2}$. We can now express now the number of edges with cost 3, and at least 4 in M_{egal}.

$$c(M_{\text{egal}}) \geq 3 \cdot \frac{\hat{x} + |M|}{2} + 4 \cdot \left(|M| - \frac{\hat{x} + |M|}{2}\right)$$
$$= 3.5|M| - 0.5\hat{x} \qquad (4)$$

Let $|M'^{(1,2)}| = z_1$. Then exactly z_1 edges in M' have cost 3. It follows from (1) that $z_1 \geq \hat{x}$. Suppose that $z_2 \leq z_1$ edges in $M'^{(1,2)}$ correspond to edges in $M_{\text{egal}}^{(1,2)}$. Recall that $|M_{\text{egal}}^{(1,2)}| = \frac{\hat{x}+|M|}{2}$. The remaining $\frac{|M|+\hat{x}}{2} - z_2$ edges in $M_{\text{egal}}^{(1,2)}$ have cost at most 4 in M'. This leaves $|M| - |M_{\text{egal}}^{(1,2)}| - (z_1 - z_2) = \frac{|M|-\hat{x}}{2} - z_1 + z_2$ edges in M_{egal} that are as yet unaccounted for; these have cost at most 5 in both M_{egal} and M'. We thus obtain:

$$c(M') \leq 3z_1 + 4\left(\frac{|M| + \hat{x}}{2} - z_2\right) + 5\left(\frac{|M| - \hat{x}}{2} - z_1 + z_2\right)$$
$$= 4.5|M| - 0.5\hat{x} - 2z_1 + z_2$$
$$\leq 4.5|M| - 1.5\hat{x} \qquad (5)$$

Combining (4) and (5) delivers the following bound.

$$\frac{c(M')}{c(M_{\text{egal}})} \leq \frac{4.5|M| - 1.5\hat{x}}{3.5|M| - 0.5\hat{x}} < \frac{9}{7}$$

The last inequality holds for every $\hat{x} > 0$.

We derived that M', the 2-approximate solution with respect to the weight function $w(uv)$ delivers a $\frac{9}{7}$-approximation in both cases.   □

Using analogous techniques we can establish similar approximation bounds for EGAL 4-SRI and EGAL 5-SRI, as follows.

Theorem 6. EGAL 4-SRI *is approximable within* $11/7$.

Theorem 7. EGAL 5-SRI *is approximable within* $13/7$.

Using a similar reasoning for each $d \geq 6$, our approach gives a c_d-approximation algorithm for EGAL d-SRI where $c_d > 2$. In these cases the 2-approximation algorithm of Feder [6,7] should be used instead.

3   Solvability and Most-Stable Matchings in d-SRTI

In this section we study the complexity and approximability of the problem of deciding whether an instance of d-SRTI admits a stable matching, and the problem of finding a most-stable matching given an instance of d-SRTI.

We begin by defining two problems that we will be studying in this section from the point of view of complexity and approximability.

Problem 3. SOLVABLE d-SRTI
Input: $\mathcal{I} = \langle G, \mathcal{O} \rangle$, *where* G *is a graph and* $\mathcal{O}$ *is a set of preference lists, each of length at most* d, *possibly involving ties.*
Question: Is $\mathcal{I}$ *solvable?*

Problem 4. MIN BP d-SRTI
Input: An instance $\mathcal{I}$ *of* d-SRTI.
Output: A matching M *in* $\mathcal{I}$ *such that* $|bp(M)| = bp(\mathcal{I})$.

We will show that SOLVABLE 3-SRTI is NP-complete and MIN BP 3-SRTI is hard to approximate. In both cases we will use a reduction from the following satisfiability problem:

Problem 5. $(2, 2)$-E3-SAT
Input: $\mathcal{I} = B$, *where* B *is a Boolean formula in CNF, in which each clause comprises exactly 3 literals and each variable appears exactly twice in unnegated and exactly twice in negated form.*
Question: Is there a truth assignment satisfying B?

$(2, 2)$-E3-SAT is NP-complete, as shown by Berman et al. [2]. We begin with the hardness of SOLVABLE 3-SRTI.

Theorem 8. SOLVABLE 3-SRTI *is NP-complete.*

Proof. Clearly SOLVABLE 3-SRTI belongs to NP. To show NP-hardness, we reduce from (2, 2)-E3-SAT as defined in Problem 5. Let B be a given instance of (2, 2)-E3-SAT, where $X = \{x_1, x_2, \ldots, x_n\}$ is the set of variables and $C = \{c_1, c_2, \ldots, c_m\}$ is the set of clauses. We form an instance $\mathcal{I} = (G, \mathcal{O})$ of 3-SRTI as follows. Graph G consists of a *variable gadget* for each x_i ($1 \le i \le n$), a *clause gadget* for each c_j ($1 \le j \le m$) and a set of *interconnecting edges* between them; these different parts of the construction, together with the preference orderings that constitute $\mathcal{O}$, are shown in Fig. 1 and will be described in more detail below.

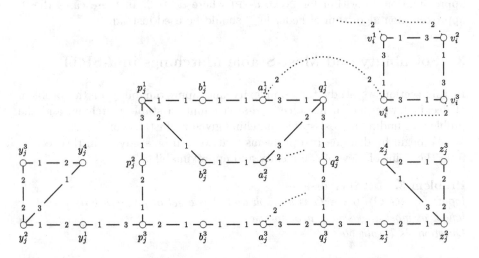

Fig. 1. Clause and variable gadgets for 3-SRTI. The dotted edges are the interconnecting edges. The notation used for edge $a_j^1 v_i^4$ implies that the first literal of the corresponding clause c_j is the second occurrence of the corresponding variable x_i in negated form.

When constructing G, we will keep track of the order of the three literals in each clause of B and the order of the two unnegated and two negated occurrences of each variable in B. Each of these four occurrences of each variable is represented by an interconnecting edge.

A variable gadget for a variable x_i ($1 \le i \le n$) of B comprises the 4-cycle $\langle v_i^1, v_i^2, v_i^3, v_i^4 \rangle$ with cyclic preferences. Each of these four vertices is incident to an interconnecting edge. These edges end at specific vertices of clause gadgets. The clause gadget for a clause c_j ($1 \le j \le m$) contains 20 vertices, three of which correspond to the literals in c_j; these vertices are also incident to an interconnecting edge.

Due to the properties of (2, 2)-E3-SAT, x_i occurs twice in unnegated form, say in clauses c_j and c_k of B. Its first appearance, as the rth literal of c_j ($1 \le r \le 3$), is represented by the interconnecting edge between vertex v_i^1 in the variable gadget corresponding to x_i and vertex a_j^r in the clause gadget corresponding to c_j. Similarly the second occurrence of x_i, say as the sth literal of c_k ($1 \le s \le 3$)

is represented by the interconnecting edge between v_i^3 and a_k^s. The same variable x_i also appears twice in negated form. Appropriate a-vertices in the gadgets representing those clauses are connected to v_i^2 and v_i^4. We remark that this construction involves a gadget similar to one presented by Biró et al. [3] in their proof of the NP-hardness of MIN BP 3-SRI.

In [5, Appendix B] we prove that there is a truth assignment satisfying B if and only if there is a stable matching M in $\mathcal{I}$. □

Our construction shows that the complexity result holds even if the preference lists are either strictly ordered or consist of a single tie of length two. Moreover, Theorem 8 also immediately implies the following result.

Corollary 9. MIN BP 3-SRTI *is* NP-*hard.*

The following result strengthens Corollary 9.

Theorem 10. MIN BP 3-SRTI *is not approximable within* $n^{1-\varepsilon}$, *for any* $\varepsilon > 0$, *unless* P= NP, *where* n *is the number of agents.*

Proof (sketch). The core idea of our proof is to gather several copies of the 3-SRTI instance created in the proof of Theorem 8, together with a small unsolvable 3-SRTI instance. By doing so, we create a MIN BP 3-SRTI instance $\mathcal{I}$ in which $bp(\mathcal{I})$ is large if the Boolean formula B (originally given as an instance of $(2, 2)$-E3-SAT) is not satisfiable, and $bp(\mathcal{I}) = 1$ otherwise. Therefore, finding a good approximation for $\mathcal{I}$ will imply a polynomial-time algorithm to decide the satisfiability of B. □

To complete the study of cases of MIN BP d-SRTI, we establish a positive result for instances with degree at most 2.

Theorem 11. MIN BP 2-SRTI *is solvable in* $\mathcal{O}(|V|)$ *time.*

Proof. For an instance $\mathcal{I}$ of MIN BP 2-SRTI, clearly every component of the underlying graph G is a path or cycle. We claim that $bp(\mathcal{I})$ equals the number of *odd parties* in G, where an *odd party* is a cycle $C = \langle v_1, v_2, ..., v_k \rangle$ of odd length, such that v_i strictly prefers v_{i+1} to v_{i-1} (addition and subtraction are taken modulo k).

Since an odd party never admits a stable matching, $bp(\mathcal{I})$ is bounded below by the number of odd parties [21]. This bound is tight: by taking an arbitrary maximum matching in an odd party component, a most-stable matching is already reached. Now we show that a stable matching M can be constructed in all other components.

Each component that is not an odd cycle is therefore a bipartite subgraph (indeed either a path or an even cycle). Such a subgraph therefore gives rise to the restriction of SRTI called the *Stable Marriage problem with Ties and Incomplete lists* (SMTI). An instance of SMTI always admits a stable solution and it can be found in linear time [19]. Thus these components contribute no blocking edge.

Regarding odd-length cycles that are not odd parties, we will show that there is at least one vertex not strictly preferred by either of its adjacent vertices.

Leaving this vertex uncovered and adding a perfect matching in the rest of the cycle results in a stable matching.

Assume that every vertex along a cycle C_k (where k is an odd number) is strictly preferred by at least one of its neighbours. Since each of the k vertices is strictly preferred by at least one vertex, and a vertex v can prefer at most one other vertex strictly, every vertex along C_k has a strictly ordered preference list. Now every vertex can point at its unique first-choice neighbour. To avoid an odd cycle, there must be a vertex pointed at by both of its neighbours. This implies that there is also a vertex v pointed at by no neighbour, and v is hence ranked second by both of its neighbours.   □

Open Questions. Theorems 4, 6 and 7 improve on the best known approximation factor for EGAL d-SRI for small d. It remains open to come up with an even better approximation or to establish an inapproximability bound matching our algorithm's guarantee. A more general direction is to investigate whether the problem of finding a minimum weight stable matching can be approximated within a factor less than 2 for instances of d-SRI for small d.

References

1. Abraham, D.J., Biró, P., Manlove, D.F.: "Almost Stable" matchings in the room-mates problem. In: Erlebach, T., Persinao, G. (eds.) WAOA 2005. LNCS, vol. 3879, pp. 1–14. Springer, Heidelberg (2006)
2. Berman, P., Karpinski, M., Scott, A.D.: Approximation hardness of short symmetric instances of MAX-3SAT. ECCC report, no. 49 (2003)
3. Biró, P., Manlove, D.F., McDermid, E.J.: "Almost stable" matchings in the room-mates problem with bounded preference lists. Theor. Comput. Sci. **432**, 10–20 (2012)
4. Biró, P., Manlove, D.F., McDermid, E.J.: Almost stable matchings in the room-mates problem with bounded preference lists. Theor. Comput. Sci. **411**, 1828–1841 (2010)
5. Cseh, Á., Irving, R.W., Manlove, D.F.: The stable roommates problem with short lists. CoRR abs/1605.04609 (2016). http://arxiv.org/abs/1605.04609
6. Feder, T.: A new fixed point approach for stable networks and stable marriages. J. Comput. Syst. Sci. **45**, 233–284 (1992)
7. Feder, T.: Network flow and 2-satisfiability. Algorithmica **11**, 291–319 (1994)
8. Gale, D., Shapley, L.S.: College admissions and the stability of marriage. Am. Math. Monthly **69**, 9–15 (1962)
9. Gusfield, D.: Three fast algorithms for four problems in stable marriage. SIAM J. Comput. **16**(1), 111–128 (1987)
10. Gusfield, D., Irving, R.W.: The Stable Marriage Problem: Structure and Algorithms. MIT Press, Cambridge (1989)
11. Gusfield, D., Pitt, L.: A bounded approximation for the minimum cost 2-SAT problem. Algorithmica **8**, 103–117 (1992)
12. Hamada, K., Iwama, K., Miyazaki, S.: An improved approximation lower bound for finding almost stable maximum matchings. Inf. Process. Lett. **109**, 1036–1040 (2009)

13. Irving, R.W.: An efficient algorithm for the "stable roommates" problem. J. Algorithms **6**, 577–595 (1985)
14. Irving, R.W., Leather, P., Gusfield, D.: An efficient algorithm for the "optimal" stable marriage. J. ACM **34**, 532–543 (1987)
15. Irving, R.W., Manlove, D.F.: The stable roommates problem with ties. J. Algorithms **43**, 85–105 (2002)
16. Kujansuu, E., Lindberg, T., Mäkinen, E.: The stable roommates problem and chess tournament pairings. Divulgaciones Matemáticas **7**, 19–28 (1999)
17. Khot, S., Regev, O.: Vertex cover might be hard to approximate to within 2-ε. J. Comput. Syst. Sci. **74**, 335–349 (2008)
18. Manlove, D.F.: Algorithmics of Matching Under Preferences. World Scientific, Singapore (2013)
19. Manlove, D.F., Irving, R.W., Iwama, K., Miyazaki, S., Morita, Y.: Hard variants of stable marriage. Theor. Comput. Sci. **276**, 261–279 (2002)
20. Ronn, E.: NP-complete stable matching problems. J. Algorithms **11**, 285–304 (1990)
21. Tan, J.J.M.: A necessary and sufficient condition for the existence of a complete stable matching. J. Algorithms **12**, 154–178 (1991)
22. Teo, C.-P., Sethuraman, J.: LP based approach to optimal stable matchings. In: Proceedings of SODA 1997, pp. 710–719. ACM-SIAM (1997)
23. Teo, C.-P., Sethuraman, J.: The geometry of fractional stable matchings and its applications. Math. Oper. Res. **23**, 874–891 (1998)

The Price of Stability
of Simple Symmetric Fractional Hedonic Games

Christos Kaklamanis, Panagiotis Kanellopoulos[✉],
and Konstantinos Papaioannou

Computer Technology Institute and Press "Diophantus"
and Department of Computer Engineering and Informatics,
University of Patras, 26504 Rio, Greece
{kakl,kanellop,papaioann}@ceid.upatras.gr

Abstract. We consider simple symmetric fractional hedonic games, in which a group of utility maximizing players have hedonic preferences over the players' set, and wish to be partitioned into clusters so that they are grouped together with players they prefer. Each player either wishes to be in the same cluster with another player (and, hence, values this agent at 1) or is indifferent (and values this player at 0). Given a cluster, the utility of each player is defined as the number of players inside the cluster that are valued at 1 divided by the cluster size, and a player will deviate to another cluster if this leads to higher utility. We are interested in Nash equilibria of such games, where no player has an incentive to unilaterally deviate to another cluster, and we focus on the notion of the price of stability. We present new and improved bounds on the price of stability both for the normal utility function and for a slightly modified one.

1 Introduction

Economic entities, be it individuals or corporations, interact frequently in the context of performing complex tasks or even in enjoying cultural activities. For example, people usually tend to like or dislike other people and, therefore, wish to socialize or distance themselves depending on the occasion. The choice about which party to attend, which restaurant to dine in, etc., usually depends also on the other participants that will be present. Such a behavior is captured by the class of *hedonic games*, where participating agents have preferences over coalitions (or groups), and, based on these preferences, they behave accordingly when selecting which group to join. In such scenarios, when an agent has decided to join a specific group, e.g., for having dinner, his utility depends only on the other agents in the same group and not on how the remaining agents have been grouped together. Such preferences are termed *hedonic preferences* and completely ignore inter-coalitional dependencies. Due to their simplicity, hedonic games can be used to model a large spectrum of activities (e.g., clustering in social networks [2], distributed task allocation for wireless agents [22], etc.).

© Springer-Verlag Berlin Heidelberg 2016
M. Gairing and R. Savani (Eds.): SAGT 2016, LNCS 9928, pp. 220–232, 2016.
DOI: 10.1007/978-3-662-53354-3_18

Hedonic games can be very expressive and admit a large class of utility functions over coalitions. For instance, given a group, we may care about the sum of utility we obtain over all members of the group, or only care about the minimum or maximum utility (again, over all members of the group). In addition, we may prefer a smaller group containing people that we value significantly over a larger group with the same set of preferred people as well as several other members that we are indifferent to; in this case, we are interested in the average utility we obtain. What constitutes an acceptable or desired solution in such games is a question that has also attracted significant attention. Clearly, a natural objective is to compute a solution that maximizes some global function over all participating agents. This solution, however, may leave several agents dissatisfied and they may not adhere to it, but, on the contrary, may choose to deviate to another group if this is to their best interest. Then, such deviations may incentivize further agents (or even groups of agents) to deviate on their own (or, respectively, in collaboration), and so on, until some group formation is reached where all agents are satisfied, if such a formation exists.

Related Work. Hedonic games (see [5] for a very recent survey) that rely on hedonic preferences were introduced by Drèze and Greenberg [12]. Bloch and Diamantoudi [8] consider a bargaining procedure of coalition formation in hedonic games and present necessary and sufficient conditions for existence of pure strategy stationary perfect equilibria. Bogomolnaia and Jackson [9] present sufficient conditions for the existence of core stable partitions in hedonic settings and also consider the weaker notion of individual stability, where no player can deviate to another cluster without either hurting itself or hurting a member of its new cluster. Peters and Elkind [20] investigate the computational complexity of stability-related questions in hedonic games. Feldman et al. [14] consider the non-cooperative version of hedonic clustering games, where they characterize Nash equilibria and provide upper and lower bounds on the price of anarchy and price of stability. Other instances of hedonic games include those studied by Branzei and Larson [10] as well as by Elkind and Wooldridge [13]. An important subclass is that of additively separable hedonic games (see [3,16,18]), where the total utility of each player is defined as the sum of utility it obtains from each player in its cluster.

In *fractional hedonic games*, the utility of each player is defined as the sum of utility it obtains from each player in its cluster divided by the cluster size. Aziz et al. [2] introduced the model, considered more general stability notions, such as core stability, and presented positive results for several classes of graphs. Then, Aziz et al. [4] consider the computational complexity of computing partitions that maximize the social welfare, defined as the sum of the players' utilities, in fractional hedonic games, without caring about stability. They consider three different notions of social welfare (i.e., utilitarian, egalitarian, and Nash welfare) and show that maximizing social welfare is NP-hard even for the subclass of *simple symmetric fractional hedonic games*, where the utility obtained from a single player can be either 0 or 1 and the utility is symmetric. On the positive side, they present polynomial time algorithms with small constant approximation ratio for the notions of utilitarian and egalitarian social welfare and the class of simple symmetric fractional hedonic games. Olsen [19], among other results, suggested an alternate utility function for fractional hedonic games, where the utility function of player i does not take i into account when averaging over the cluster size, i.e., each player is interested in the average utility obtained from all other players in the same cluster.

Bilò et al. [6,7] consider the price of anarchy and stability in fractional hedonic games. They show that when the utility function may take negative values, Nash stable outcomes are not guaranteed to exist, but if all values are nonnegative, then the partition where all players are in the same cluster is Nash stable. For the last case, they show an upper bound of $O(n)$ on the price of anarchy, which is tight even for simple symmetric fractional hedonic games. Furthermore, they show a lower bound of $\Omega(n)$ on the price of stability for games played on weighted stars and non-negative utility functions. For the price of stability in simple symmetric fractional hedonic games, they show a lower bound of 2 for general graphs, an upper bound of 4 for triangle-free graphs, and almost tight bounds for the case of bipartite graphs. In particular, they present an upper bound of $6(3 - 2\sqrt{2}) \approx 1.0294$ and a lower bound of 1.003. In addition, Bilò et al. [7] observe that their upper bounds still hold for the utility function defined by Olsen. Further notions of stability in fractional hedonic games have been investigated by Brandl et al. [11].

Our Contribution. We present new and improved bounds on the price of stability for the class of simple symmetric fractional hedonic games. We improve upon the lower bound of [7] and show a lower bound of $1 + \sqrt{6}/2 \approx 2.224$ for general graphs. The construction we use in the proof admits an optimal partition consisting of cliques of different sizes, while the only Nash stable partition is the grand coalition, where all players form a single cluster.

Then, we consider games played on graphs of girth at least 5, i.e., graphs without cycles of size 3 and 4. We prove that the price of stability, for this class of graphs, is 1. This result complements a result of Bilò et al. [7] that there exists a bipartite graph with price of stability at least 1.003. Since bipartite graphs have no cycles of length 3 but may have cycles of length 4, we obtain a clear separation of which girth values lead to price of stability equal to 1.

Our final result concerns the utility function defined by Olsen [19], where the average utility is computed with respect to the cluster size minus 1, i.e., each player only considers the average utility it obtains by the nodes it is grouped together with. We show that, under this utility function, the price of stability of simple symmetric fractional hedonic games is 1; the previously known bounds were those obtained in [7] for the standard utility function.

Roadmap. The remainder of the paper is structured as follows. We begin, in Sect. 2, by formally introducing the class of simple symmetric fractional hedonic games and presenting the necessary definitions. Then, in Sect. 3, we present the results on the price of stability and we conclude with open problems in Sect. 4.

2   Preliminaries

A *fractional hedonic game* is a non-cooperative strategic game played by a set N of n utility maximizing players. A *partition* (or *clustering*) of the game consists of a set $\mathcal{C} = \{C_1, C_2, \dots\}$ of clusters such that $\cup_i C_i = N$, and $C_i \cap C_j = \emptyset$ for any pair $i \neq j$, i.e., each player belongs to exactly one cluster. We let $C(u)$ denote the cluster that player u belongs to.

Each player i has a utility function $u_i : N \to \mathbb{R}$ that denotes how much player i values each of the remaining players. We are interested in the class of *simple symmetric fractional hedonic games*, where $u_i(j)$ is either 0 or 1 and, furthermore, $u_i(j) = u_j(i)$. Also, for any player i, $u_i(i) = 0$. Simple symmetric fractional hedonic games admit a graph representation by considering a connected[1] undirected graph $G = (V, E)$ with $|V| = n$. Each node $u \in V$ corresponds to a player and an edge $(i, j) \in E$ denotes the fact that player i values player j at 1 and vice versa. If, otherwise, edge $(i, j) \notin E$ then player i values j at 0 and vice versa. Note that edges are undirected and the weight of each edge is 1. Let $\deg_G(i)$ denote the degree of node i in graph G and let $\deg_C(i)$ denote the number of neighbors of node i that belong to cluster C. Given a partition $\mathcal{C}$, the utility of player i that is in cluster $C(i)$ is defined as

$$u_i(C(i)) = \frac{\sum_{j \in C(i)} u_i(j)}{|C(i)|} = \frac{\deg_{C(i)}(i)}{|C(i)|}.$$

Clearly, for any player i and any partition $\mathcal{C}$ it holds that $0 \leq u_i(C(i)) \leq \frac{n-1}{n}$.

The *social welfare* $\mathrm{SW}(\mathcal{C})$ of partition $\mathcal{C}$ is defined as the sum of the players' utility, i.e., $SW(\mathcal{C}) = \sum_i u_i(C(i))$. An equivalent way to define the social welfare is by taking into account the number of edges $E(C)$ inside each cluster C. Hence, we obtain that

$$\mathrm{SW}(\mathcal{C}) = \sum_{C \in \mathcal{C}} \frac{2E(C)}{|C|}.$$

We denote by $\mathcal{C}^*$ the partition that maximizes the social welfare.

Since each player is utility maximizing, given a partition $\mathcal{C}$, player i may deviate from its current cluster $C(i)$ in $\mathcal{C}$ and join another cluster C', if it holds that $u_i(C(i)) < u_i(C' \cup i)$, i.e., whenever $\frac{\deg_{C(i)}(i)}{|C(i)|} < \frac{\deg_{C'}(i)}{|C'|+1}$. A player i is *Nash stable* if there is no cluster $C' \neq C(i)$ such that its utility improves by deviating to C', i.e., for any cluster C' it holds that $\frac{\deg_{C(i)}(i)}{|C(i)|} \geq \frac{\deg_{C'}(i)}{|C'|+1}$. A cluster is *Nash stable* if all players in the cluster are Nash stable. A partition is a *Nash stable partition* if all clusters are Nash stable.

The *price of stability* PoS (introduced in [1]) denotes the best-case performance deterioration arising from the requirement that the resulting partition is Nash stable. Given a graph G, the corresponding fractional hedonic game Γ_G and its set of Nash stable partitions $\mathcal{C}_s$, the price of stability for the game Γ_G

---

[1] Our upper bounds also hold for disconnected graphs by considering each component separately.

is formally defined as $\text{PoS}(\Gamma_G) = \min_{C \in \mathcal{C}_s} \frac{\text{SW}(C^*)}{\text{SW}(C)}$. Similarly, the price of stability for the class of simple symmetric fractional hedonic games is defined as $\text{PoS} = \max_G \text{PoS}(\Gamma_G)$.

A variant of fractional hedonic games was introduced by Olsen [19], where the single difference is that the utility of each player i is now defined as

$$u_i'(C(i)) = \begin{cases} \frac{\deg_{C(i)}(i)}{|C|-1}, & \text{if} |C| > 1; \\ 0, & \text{otherwise.} \end{cases}$$

3   Price of Stability of Fractional Hedonic Games

This section contains the results on the price of stability of simple symmetric fractional hedonic games. We begin by presenting an improved lower bound of $1 + \sqrt{6}/2 \approx 2.224$. Then, we consider the case where the game is played on graphs of girth at least 5, i.e., there are no triangles and no cycles of length 4. For this case, we prove that there exists an optimal partition that is also Nash stable, i.e., the price of stability is 1. We conclude by considering a slightly different utility function (defined by Olsen [19]) where each agent averages the value obtained over the cluster size minus one.

3.1   A Lower Bound for General Graphs

Our construction extends in a non-trivial way the graph used in the lower bound of [7]. The proof relies on showing that the grand coalition is the only Nash stable partition while the optimal partition contains cliques of different sizes. We now present the main result of this section.

Theorem 1. *The price of stability of simple symmetric fractional hedonic games is at least $1 + \frac{\sqrt{6}}{2} - \epsilon$ for $\epsilon > 0$.*

Proof (Sketch). Let α be a positive integer. Consider the following graph G that is also presented in Fig. 1. It consists of $\alpha + 2$ cliques K^κ where $1 \leq \kappa \leq \alpha + 2$ and some additional nodes and edges to be detailed later. Clique K^1 contains $4(\sqrt{6}+1)\alpha^2$ nodes[2], clique K^2 contains $4\alpha^2 + 2$ nodes, while each remaining clique contains 4α nodes. There exist $4(\sqrt{6}+1)\alpha^2$ additional nodes where each of them has degree 1 and is connected to a node in K^1 so that no pair of additional nodes shares a neighbor. The total number of nodes is $n = 8(2 + \sqrt{6})\alpha^2 + 2$. There exist additional edges as follows: each node in K^2 is connected to any node in K^κ, for $\kappa \in [1, \ldots, \alpha + 2]$. The total number of edges in G is

$$E(G) = 4(\sqrt{6}+1)\alpha^2 + \frac{4(\sqrt{6}+1)\alpha^2(4(\sqrt{6}+1)\alpha^2 - 1)}{2} + \frac{(4\alpha^2 + 2)(4\alpha^2 + 1)}{2}$$

$$+ \alpha \frac{4\alpha(4\alpha - 1)}{2} + (4\alpha^2 + 2)(4(\sqrt{6}+1)\alpha^2 + 4\alpha^2)$$

$$= 32(3 + \sqrt{6})\alpha^4 + 8\alpha^3 + 2(11 + 5\sqrt{6})\alpha^2 + 1,$$

---

[2] In fact, $|K^1|$ should be either $\lceil 4(\sqrt{6} + 1)\alpha^2 \rceil$ or $\lfloor 4(\sqrt{6} + 1)\alpha^2 \rfloor$ but the proof still follows in the same way. We set $|K^1| = 4(\sqrt{6}+1)\alpha^2$ to keep the presentation cleaner.

where, in the first equality, the first term is due to the edges connecting the additional nodes to the nodes in K^1, the second term is due to the edges inside K^1, the third term is due to the edges inside K^2, the fourth term is due to the edges inside the remaining cliques, while the last term is due to edges connecting nodes of K^2 to nodes in other cliques.

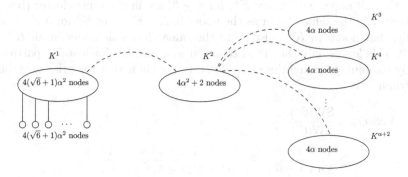

Fig. 1. The graph G used in the lower bound. Each bubble is a clique and dashed lines represent the edges connecting any node in K^2 to any node at another clique.

Consider the partition $\mathcal{C}$ where each node in K^1 forms a cluster with its additional neighbor, nodes in K^2 and K^3 form a single cluster together, while each K^κ, for $\kappa \geq 4$, is a cluster. Then, the social welfare is

$$SW(\mathcal{C}) = \sum_{C \in \mathcal{C}} \frac{2E(C)}{|C|}$$
$$= |K^1| + (|K^2| + |K^3| - 1) + \sum_{\kappa=4}^{\alpha+2} (|K^\kappa| - 1)$$
$$= 4(\sqrt{6} + 1)\alpha^2 + 4\alpha^2 + 4\alpha + 1 + (\alpha - 1)(4\alpha - 1)$$
$$= 4(3 + \sqrt{6})\alpha^2 - \alpha + 2.$$

Clearly, for the optimal partition $\mathcal{C}^*$ it holds that $SW(\mathcal{C}^*) \geq SW(\mathcal{C})$, hence

$$SW(\mathcal{C}^*) \geq 4(3 + \sqrt{6})\alpha^2 - \alpha + 2. \tag{1}$$

Consider now the partition $\mathcal{C}'$ where all nodes form a single cluster, i.e., the grand coalition. Its social welfare is

$$SW(\mathcal{C}') = \frac{2E(G)}{n}$$
$$= \frac{32(3 + \sqrt{6})\alpha^4 + 8\alpha^3 + 2(11 + 5\sqrt{6})\alpha^2 + 1}{4(2 + \sqrt{6})\alpha^2 + 1}. \tag{2}$$

We can show that $\mathcal{C}'$ is the only Nash stable partition using the following approach; due to lack of space, details are omitted. First, note that, in any Nash stable partition, any additional node is in the same cluster as its neighbor in K^1; otherwise, its utility would be 0. Next, we can show that all nodes in K^2 must be in the same cluster. Then, we argue that all nodes in K^1 and the additional nodes must be in the same cluster as those in K^2. Our next step would be to show that all nodes within any K^κ, for $\kappa \geq 3$, are in the same cluster (but not necessarily in the same cluster as the nodes in K^1, K^2 or in K^λ for $\lambda \neq \kappa$), and, finally, that nodes in K^2 must be in the same cluster as nodes in all K^κ, for $\kappa \geq 3$. This leaves only the grand coalition as a possible Nash stable partition.

By combining inequalities (1) and (2), we conclude that the price of stability for graph G is

$$
\begin{aligned}
\mathrm{PoS}(G) &= \frac{\mathrm{SW}(\mathcal{C}^*)}{\mathrm{SW}(\mathcal{C}')} \\
&\geq \frac{(4(3+\sqrt{6})\alpha^2 - \alpha + 2)(4(2+\sqrt{6})\alpha^2 + 1)}{32(3+\sqrt{6})\alpha^4 + 8\alpha^3 + 2(11+5\sqrt{6})\alpha^2 + 1} \\
&= \frac{16(12+5\sqrt{6})\alpha^4 - 4(2+\sqrt{6})\alpha^3 + 4(7+3\sqrt{6})\alpha^2 - \alpha + 2}{32(3+\sqrt{6})\alpha^4 + 8\alpha^3 + 2(11+5\sqrt{6})\alpha^2 + 1} \\
&\geq 1 + \frac{\sqrt{6}}{2} - \epsilon
\end{aligned}
$$

as α tends to infinity, where ϵ is an arbitrarily small positive number.   □

3.2   Graphs of Girth at Least 5

We now consider the class of triangle-free and quadrilateral-free graphs, i.e., when there are no cycles of length 3 or 4. We show that there exists an optimal partition where all clusters are stars and then argue that it must also be Nash stable; this implies a price of stability of 1.

In our proof we exploit the following lemma that upper-bounds the number of edges in a quadrilateral-free graph.

Lemma 1 (Due to [17, 21], as mentioned in [15]). *The maximum number of edges in a quadrilateral-free graph with $n \geq 4$ nodes is $f(n) \leq \frac{1}{4}n(1+\sqrt{4n-3})$.*

We now present a key lemma that details the structure of at least one optimal partition. We will then exploit this result in order to prove the upper bound of 1 on the price of stability.

Lemma 2. *There exists an optimal partition where all clusters are stars.*

Proof. Consider an optimal partition $\mathcal{C}^*$ and let C be a cluster with at least two nodes of degree greater than one. Let i and j be two such nodes. We split C in two clusters $C(i)$ and $C(j)$ so that $i \in C(i)$, $j \in C(j)$ and, furthermore, $C(i)$ is either a cluster with only two nodes or a star such that there is no edge between a leaf in $C(i)$ and a node in $C(j)$. In addition, we require that $C(j)$ is still connected. Note that there always exists a node i with degree at least 2 so that these constraints are satisfied. E.g., we can always identify a star whose removal does not increase the number of connected components and remove any leaf that is connected to $C(j)$. Also, let $c_i = |C(i)|$ and $c_j = |C(j)|$.

The total utility of the players in C is $u(C) = \frac{2e_i + 2e_j + 2e_{ij}}{c_i + c_j}$, where e_i (respectively, e_j) is the number of edges among nodes that are in $C(i)$ (respectively, in $C(j)$) and e_{ij} is the number of edges where one endpoint is in $C(i)$ and the other is in $C(j)$. Similarly, the total utility of players in $C(i)$ is $u(C(i)) = \frac{2e_i}{c_i} = \frac{2c_i - 2}{c_i}$, since $C(i)$ is a star, and the total utility of players in $C(j)$ is $u(C(j)) = \frac{2e_j}{c_j}$. It suffices to prove that $u(C(i)) + u(C(j)) \geq u(C)$, i.e.,

$$\frac{2c_i - 2}{c_i} + \frac{2e_j}{c_j} - \frac{2c_i - 2 + 2e_j + 2e_{ij}}{c_i + c_j} \geq 0. \tag{3}$$

The left hand side of (3) becomes

$$2c_i c_j(c_i + c_j) - 2c_i c_j - 2c_j^2 + 2e_j c_i(c_i + c_j) - 2c_i^2 c_j - 2c_i c_j + 2e_j c_i c_j + 2e_{ij} c_i c_j,$$

and subsequently

$$2c_i c_j^2 - 2c_j^2 + 2e_j c_i^2 - 2e_{ij} c_i c_j.$$

Since $e_j \geq c_j - 1$, in order to prove (3) it suffices to show that

$$c_i c_j^2 + c_i^2 c_j - e_{ij} c_i c_j - c_i^2 - c_j^2 \geq 0. \tag{4}$$

We now distinguish among two cases, depending on whether $e_{ij} \leq \frac{c_i + c_j}{2}$ or not. If $e_{ij} \leq \frac{c_i + c_j}{2}$, then by substituting it in (4) we obtain $\frac{c_i c_j^2}{2} + \frac{c_i^2 c_j}{2} - c_i^2 - c_j^2$ which is clearly nonnegative since $c_i, c_j \geq 2$. So, (4) holds, and, subsequently, (3) also holds and, then, splitting C to $C(i)$ and $C(j)$ does not decrease the social welfare while increasing the number of star clusters by at least 1.

Otherwise, when $e_{ij} > \frac{c_i + c_j}{2}$, it has to be the case that $C(i)$ is a cluster of size 2 where both nodes are connected to nodes in $C(j)$. To see that, observe that if $C(i)$ was a star with at least 2 leaves, then since, by the discussion above, there are no edges connecting the leaves to nodes in $C(j)$, all the e_{ij} edges are adjacent to the star center. Since the endpoints of these e_{ij} edges in $C(j)$ must have distance at least 3 among themselves, as otherwise the girth would be less than 5, it cannot be that $e_{ij} > \frac{c_i + c_j}{2}$. It remains, therefore, to consider the case where $C(i)$ contains exactly two nodes and there are more than $\frac{2 + c_j}{2}$ edges connecting them with nodes in $C(j)$. First, observe that these e_{ij} endpoints are at distance at least 2 from each other, as otherwise the girth would be less than 5. Similarly, all endpoints connected to the same node in $C(i)$ must be at distance at least 3 from each other. Then, we can decompose $C(j)$ into tree clusters so that each tree cluster contains at most one endpoint connected to the same node in $C(i)$ and at most two endpoints in total. In the worst case, this decomposition leads to a social welfare of at least $\frac{c_j+2}{3}$, by considering a clustering where each cluster contains exactly two of the e_{ij} endpoints and has, therefore, utility at least $4/3$. Hence, the total social welfare of $C(i)$ and the tree decomposition of $C(j)$ is at least $1 + \frac{c_j+2}{3}$, i.e., $1 + \frac{n}{3}$, since $c_i + c_j = n$ and $c_i = 2$. The argument concludes by observing that maximum social welfare in the original cluster C is, due to Lemma 1, $SW(C) = \frac{2E(C)}{|C|} \leq \frac{1+\sqrt{4n-3}}{2}$, and since $\frac{1+\sqrt{4n-3}}{2} < 1 + \frac{n}{3}$ for any $n \geq 4$.

To conclude, we have shown that a non-star cluster C can be split into two clusters $C(i)$ and $C(j)$ where $C(i)$ is a star, without decreasing the social welfare. By iterating the process, we reach an optimal partition where all clusters are stars. □

The following result (Lemma 3 in [7]) states that in a triangle-free graph, the partition, that maximizes the social welfare among partitions consisting only of stars, is stable.

Lemma 3 (Due to [7]). *Let G be a triangle-free graph, then any optimal star clustering is stable.*

By combining Lemmas 2 and 3, we obtain the main result of this section.

Theorem 2. *The price of stability of simple symmetric fractional hedonic games on graphs of girth at least 5 is 1.*

3.3 Olsen's Utility Function

In this section we consider the alternate utility function considered by Olsen [19], i.e., the utility of a player i belonging to cluster $C(i)$ is $u_i'(C(i)) = \frac{\deg_{C(i)}(i)}{|C(i)|-1}$. We show that, under this utility function, the price of stability of simple symmetric fractional hedonic games is 1 by arguing about the structure of the optimal partition. In particular, we show that there exists an optimal partition satisfying a desirable structure, and then we argue that this partition is Nash stable. We begin with a technical lemma.

Lemma 4. *For any integers x, y, z such that $1 \leq x \leq y \leq z - 1$, it holds that $z^2 - 3z + xy - x^2 + x + y - yz + 2 \geq 0$.*

Proof. Fix z and let $f(x,y) = -x^2 + xy + x + (1-z)y + z^2 - 3z + 2$. It suffices to prove that $f(x,y) \geq 0$ for any $1 \leq x \leq y \leq z-1$. The derivative with respect to y is $f_y'(x,y) = x+1-z$. Note that $f_y'(x,y) < 0$ whenever $x < z-1$ and $f_y'(x,y) = 0$ only when $x = z-1$. In both cases, the value $y = z-1$ minimizes $f(x,y)$. Then, $f(x, z-1) = -x^2 + zx + 1 - z$ with $f(1, z-1) = 0$ and $f(z-1, z-1) = 0$. The proof follows since $f(x, z-1)$ is increasing up to $x = z/2$ and then becomes decreasing. □

The next lemma specifies the structure of an optimal partition. In particular, we show that there exists an optimal partition where each cluster C is either a singleton cluster (when $|C| = 1$), a path of two nodes (when $|C| = 2$), a star with 2 leaves or a triangle (when $|C| = 3$), or a star (whenever $|C| \geq 4$).

Lemma 5. *There exists an optimal partition C^* where each cluster C with $|C| \geq$ 4 is a star.*

Proof. Consider a cluster C with k nodes, with $k \geq 4$, that is not a star. Let $\ell = E(C)$ be the number of edges in C. Then, the social welfare is $\text{SW}(C) = \frac{2\ell}{k-1}$.

Let i be a node in C with the smallest degree and let $x = \deg(i)$; in case of ties, we pick i arbitrarily. Let j be the neighbor of i with the smallest degree among all i's neighbors and let $y = \deg(j)$; again, we pick j arbitrarily in case of ties. We argue that we can split C into two clusters, i.e., $C_1 = \{i, j\}$ and $C_2 = C \setminus C_1$ without decreasing the social welfare. The total social welfare of the two clusters is $\text{SW}' = \text{SW}(C_1) + \text{SW}(C_2) = 2 + \frac{2\ell'}{k-3}$, where ℓ' is the number of edges in cluster C_2. It holds that $\ell = \ell' + x + y - 1$, as i (respectively, j) has $x - 1$ (respectively, $y - 1$) neighbors in C_2 while the edge (i, j) also exists in C but not in C_2.

We now provide a lower bound on ℓ' based on x, y and k. By the definitions of x and y, it holds that C contains x nodes with degree at least y and $k - x$ nodes with degree at least x. Hence, we obtain that $\ell \geq \frac{xy + (k-x)x}{2}$. Since $\ell' = \ell - x - y + 1$, we obtain that

$$\ell' \geq \frac{kx + xy - x^2 - 2x - 2y + 2}{2}. \tag{5}$$

It suffices to prove that $\text{SW}' \geq \text{SW}(C)$. We have

$$
\begin{aligned}
\text{SW}' - \text{SW}(C) &= \frac{2k + 2\ell' - 6}{k - 3} - \frac{2\ell' + 2x + 2y - 2}{k - 1} \\
&= \frac{2k^2 - 6k + 4\ell' - 2kx - 2ky + 6x + 6y}{k^2 - 4k + 3} \\
&\geq \frac{2k^2 - 6k + 2kx + 2xy - 2x^2 - 4x - 4y + 4 - 2kx - 2ky + 6x + 6y}{k^2 - 4k + 3} \\
&= \frac{2k^2 - 6k + 2xy - 2x^2 + 2x + 2y - 2ky + 4}{k^2 - 4k + 3} \\
&\geq 0.
\end{aligned}
$$

The first inequality follows by using (5), while the last inequality holds due to Lemma 4 (by setting $z = k$) and since $k \geq 4$.

By repeating this process as long as there exists a non-star cluster of size at least 4, we obtain an optimal solution with the desired properties. $\qquad \square$

We now show that there exists an optimal partition that is Nash stable, hence the price of stability is 1.

Theorem 3. *The price of stability of simple symmetric fractional hedonic games is 1 when the utility function of player i is defined as $u_i'(C(i)) = \frac{\deg_{C(i)}(i)}{|C(i)|-1}$.*

Proof. Consider an optimal partition satisfying the properties of Lemma 5. Clearly, any node in a triangle and any node that is a root in a star is satisfied since its utility is 1. The only players that might wish to deviate are either nodes in singleton clusters, or leaves in a star. Observe that there cannot be an edge connecting two nodes from the set of leaves and singletons, otherwise they would form a new cluster and the social welfare would strictly increase; this contradicts our assumption that we begin from an optimal partition. Similarly, there cannot be an edge connecting a leaf or singleton node i to a node j that belongs to a cluster C forming a triangle, as then the social welfare would strictly increase by creating cluster $\{i,j\}$ and reducing C to $C \setminus \{j\}$.

We first let all singleton clusters deviate and join their preferred star. Since the social welfare of any star cluster C with k nodes is $\mathrm{SW}(C) = \frac{2(k-1)}{(k-1)}$, i.e., $\mathrm{SW}(C) = 2$ irrespective of the number of leaves, these deviations do not decrease the social welfare. Then, all possible subsequent deviating moves (which can be made only by leaves) lead to partitions where the number of triangles remains the same, the number of stars remains also the same, but the structure of these stars may change as the leaves deviate. Observe that a deviating move of node i from star $C(i)$ to another star C, requires that $|C(i)| > |C| + 1$ and strictly decreases the maximum size among these two star clusters, i.e., $|C(i)|$ in this example. Therefore, by considering the lexicographic order π of all star-clusters in the partition $\mathcal{C}$ based on the number of nodes, from the minimum to the maximum, we observe that any deviating move that leads to partition $\mathcal{C}'$ satisfies $\pi(\mathcal{C}) < \pi(\mathcal{C}')$ and, hence, this process is guaranteed to end. Furthermore, any deviating move does not decrease the social welfare, as the new clusters remain stars. This concludes the proof of the theorem.   □

4   Conclusions

We have presented new bounds on the price of stability of simple symmetric fractional hedonic games. The most important open question concerns the upper bound for general graphs. We conjecture that the price of stability is constant but the proof of such a claim remains elusive. Interestingly, for graphs of girth at least 5, we show that there exists an optimal partition that is stable. Note that for the remaining case, for graphs of girth at least 4, Bilò et al. [7] have shown a constant upper bound of 4 and a lower bound of approximately 1.003; they also show an improved upper bound of approximately 1.03 for the case of bipartite graphs. Whether triangle-free graphs indeed behave differently than bipartite graphs is another natural question; we have not been able to find a lower bound for non-bipartite triangle-free graphs with price of stability greater than 1.03.

References

1. Anshelevich, E., Dasgupta, A., Kleinberg, J.M., Tardos, E., Wexler, T., Roughgarden, T.: The price of stability for network design with fair cost allocation. SIAM J. Comput. **38**(4), 1602–1623 (2008)
2. Aziz, H., Brandt, F., Harrenstein, P.: Fractional hedonic games. In: Proceedings of the 13th International Conference on Autonomous Agents and Multiagent Systems (AAMAS), pp. 5–12 (2014)
3. Aziz, H., Brandt, F., Seedig, H.: Computing desirable partitions in additively separable hedonic games. Artif. Intell. **195**, 316–334 (2013)
4. Aziz, H., Gaspers, S., Gudmundsson, J., Mestre, J., Täubig, H.: Welfare maximization in fractional hedonic games. In: Proceedings of the 24th International Joint Conference on Artificial Intelligence (IJCAI), pp. 461–467 (2015)
5. Aziz, H., Savani, R.: Hedonic games. In: Brandt, F., Conitzer, V., Endriss, U., Lang, J., Procaccia, A. (eds.) Handbook of Computational Social Choice. Cambridge University Press, Cambridge (2016)
6. Bilò, V., Fanelli, A., Flammini, M., Monaco, G., Moscardelli, L.: Nash stability in fractional hedonic games. In: Liu, T.-Y., Qi, Q., Ye, Y. (eds.) WINE 2014. LNCS, vol. 8877, pp. 486–491. Springer, Heidelberg (2014)
7. Bilò, V., Fanelli, A., Flammini, M., Monaco, G., Moscardelli, L.: On the price of stability of fractional hedonic games. In: Proceedings of the 14th International Conference on Autonomous Agents and Multiagent Systems (AAMAS), pp. 1239–1247 (2015)
8. Bloch, F., Diamantoudi, E.: Noncooperative formation of coalitions in hedonic games. Int. J. Game Theory **40**(2), 262–280 (2011)
9. Bogomolnaia, A., Jackson, M.O.: The stability of hedonic coalition structures. Games Econ. Behav. **38**(2), 201–230 (2002)
10. Branzei, S., Larson, K.: Social distance games. In: Proceedings of the 22nd International Joint Conference on Artificial Intelligence (IJCAI), pp. 273–279 (2011)
11. Brandl, F., Brandt, F., Strobel, M.: Fractional hedonic games: individual and group stability. In: Proceedings of the 14th International Conference on Autonomous Agents and Multi-Agent Systems (AAMAS), pp. 1219–1227 (2015)
12. Drèze, J.H., Greenberg, J.: Hedonic coalitions: optimality and stability. Econometrica **48**(4), 987–1003 (1980)
13. Elkind, E., Wooldridge, M.: Hedonic coalition nets. In: Proceedings of the 8th International Conference on Autonomous Agents and Multi-Agent Systems (AAMAS), pp. 417–424 (2009)
14. Feldman, M., Lewin-Eytan, L., Naor, J.: Hedonic clustering games. ACM Trans. Parallel Comput. **2**(1), Article 4 (2015)
15. Furedi, Z.: Graphs without quadrilaterals. J. Comb. Theory, Ser. B **34**, 187–190 (1983)
16. Gairing, M., Savani, S.: Computing stable outcomes in hedonic games. In: Proceedings of the 3rd International Symposium on Algorithmic Game Theory (SAGT), pp. 174–184 (2010)
17. Kővári, T., Sós, V.T., Turán, P.: On a problem of K. Zarankiewicz. Colloquium Mathematicae **3**, 50–57 (1954)
18. Olsen, M.: Nash stability in additively separable hedonic games and community structures. Theory Comput. Syst. **45**(4), 917–925 (2009)
19. Olsen, M.: On defining and computing communities. In: Proceedings of the 18th Computing: Australasian Theory Symposium (CATS), pp. 97–102 (2012)

20. Peters, D., Elkind, E.: Simple causes of complexity in hedonic games. In: Proceedings of the 24th International Joint Conference on Artificial Intelligence (IJCAI), pp. 617–623 (2015)
21. Reiman, I.: Über ein problem von K. Zarankiewicz. Acta Mathematica Academiae Scientiarium Hungaricae **9**, 269–278 (1958)
22. Saad, W., Han, Z., Basar, T., Debbah, M., Hjorungnes, A.: Hedonic coalition formation for distributed task allocation among wireless agents. IEEE Trans. Mob. Comput. **10**(9), 1327–1344 (2011)

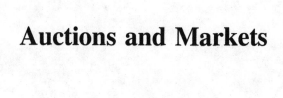

Auctions and Markets

Truthfulness and Approximation with Value-Maximizing Bidders

Salman Fadaei and Martin Bichler[✉]

Department of Informatics, Technical University of Munich, Munich, Germany
salman.fadaei@gmail.com, bichler@in.tum.de

Abstract. In many markets bidders want to maximize value rather than payoff. This is different to the quasi-linear utility functions, and leads to different strategies and outcomes. We refer to bidders who maximize value as *value bidders*. While simple single-object auction formats are truthful for value bidders, standard multi-object auction formats allow for manipulation. It is straightforward to show that there cannot be a truthful and revenue-maximizing deterministic auction mechanism with value bidders and general valuations. Using approximation as a means to achieve truthfulness, we study truthful approximation mechanisms for value bidders. We show that the approximation ratio that can be achieved with a deterministic and truthful approximation mechanism with n bidders and m items cannot be higher than $1/n$ for general valuations. For randomized approximation mechanisms there is a framework with a ratio of $O(\frac{\sqrt{m}}{\epsilon^3})$ with probability at least $1 - \epsilon$, for $0 < \epsilon < 1$.

Keywords: Value bidders · Revenue · Truthfulness · Approximation mechanisms

1 Introduction

In auction theory, bidders are typically modeled as payoff-maximizing individuals using a quasilinear utility function. Under these utility functions the Vickrey-Clarke-Groves mechanism is the unique mechanism to obtain maximum welfare in dominant strategies. Sometimes, however, payoff-maximization might just not be the right assumption and bidders rather maximize value subject to a budget constraint.

In *display ad auctions* individual user impressions on a web site are auctioned off. Advertising buyers bid on an impression and, if the bid is won, the buyer's ad is instantly displayed on the publisher's site. Demand-side platforms (DSPs) are intermediaries, who provide the technology to bid for advertisers on such advertising exchanges. A number of papers describe bidding strategies and heuristics in display ad auctions. Zhang [14] gives an up-to-date overview. In all

---

M. Bichler—A full version of this paper is available at http://arxiv.org/abs/1607.03821. The financial support from the Deutsche Forschungsgemeinschaft (DFG) (BI 1057/1-4) is gratefully acknowledged.

© Springer-Verlag Berlin Heidelberg 2016
M. Gairing and R. Savani (Eds.): SAGT 2016, LNCS 9928, pp. 235–246, 2016.
DOI: 10.1007/978-3-662-53354-3_19

of these papers the task of the DSP or advertiser is to maximize the values of impressions typically subject to a budget constraint for a campaign.

Value maximization subject to a budget is not limited to display ad auctions. Private individuals often determine a budget before making a purchase, and then buy the best item or set of items (e.g., cars, real-estate) that meets the budget. Actually, in classical micro-economic consumer choice theory, consumers select a package of objects that maximizes value subject to a budget constraint, they don't maximize payoff. Maximizing value subject to a budget constraint is also wide-spread in business due to principal-agent relationships. For example, in spectrum auctions, national telecoms have different preferences for different packages of spectrum licenses based on the corresponding net present values of business cases. These billion dollar net present values exceed the financial capabilities of the local telecom by far, but not those of its stakeholder, a multinational, which has mainly long run strategic incentives of operating in the local market. Thus, the stakeholder provides the local telecom with allowances for individual packages based on the underlying net present value. The local management then tries to win the most valuable package within the allowances provided by the stakeholder.

In this paper, we analyze truthful approximation mechanisms for value maximizing bidders. One can think of several market types with value bidders depending on the nature of the preferences for one or more objects and of the budgets: single-object markets, assignment markets, and combinatorial markets, where bidders have preferences for and can bid on packages of objects.

We discuss truthful mechanisms for a number of these environments, but focus on approximation mechanisms for combinatorial auction markets (with a fully expressive XOR bid language). Combinatorial auction markets allow for general preferences including substitutes and complements and the efficiency of the mechanism is not limited just due to restrictions in the expressiveness of the bid language.

1.1 Our Results

First, we analyze a truthful Pareto-optimal mechanism for markets with value bidders. We show that such a mechanism exists. Then we study truthful revenue maximizing mechanisms. We focus on revenue rather than welfare. Social welfare is difficult to analyze in environments where bidders have values and budget constraints. We show that for single-minded and single-valued value bidders there are simple truthful mechanisms that maximize revenue, but that this is not possible for multi-minded value bidders. Next, we explore truthful approximation mechanisms. We maximize revenue without considering incentives, and refer to this as optimal revenue. We will then say that a strategy-proof mechanism returns (at least) a ratio α of the optimal if it's revenue is always greater than or equal to α times the optimal revenue. Our first main result concerns deterministic approximation mechanisms for multi-unit package auctions.

Theorem. *The best revenue ratio achievable by a deterministic and truthful mechanism with value bidders in a market with n bidders and m homogeneous items is $\frac{1}{n}$, for any $n \geq 2$ and $m \geq 2$.*

The theorem has a straightforward extension to combinatorial markets. In quasi-linear mechanism design, randomization is often a remedy to achieve higher approximation ratios. Approximation mechanisms for quasi-linear bidders do typically not lead to strategy-proofness with value bidders. However, there is a recent contribution by Dobzinski et al. [4], which is also truthful for value bidders with a simple change of the payment rule.

Theorem. *There exists a polynomial-time randomized mechanism for value bidders which is universally truthful and guarantees an approximation ratio of $O(\frac{\sqrt{m}}{\epsilon^3})$ with probability at least $1 - \epsilon$, for $0 < \epsilon < 1$.*

1.2 Related Literature

The Gibbard-Satterthwaite theorem describes one of the most celebrated results in social choice theory. Gibbard [6] proved that any non-dictatorial voting scheme with at least three possible outcomes is not strategy-proof. Satterthwaite [13] showed that if a committee is choosing among at least three alternatives, then every strategy-proof voting procedure is dictatorial. There have been a number of papers on multi-unit assignment problems without money, showing that the only strategy-proof and Pareto-optimal mechanisms are serial dictatorships.

Closest to our assumptions is the model analyzed by Feldman [5] in which bidders have an overall budget and a value for ad slots in sponsored search and they want to maximize the number of clicks given their budget. They also argue that a bidder is incentivized to spend the entire budget to maximize exposure or the number of clicks in the market. Maximizing the total value of clicks is actually well-motivated also in more recent literature on bidding heuristics for display ad auctions [14].

2 Preliminaries and Notations

In a combinatorial market we have m non-homogeneous items, J, one seller 0, and n bidders, I (I_0 includes the seller). Each bidder $i \in I$ has a valuation $v_i(a_i)$ for any package $a_i \subseteq a$ assigned to an agent i and allocation $a \in A$, where A describes the set of all allocations. For brevity, we will drop the subscript in a_i and write $v_i(a)$, even though the bidder is only interested in his own allocation and not the allocation overall. A feasible allocation of bundles of items to bidders is described as $a = \bigcup_{i \in I} a_i$ with $\bigcap_{i \in I} a_i = \emptyset$ and $a \in A$. When we discuss combinatorial auctions with heterogeneous items, we assume general cardinal valuations and allow for substitutes and complements and free disposal $(v_i(S) \leq v_i(T), S \subseteq T \subseteq J)$. In contrast to mechanism design with quasi-linear utility functions where utility is defined as valuation minus price of a bundle, $u_i(a) = v_i(a) - p_i(a)$, we assume that these bidders have no value for residual budget or payoff, but they want to win their highest-valued package subject to some budget constraint b_i.

There are different ways, how budget can be considered. In combinatorial markets the value of larger packages of objects a can exceed the overall budget constraint $v_i(a) > b_i(a)$. This can also happen in single object auctions. In this paper we are concerned with maximizing revenue, and we consider the willingness-to-pay for a package rather than the true value and trim the valuation of such packages to the amount of the overall budget constraint b_i. For example, a telecom in a spectrum auction market might have a very high value for the package of all spectrum licenses in the market, but it has only one million dollar budget available. Therefore, we assume that his value, i.e., willingness-to-pay, for the package of all licenses is $v_i = \$1$ million. In summary, the value bidders' utility function is $u_i(a) = v_i(a)$ if $p_i(a) \leq v_i(a)$, and $u_i(a) = -\infty$ otherwise. This means, utility is non-transferable between the value bidders in our model. The values v_i are assumed to be monotone non-decreasing (free disposal) and they are normalized with $v_i(\emptyset) = 0$, $\forall i \in I$.

Note that value bidders would not bid beyond their valuation $v_i(a)$, even if their overall budget b_i is not binding. For example, an advertiser on a digital advertising exchange does not want to have an allocation which is within his overall budget b_i, but where he has to pay more for every impression than what the net present value for these impressions is. In spectrum auctions, principals typically determine a budget for different packages, which is based on the net present value of the licenses in the package. The management needs to consider these limits and cannot bid beyond.

We consider an offline environment, where we can match bidders to objects in a single step. We will start discussing the simplest variant of a combinatorial auction market in our paper: the multi-unit package auction. In a multi-unit auction we have m identical units of an item. We use the notation of an $n \times m$ auction to point to a multi-unit auction with n bidders and m identical items or units. In a multi-unit auction we use the notation $(s_1, s_2, \ldots, s_n)$ to denote an allocation in which s_i units are assigned to bidder i. We focus on package auctions, because without package bids, bidders cannot express their preferences for complements or substitutes, which can lead to arbitrarily low revenue with general valuations. Moreover, in a multi-unit market we use $v_i(s)$ to describe the valuation of bidder i on s units. When an auctioneer presents a bidder with a package s, and the bidder responds with his value $v_i(s)$, we will also refer to this as a *value query*. The optimization goal is to find an allocation of objects to the bidders, where bidder i gets s_i units, with $\sum_i s_i \leq m$, that maximizes $\sum_i v_i(s_i)$.

Definition 1. *A (direct revelation) mechanism is a social choice function $f : V_1 \times \ldots \times V_n \to A$ and a vector of payment functions $p_1, \ldots, p_n$, where $p_i : V_1 \times \ldots \times V_n \to \Re$ is the amount that player i pays.*

$V_i \subseteq \Re^A$ describes the set of possible valuation functions for bidder i. We sometimes refer to the social choice function as the allocation rule of a mechanism.

Definition 2. *A mechanism* $(f, p_1, ..., p_n)$ *is called incentive compatible if for every bidder* i, *every* $v_1 \in V_1, ..., v_n \in V_n$ *and every* $v_i' \in V_i$, *if we denote* $a = f(v_i, v_{-i})$ *and* $a' = f(v_i', v_{-i})$, *then* $u_i(a) \geq u_i(a')$.

We will also talk about a *truthful* or *strategy-proof* mechanism in this context when truthtelling is a dominant strategy.

Desirable goals in mechanism design are *Pareto optimality*, the maximization of *social welfare*, and *revenue*. Utilitarian social welfare functions add up the value of each individual in order to obtain society's overall welfare. As indicated earlier, the notion of social welfare is difficult if bidders have budget constraints. In this paper we focus on maximizing the auctioneers' revenue. In other words, $v_i(a)$ is the willingness-to-pay of a bidder i for a package a (eventually trimmed to b_i), and the auctioneer wants to maximize the sum of these values in the allocation: $f_R = \max_{a \in A} \sum_{i \in I_0} v_i(a)$. Assuming that every bidder is telling the truth, f_R states the optimal revenue or is optimal, for short.

The other desirable goal in mechanism design is Pareto-optimality.

Definition 3. *A pair of allocation and payments* $(a, p_1, ..., p_n)$ *is Pareto-optimal if for no other pair* $(a', p_1', ..., p_2')$ *are all bidders and seller better off,* $u_i(a') \geq u_i(a)$, *including the seller* $\sum_{i \in I} p_i' \geq \sum_{i \in I} p_i$, *with at least one of the inequalities strict.*

A quasi-linear mechanism is Pareto-optimal if in equilibrium it selects an allocation or choice a such that $\forall i \forall a', \sum_i v_i(a) \geq \sum_i v_i(a')$. Therefore, an allocation that solves the social welfare maximization problem is Pareto optimal. In a non-quasi-linear environment with value bidders, maximizing social welfare is a sufficient, but not a necessary condition for Pareto-optimal allocations with value bidders.

Example 1. To see this consider an example with bidder 1 interested in item A for \$8 and B for \$5, and a bidder 2 with a value of \$7 for A and \$6 for B. If the auctioneer allocates B to bidder 1 and A to bidder 2 with the price equal to their bid, then the utility of the auctioneer and the two bidders would be $(12, 5, 7)$ rather than $(14, 8, 6)$ in the social welfare maximizing allocation. Both allocations are Pareto-optimal, though.

A mechanism is *individually rational* if bidders always get nonnegative utility. Most of our analysis focuses on individually rational and truthful approximation mechanisms for value bidders. The algorithmic problem of finding the optimal social welfare for general valuations in combinatorial auctions is $O(\sqrt{m})$ [7], which is a natural upper bound on the approximation factor of truthful approximation mechanisms. For quasi-linear bidders randomized approximation mechanisms with the same approximation ratio have been found [4,10]. However, the best deterministic truthful approximation guarantee known for general combinatorial auctions is $O(\frac{m}{\sqrt{\log m}})$ [8].

The algorithmic problem of allocating multiple units of an item to multiminded bidders reduces to the knapsack problem, for which a simple greedy algorithm proves an approximation ratio of 2 [10]. Just like for the knapsack problem, the algorithmic allocation problem can be approximated arbitrarily well and has an FPTAS: approximation ratio of $1 + \epsilon$ obtained in time that is polynomial in n, $\log m$, and ϵ^{-1}. For quasi-linear bidders, the framework by [10] can be used such that any approximation algorithm witnessing an LP integrality gap can be transformed into an algorithm that is truthful in expectation. For the multi-unit auction problem the integrality gap is 2 and, hence, the framework of Lavi and Swamy gives a 2-approximation. [3] presented an FPTAS for multi-unit auctions that is truthful in expectation for quasi-linear bidders.

3   Pareto-Optimality and Revenue Maximization

We will first analyze if revenue maximization and Pareto efficiency can be implemented in dominant strategies with value bidders. Due to the revelation principle, we limit ourselves to direct revelation mechanisms.

3.1   Truthful and Pareto-Optimal Mechanisms

Because our setting is similar to the setting without money, achieving strategy-proof and Pareto-optimal mechanisms might seem impossible at first sight. However, payments are available and provide an escape route from the many impossibility results in mechanism design without money. In the following we show that there exists a Pareto-optimal and strategy-proof mechanism for the problem, which is a simple greedy algorithm similar in spirit to [12].

Definition 4 (PO auction). *Given a set of items J and bidders I, find the $i \in I$ and $S \subseteq J$ with the highest $v_i(S)$. Allocate S to i at the price equal to $v_i(S)$, and recurse on $(J \setminus S)$ and $(I \setminus \{i\})$.*

For the following theorem, we assume that bidders have strict valuations. That is, for all bundles S, and T in the bidder i's demand set, we have that $v_i(S) \neq v_i(T)$, for all bidders $i \in I$. This is a reasonable assumption for example in spectrum auctions, where it is unlikely that two different packages have the same value. We will discuss truthful mechanisms for general valuations including ties in Sect. 4.

Theorem 1. *The PO auction is a deterministic, strategy-proof, and Pareto-optimal mechanism for value bidders with strict valuations.*

All proofs can be found in the a long version of the paper. It is easy to see that the mechanism achieves the best possible revenue in multi-unit markets with linear valuations. However, the revenue of the mechanism in general can be as low as $\frac{1}{n}$ of the optimal revenue (see Example 2).

Example 2. Suppose in a multi-unit market, bidders have valuations $v_i(1) = x + \epsilon, v_i(2) = x + 2\epsilon, \ldots, v_i(m) = x + m \cdot \epsilon$, for any $\epsilon > 0$, and $i \in I$. With these valuations, the mechanism will return revenue of $x + m \cdot \epsilon$. But the optimal revenue can be higher than $n \cdot x$. Thus, the mechanism returns a result with a revenue lower than $\frac{1}{n}$ of the optimal revenue.

Note that if the mechanism chooses the revenue-maximizing allocation based on the bids it would not be strategy-proof any more because bidders could shade lower-valued bids in order to get a higher valued package, as we will discuss in the next section. We will discuss the goal of revenue maximizing mechanisms next.

3.2   Truthful Mechanisms Maximizing Revenue

Let's first consider the trivial case of a single item only. A direct pay-as-bid revelation simply gives the item to the highest bidder. Reporting the truth is a weakly dominant strategy for this mechanism. All bidders will report $v_i(a)$, since they pay what they bid but have no value for payoff. The losing bidders cannot gain from decreasing their bid, but would risk making a loss if they bid beyond their valuation. The same result can be extended to *single-minded* and *single-valued* value bidders.

Definition 5 *[12]. A valuation v is called single-minded if there exists a bundle of items S^* and a value $v^* \in \mathbb{R}$ such that $v(S) = v^*$ for all $S \supseteq S^*$ and $v(S) = 0$ for all other S. A single-minded bid is the pair (S^*, v^*).*

Definition 6 *[1]. A bidder i is a single-valued (multi-minded) value bidder if there exists a real value $v_i > 0$ such that for any bundle $S \in J$, $v_i(S) \in \{0, v_i\}$ and $v_i(T) = v_i$ for all $T \supseteq S$ if $v_i(S) = v_i$. Both the bidder's value and his collection of desired bundles are assumed to be private information, known only to the bidder himself.*

Theorem 2. *A mechanism $(f, p_1, \ldots, p_n)$ with single-valued or single-minded value bidders is incentive compatible if and only if the following conditions hold:*

1. f is monotone in every v_i.
2. Every winning bidder pays his bid.

Notice that with single-minded or single-valued value bidders the winner determination problem to select the revenue-maximizing allocation is still NP-hard [11]. Theorem 2 allows us to use the existing monotone approximation algorithms for the allocation problem in markets with single-minded or single-valued value bidders which are such that the computational hardness cannot be ignored. With a pay-as-bid payment rule such single-parameter value bidders would not have an incentive to deviate from truthful bidding. Unfortunately, these positive results do not carry over to multi-minded value bidders with general valuations.

Theorem 3. *There is no strategy-proof and revenue maximizing auction mechanism for general value bidders and more than one object for sale.*

In the quasi-linear setting, assignment markets allow even for strategy-proof ascending auctions [2]. Unfortunately, the negative result in Theorem 3 even holds for assignment markets, where value bidders can only win at most one from multiple items.

Corollary 1. *There is no strategy-proof and revenue maximizing auction mechanism for value bidders in assignment markets.*

Proof. Suppose there is a market with two items A and B and two bidders. Bidder 1 has a value of x for item B, while bidder 2 has a value of x for item A and a value of $x + \epsilon$ for item B. Bidder 2 can increase his utility by bidding 0 for item A, which would make him win item B and lead to a revenue of $\frac{1}{2}$ of the optimal.

One escape route from these negative results on strategy-proof and revenue maximizing auctions is to give up on optimal solutions and restrict attention to approximation mechanisms to achieve strategy-proofness. In other words, we try to keep strategy-proofness at the expense of optimal revenue. There is a growing literature on approximation mechanisms for quasi-linear bidders [9] and it is a natural question to understand approximation mechanisms for value bidders.

4   Deterministic Mechanisms for General Value Bidders

In this section, we characterize properties of truthful revenue maximizing mechanisms for value bidders in general. This allows us to answer the question whether an approximation ratio better than $\frac{1}{n}$ is possible for markets with n bidders. This is what we can achieve when bidders are only allowed to submit bids on the grand bundle (the bundle of all items). We try to analyze this by characterizing the allocation rule of all possible truthful mechanisms. We present our negative results for multi-unit auctions which are a subset of combinatorial auctions.

Generally speaking, with value bidders at most one value query for each bidder can be verified by a truthful mechanism. This severely restricts the possibility of designing deterministic truthful mechanisms. One observation from the golden ratio mechanism and the randomized 2×2 mechanism is that only a single value query for one package S is used by the auctioneer, i.e., only a single value $v_i(S)$ of each bidder is considered and then the bidder can either win this package or a lower valued one. An example for another allocation rule f in a 3×4 market which satisfies the same properties is: "if $v_1(4) > v_2(2) + v_3(2)$ then $(4, 0, 0)$ else $(0, 2, 2)$." In this example, each bidder can only win the package, which is evaluated in the condition of the allocation rule or the empty set. No bidder has an incentive to lie in the value query, because this package is assigned to the bidder.

4.1   Allocation Rule Revisited

We restrict our attention to those valuations of bidders which make the allocation rule assign all units to only one bidder. These valuations play an essential role in defining the outcome of the allocation rule and we discuss this further in this section. We first define two new notations. Let $v = (v_1, \ldots, v_n) \in V_1 \times \ldots \times V_n$ point to an arbitrary set of valuations. We denote by $(i \hookleftarrow s)$ the assignment in which s units are assigned to bidder i. Given an allocation rule $f : V_1 \times \ldots \times V_n \to A$, we define a new function

$$F_i : V_1 \times \ldots \times V_n \to \{true, false\}$$
$$F_i(v) = \begin{cases} true & \text{if } f(v) = (i \hookleftarrow m), \\ false & \text{otherwise.} \end{cases}$$

Intuitively speaking, F_i determines whether a set of bidders' valuations will result in the assignment of m units to bidder i. Notice, $F_i(\cdot)$'s are disjunctive, i.e. $F_i(v)$ & $F_j(v) = false, \forall i, j \in I$, because the grand bundle cannot be assigned to more than one bidder, simultaneously. Now, using F_i's we redefine the allocation rule as follows.

$$f^* : V_1 \times \ldots \times V_n \to A$$
$$f^*(v) = \begin{cases} (i \hookleftarrow m) & \text{if } F_i(v) = true, \forall i \in I, \\ f(v) & \text{otherwise.} \end{cases}$$

Example 3. Consider the following allocation rule for the 3×4 market.

1. If $v_1(4) > \max(v_2(4), v_3(4)) \wedge (v_1(4) > v_2(2) + v_3(2))$ then $(4, 0, 0)$.
2. If $v_2(4) > \max(v_1(4), v_3(4)) \wedge (v_2(4) > v_1(2) + v_3(2))$ then $(0, 4, 0)$.
3. If $v_3(4) > \max(v_1(4), v_2(4))$.
4. If $v_1(4) > \max(v_2(4), v_3(4))$ then $(0, 2, 2)$.
5. If $v_2(4) > \max(v_1(4), v_3(4))$ then $(2, 0, 2)$.

In this allocation rule, the functions $F_i(\cdot)$'s are defined as follows. $F_1(v) \equiv v_1(4) > \max(v_2(4), v_3(4)) \wedge (v_1(4) > v_2(2) + v_3(2))$, $F_2(v) \equiv v_2(4) > \max(v_1(4), v_3(4)) \wedge (v_2(4) > v_1(2) + v_3(2))$, and $F_3(v) \equiv v_3(4) > \max(v_1(4), v_2(4))$. These functions are disjunctive and all might become false simultaneously.

This new rewriting of the allocation rule will be useful for deriving a general result in mechanism design for value bidders as will be shown in the following.

It is easy to observe that the two definitions of the allocation rule are equivalent.

Lemma 1. *For any set of valuations $v \in V_1 \times \ldots \times V_n$, we have that $f^*(v) = f(v)$.*

That f^* and f are equivalent, lets us focus on f^* and try to find conditions which f^* must satisfy in order to achieve a good revenue as well as obtaining truthfulness.

4.2 Properties of $F_i(\cdot)$

The domain of function $F_i(\cdot)$ is the set of all valuations. Thus, one might guess that in computing $F_i(\cdot)$, the valuations of bidders for any bundle $j \leq m$ might be queried. Yet, in what follows we present multiple lemmata which show that this is not the case and restrict the arguments of $F_i(\cdot)$ to only valuations of bidders for m units: $v_j(m) \; \forall j \in I$.

We first look at the lemma which describes properties of mechanisms, which avoid low revenue.

Lemma 2. *In order to avoid arbitrarily low revenues of less than $\frac{1}{n}$, arguments of function $F_i(\cdot)$ must include all value queries $v_j(m)$, $\forall j \in I$, i.e. $F_i(\cdot)$ has to be a function of all $v_j(m)$'s.*

The proofs for the following lemmata can be found in the long version of the paper. The second lemma holds only for value bidders. This lemma takes into account the truthfulness of mechanism.

Lemma 3. *In order to obtain truthfulness, no valuation from bidder i other than $v_i(m)$ can be queried in computing $F_i(\cdot)$.*

Lemma 4. *If two bidders are equal in all valuations except for the grand bundle, then in case the mechanism wishes to assign the grand bundle to one of them, it must assign the grand bundle to the stronger bidder, otherwise the mechanism will be neither revenue maximizing nor truthful.*

The next lemma proves that, by taking into consideration the truthfulness, if all units are assigned to a bidder, that bidder has to be the strongest on the grand bundle.

Lemma 5. *If $F_i(v) = true$ then $v_i(m) > \max_{j \neq i} v_j(m)$, $\forall i, j \in I$, otherwise the mechanism is not truthful.*

Considering Lemma 5, we can be more specific about the function $F_i(\cdot)$ as the following lemma states.

Lemma 6. *The rewriting of function $F_i(\cdot)$ as the following is without loss of generality. $F_i(v) = (v_i(m) > \max_{j \neq i} v_j(m)) \wedge F_i'(v)$, where*

$$F_i' : V_1 \times \ldots \times V_n \to \{true, false\}$$
$$F_i'(v) = \begin{cases} true & \text{if } F_i(v) = true, \\ false & \text{if } F_i(v) = false \wedge v_i(m) > \max_{j \neq i} v_j(m), \\ don'tcare & v_i(m) < \max_{j \neq i} v_j(m). \end{cases}$$

In the next lemma, we show that when a bidder has the highest valuation for the grand bundle, he will get the all units. In order to show this, we draw on a property of social choice functions, namely anonymity. Anonymity requires that the outcome of a social choice function is unaffected when agents are renamed.

Lemma 7. *If $v_i(m) > \max_{j \neq i} v_j(m)$ then $F_i(v) = true$, $\forall i, j \in I$, otherwise the mechanism is not truthful.*

Example 4. Consider the allocation rule given in Example 3. Obviously, the allocation rule is not anonymous since the outcome depends on the label of the bidders. If bidder 3 has the highest valuation for the grand bundle, he will get it but for bidder 1 and 2 this does not hold.

In addition, the mechanism is not truthful. Consider a case in which $v_1(4) > v_2(4) > v_3(4)$, $v_1(4) < v_2(2) + v_3(2)$, and $v_2(4) < v_1(4) + v_3(2)$. The outcome of the mechanism in this case will be $(0, 2, 2)$. But, bidder 1 can be better off by bidding $v'_1(4) < v_2(4)$, for which the outcome will be $(2, 0, 2)$.

Theorem 4. *The best revenue ratio achievable by a deterministic and truthful mechanism with value bidders in an $n \times m$ market is $\frac{1}{n}$.*

Proof. According to Lemma 7, the allocation rule assigns everything to the bidder with the highest bid for the grand bundle. This assignment has a worst-case approximation ratio $\frac{1}{n}$.

Theorem 4 is easily extensible to combinatorial markets as the following corollary states. Overall, there is a gap between the best approximation ratio of an approximation algorithm to the set packing problem $(O(\sqrt{m})$, and the best truthful approximation mechanism.

Corollary 2. *The best revenue ratio achievable by a deterministic and truthful mechanism with value bidders in a combinatorial market is $\frac{1}{n}$.*

5   A Randomized Mechanism for General Value Bidders

Interestingly, the general randomized framework designed by Dobzinski et al. [4] for mechanism design in quasi-linear settings, can easily be adapted for value bidders. We will refer to this framework as U. Therefore, we can also achieve a randomized mechanism for combinatorial auctions with value bidders, which is truthful in a universal sense.

The framework tries to distinguish two cases: either there is a dominant bidder such that allocating all items to him is a good approximation to the revenue, or there is no such bidder. In the first case all items will be assigned to a bidder. In the second case, a fixed-price auction is performed, which achieves a good approximation. In a first phase of the auction bidders are partitioned randomly in three sets. One of these sets is then used to gather statistics in phase II, which allow to set a reserve price in the second-price auction of phase III, which only allocates all items to one of the bidders. If the reserve price is not met in phase III, then the items are sold in the fixed-price auction in phase IV.

Theorem 5. *A randomized mechanism according to framework U for value bidders is universally truthful and runs in polynomial time. It guarantees an approximation ratio of $O(\frac{\sqrt{m}}{\epsilon^3})$ with probability at least $1 - \epsilon$.*

A randomized mechanism according to framework U is also tight as it obtains an $O(\sqrt{m})$-approximation of the optimal revenue for general bidder valuations. This also shows, that there exists a gap between the power of randomized versus deterministic mechanisms. Whether such a gap exists for quasi-linear mechanism design is an open problem [9].

References

1. Babaioff, M., Lavi, R., Pavlov, E.: Single-value combinatorial auctions and algorithmic implementation in undominated strategies. J. ACM **56**(1), 1–32 (2009)
2. Demange, G., Gale, D., Sotomayor, M.: Multi-item auctions. J. Polit. Econ. **94**, 863–872 (1986)
3. Dobzinski, S., Dughmi, S.: On the power of randomization in algorithmic mechanism design. In: 50th Annual IEEE Symposium on Foundations of Computer Science, FOCS 2009, pp. 505–514. IEEE (2009)
4. Dobzinski, S., Nisan, N., Schapira, M.: Truthful randomized mechanisms for combinatorial auctions. J. Comput. Syst. Sci. **78**(1), 15–25 (2012)
5. Feldman, J., Muthukrishnan, S.M., Nikolova, E., Pál, M.: A truthful mechanism for offline ad slot scheduling. In: Monien, B., Schroeder, U.-P. (eds.) SAGT 2008. LNCS, vol. 4997, pp. 182–193. Springer, Heidelberg (2008)
6. Gibbard, A.: Manipulation of voting schemes: a general result. Econometrica **41**, 587–601 (1973)
7. Halldorsson, M., Kratochvil, J., Telle, J.: Independent sets with domination constraints. Discrete Appl. Math. **99**(1–3), 39–54 (2000)
8. Holzman, R., Kfir-Dahav, N., Monderer, D., Tennenholtz, M.: Bundling equilibrium in combinatorial auctions. Games Econ. Behav. **47**(1), 104–123 (2004)
9. Lavi, R.: Computationally efficient approximation mechanisms. In: Algorithmic Game Theory, pp. 301–329 (2007)
10. Lavi, R., Swamy, C.: Truthful and near-optimal mechanism design via linear programming. J. ACM (JACM) **58**(6), 25 (2011)
11. Lehmann, D., Mueller, R., Sandholm, T.: The winner determination problem. In: Cramton, P., Shoham, Y., Steinberg, R. (eds.) Combinatorial Auctions. MIT Press, Cambridge (2006)
12. Lehmann, D., Oćallaghan, L.I., Shoham, Y.: Truth revelation in approximately efficient combinatorial auctions. J. ACM (JACM) **49**(5), 577–602 (2002)
13. Satterthwaite, M.A.: Strategy-proofness and arrow's conditions. existence and correspondence theorems for voting procedures and social welfare functions. J. Econ. Theory **10**(2), 187–217 (1975)
14. Zhang, W., Yuan, S., Wang, J.: Optimal real-time bidding for display advertising. In: Proceedings of the 20th ACM SIGKDD International Conference on Knowledge Discovery and Data Mining, KDD 2014, pp. 1077–1086. ACM, New York (2014)

Envy-Free Revenue Approximation
for Asymmetric Buyers with Budgets

Evangelos Markakis[1] and Orestis Telelis[2(✉)]

[1] Department of Informatics, Athens University of Economics and Business,
Athens, Greece
markakis@gmail.com
[2] Department of Digital Systems, University of Piraeus, Piraeus, Greece
telelis@gmail.com

Abstract. We study the computation of revenue-maximizing envy-free outcomes in a monopoly market with budgeted buyers. Departing from previous works, we focus on buyers with asymmetric combinatorial valuation functions over subsets of items. We first establish a hardness result showing that, even with two identical additive buyers, the problem is inapproximable. In an attempt to identify tractable families of the problem's instances, we introduce the notion of *budget compatible* buyers, placing a restriction on the budget of each buyer in terms of his valuation function. Under this assumption, we establish approximation upper bounds for buyers with submodular valuations over preference subsets as well as for buyers with identical subadditive valuation functions. Finally, we also analyze an algorithm for arbitrary additive valuation functions, which yields a constant factor approximation for a constant number of buyers. We conclude with several intriguing open questions regarding budgeted buyers with asymmetric valuation functions.

1 Introduction

We study a pricing problem in a monopoly market involving n buyers and m distinct goods (items) on sale. Each buyer has preferences over subsets of items, expressed through a combinatorial valuation function; each buyer is also associated with a scalar budget, that constrains his monetary capacity. The budget constitutes an exogenous constraint on liquidity expressing the fact that, no matter how high any buyer may value a subset of items, the amount of money that he can readily spend on it may be significantly lower. Under this setting, our goal is to produce an allocation of items to the buyers along with corresponding payments, so as to maximize the achieved revenue. We study this problem subject to *(i)* the buyers' budget constraints, *(ii)* individual rationality and *(iii)* envy-freeness among the buyers. Individual rationality ensures that the utility of every buyer should be non-negative, in any feasible outcome, and envy-freeness imposes that no buyer may increase his utility by acquiring the allocation of another buyer at the price paid by that buyer. Given that the problem is generally **NP**-hard, we wish to investigate both algorithmic and hardness results on approximating the optimal revenue.

© Springer-Verlag Berlin Heidelberg 2016
M. Gairing and R. Savani (Eds.): SAGT 2016, LNCS 9928, pp. 247–259, 2016.
DOI: 10.1007/978-3-662-53354-3_20

The study of envy-free pricing for revenue maximization has received considerable attention in the recent literature. It was initialized by the work of Guruswami *et al.* [15], and has been the subject of various follow-up works [1, 3–6, 9, 19], with or without budgets. Several factors have made this family of problems gain popularity. First, envy-freeness is one of the dominant and established solution concepts in the fair division literature, ever since the works of [12, 20]. Furthermore, envy-freeness along with individual rationality can be seen as a relaxation of the Walrasian equilibrium concept [14]. This relaxation aims at remedying the fact that Walrasian equilibria do not always exist for arbitrary valuation functions. Finally, in certain cases, envy-freeness can also be viewed as a relaxation of incentive compatibility, see e.g., [16], where envy-free outcomes are used as a benchmark for prior-free mechanisms.

Despite the plethora of works on envy-free pricing, results on the case of buyers with budgets have been quite elusive. Recent works on budgeted buyers concern primarily symmetric (multi-unit) valuation functions [9], with the exception of [6], wherein valuation functions with limited asymmetry are also studied. In contrast, our focus is exclusively on buyers with *asymmetric* combinatorial valuation functions. We find that asymmetric valuation functions in conjunction with budget constraints induce severe computational hardness. This motivates the introduction of assumptions on the problem's parameters, so as to identify families of instances that are computationally more benign. We introduce formally such a restricted class of instances, by conditioning on a relation between the buyers' valuation functions and their budgets. For this class of instances, we manage to obtain positive approximation results, through adaptations of familiar techniques from the literature on combinatorial auctions.

Contribution. Our first result is a negative one; we show that envy-free revenue optimization for two budgeted buyers with identical additive valuation functions is inapproximable in polynomial time (Sect. 3). This result motivates the introduction of the notion of *budget compatible buyers*; for instances of the problem that we consider subsequently, we require that each buyer has enough budget to "cover" his value for specific subsets of items that we refer to as *minimally valuable*. First we observe that, under the assumption of budget compatible buyers, certain approximation results from the literature on social welfare and revenue maximization (for buyers without budgets) [15, 18] transfer to our problem of envy-free revenue maximization for budgeted buyers.

Subsequently, in Sect. 4, we consider budgeted buyers with identical submodular valuation functions over preference subsets, as well as identical subadditive valuation functions. In the former case, we analyze an adaptation of a well known greedy algorithm from [17] for social welfare maximization, achieving a 4-approximation of the optimal envy-free revenue. For the case of identical subadditive budgeted buyers, we modify an algorithm from [8] for subadditive welfare maximization, and show that the resulting algorithm achieves an optimal $O(\sqrt{m})$-approximation. For the special case of two buyers with arbitrary subadditive valuation functions we also justify a 4-approximation. Finally, in Sect. 5, we study the setting of budgeted buyers with arbitrary additive valuation functions.

We analyze a simple greedy-like algorithm that achieves a constant approximation of the optimum envy-free revenue, for a constant number of buyers. We also highlight that the case of additive valuation functions with budget compatible buyers still remains an interesting challenge.

1.1 Related Work

The settings studied by Feldman et al. [9] and Colini-Baldeschi et al. [6] are the closest to ours. Feldman et al. [9] studied envy-free revenue maximization in a limited supply multi-unit auction setting with budgeted buyers, having linear valuation functions over the number of items they receive. They devised a uniform price mechanism that sells all (identical) items at the same price; the payments of buyers are then proportional to the number of items they receive. This mechanism was shown to output an envy-free outcome and approximate the optimum envy-free revenue within factor 2. The authors also established the problem's NP-hardness. Colini-Baldeschi et al. [6] studied a setting with budgeted buyers and multiple distinct items, each available in a limited number of identical copies. Each buyer's valuation function was defined to be linear over copies of items belonging in a buyer-specific *preference subset*, thus, exhibiting some asymmetry. For this setting the authors proved an approximation lower bound of $\Omega((\min\{m, n\})^{1/2})$ for the optimum envy-free revenue, and showed how a $\Theta(m)$-approximation can be obtained by exploiting the results of [9]. For unit-demand buyers, they analyzed a best-possible $O(\log n)$-approximation algorithm. For a multi-unit setting with a single item in limited supply of identical copies and budgeted buyers with linear valuation functions, the authors analyzed a fully polynomial-time approximation scheme, for approximating the optimum envy-free uniform-price revenue.

Guruswami et al. [15] initialized the study of envy-free revenue maximization algorithms (for buyers without budgets). Envy-freeness of an outcome in this work and several works that followed [1–5,19] ensures that every buyer receives his most preferred subset for given *prices on the items*. The authors in [15] analyzed approximation algorithms for Unit-Demand and Single-Minded buyers; they also proved computational inapproximability results for these two cases of buyers. The results on Single-Minded buyers were further improved and extended in a number of subsequent works including [1,3,5]. Fiat and Wingarten [11] also studied envy-free revenue maximization for Single-Minded buyers, but with the same definition of envy-freeness as the one we use. While most of the works on envy-free revenue maximization (without budgets) focused on settings of multiple distinct items, the recent work of Monaco et al. [19] concerns a multi-unit setting of a single good in limited supply and buyers with symmetric valuation functions. Moreover, the authors study both kinds of envy-freeness, the one that we consider here and the one studied by [15] and follow-up works.

Let us note that there has been considerable work on the design of incentive compatible mechanisms in settings involving budgeted buyers, starting with the seminal work of Dobzinski, Lavi and Nisan [7] and followed by several works including, e.g., [10,13].

2   Definitions and Preliminaries

We consider a market involving a set $N = \{1, \ldots, n\}$ of $n = |N|$ buyers and a set $M = \{1, \ldots, m\}$ of $m = |M|$ items. Each buyer $i \in N$ is associated with a budget $b_i > 0$ and with a *monotone* non-decreasing valuation function, $v_i : 2^M \mapsto \mathbb{R}^+$, satisfying $v_i(\emptyset) = 0$. For every subset of items $X \subseteq M$, $v_i(X)$ denotes the intrinsic value of buyer i for acquiring X; in effect, $v_i(X)$ corresponds to the maximum monetary amount that i is *willing to spend* for the purchase of X. The budget b_i constrains the maximum monetary amount that i *can actually spend* for any allocation of items, regardless of his value for this allocation.

In this setting we are interested in outcomes $(\mathbf{X}, \mathbf{p})$ consisting of an allocation $\mathbf{X} = (X_1, \ldots, X_n)$ of subsets of items to the buyers, along with payments $\mathbf{p} = (p_1, \ldots, p_n)$ that the buyers are required to issue for their allocated subsets. In particular, we study the computation of outcomes $(\mathbf{X}, \mathbf{p})$ that approximately maximize the raised revenue $\mathcal{R}(\mathbf{X}, \mathbf{p}) = \sum_i p_i$, while respecting the buyers' budgets, i.e., $p_i \leq b_i$ for every $i \in N$, and satisfying the constraints of *Individual Rationality* and *Envy Freeness*, imposed on the buyers' utilities. The utility of a buyer i for an outcome $(\mathbf{X}, \mathbf{p})$ is defined as:

$$u_i(\mathbf{X}, \mathbf{p}) \equiv u_i(X_i, p_i) = \begin{cases} v_i(X_i) - p_i, & \text{if } p_i \leq b_i \\ -\infty & \text{otherwise} \end{cases} \tag{1}$$

Individual Rationality requires that $u_i(X_i, p_i) \geq 0$ for every buyer $i \in N$. An outcome $(\mathbf{X}, \mathbf{p})$ is *envy-free* if it satisfies:

$$u_i(X_i, p_i) \geq u_i(X_{i'}, p_{i'}), \quad \text{for every two } i, i' \in N \tag{2}$$

Thus, i *does not envy* i' if his utility would not increase if he received the allocation of i' at the price payed by i'. By definition of $u_i(\cdot, \cdot)$ in (1), we observe that (2) holds trivially if $b_i < p_{i'}$, i.e., if i *cannot afford* to pay $p_{i'}$.

The *social welfare* of an allocation $\mathbf{X}$ is denoted by $SW(\mathbf{X})$ and defined as the sum of the buyers' values for it, i.e., $SW(\mathbf{X}) = \sum_i v_i(X_i)$. For any subset $N' \subseteq N$, we will use the notation $SW_{N'}(\mathbf{X})$ for $\sum_{i \in N'} v_i(X_i)$. We denote the optimum envy-free revenue of the instance in context by $\mathcal{R}^*$; we use $\mathcal{R}^*_{N'}$ for the optimum envy-free revenue of the instance restricted to the subset of buyers N'.

Valuation Functions. We consider the classes of *additive* (**ADD**), *submodular* (**SM**), and *subadditive* (**SA**) valuation functions for the buyers. These classes have been studied extensively in the context of combinatorial auctions.

Definition 1. *A function $v : 2^M \mapsto \mathbb{R}^+$ belongs to the class:*

- **ADD**, *if for every $X \subseteq M$, $v(X) = \sum_{j \in X} v(\{j\})$.*
- **SM**, *if for every two $X, Y \subseteq M$, $v(X) + v(Y) \geq v(X \cup Y) + v(X \cap Y)$.*
- **SA**, *if for every two $X, Y \subseteq M$, $v(X) + v(Y) \geq v(X \cup Y)$.*

It is known that **ADD** $\subset$ **SM** $\subset$ **SA**; for further information on these classes of valuation functions, we refer the reader to the seminal work of Lehmann,

Lehmann and Nisan [17]. We note that all our results are based on algorithms accessing the buyers' valuation functions through *value queries*. In the presentation our results, we will find it convenient to use the following notation, in the manipulation of valuation function values: $v(A|B) \equiv v(A \cup B) - v(B)$.

3 Budget Compatibility

Our main result in this section is that envy-free revenue approximation is intractable, even for two buyers with *identical* additive valuation functions. The proof is by reduction from the EQUAL-SUM-SUBSETS problem [21].

Theorem 1. *For any polynomial-time computable function $\rho(m)$, the optimal envy-free revenue cannot be approximated within a factor of $\rho(m)$, unless $\mathbf{P} = \mathbf{NP}$, even for 2 buyers with identical additive valuation functions on m items and equal budgets.*

To alleviate the problem's hardness, we introduce below the notion of *budget-compatible* buyers, a restriction that relates the valuation function of each buyer with his budget. In subsequent sections we consider only instances with such buyers. As we observe at the end of this section, the assumption of budget-compatible buyers already yields some simple positive results for certain classes of valuation functions.

Definition 2. *Consider a buyer with budget $b > 0$ and a non-decreasing valuation function $v : 2^M \mapsto \mathbb{R}^+$, over subsets of goods from a set M, $|M| = m \geq 1$. A subset $Y \subseteq M$ is minimally valuable for the buyer if $v(Y) > 0$ and $v(Y \setminus \{j\}) = 0$ for every $j \in Y$. The buyer is **Budget-Compatible (BC)** if $b \geq v(X)$, for every minimally valuable subset of items $X \subseteq M$.*

Let us note that for the most general class of **SA** valuation functions that we consider, the *minimally valuable subsets* are always singletons. If $v : 2^M \mapsto \mathbb{R}_+$ is a (non-decreasing) **SA** valuation function and $v(X) > 0$ for some $X \subseteq M$ with $|X| \geq 2$, there exists $j_0 \in X$ such that $v(\{j_0\}) > 0$; otherwise, $\sum_{j \in X} v(\{j\}) = 0 < v(X)$ contradicts the subadditivity of v. By the monotonicity of v, no strict superset of $\{j_0\}$ can be minimally valuable, according to Definition 2.

For functions that do not belong to the class **SA** the minimally valuable subsets can be non-singletons. Consider for example the class of *single-minded buyers* [18]; each such buyer i has a scalar value $v_i > 0$ for a particular subset M_i of items (or any superset of it) and 0 for all other subsets. Then M_i is the (unique) minimally valuable subset of i. Budget compatibility in this setting yields a (best-possible) $O(\sqrt{m})$-approximation of the optimum envy-free revenue, via the celebrated greedy algorithm of Lehmann, O'Callaghan and Shoham [18], for social welfare maximization.

Another result that follows immediately under the assumption of BC budgeted buyers is a factor $O(\log n)$-approximation of the optimal envy-free revenue for *Unit Demand* valuation functions, which make up a strict subclass of

SM valuation functions. A valuation function $v : 2^M \mapsto \mathbb{R}_+$ is *Unit Demand* if $v(X) = \max_{j \in X} v(\{j\})$. For Unit Demand buyers (without budgets), Guruswami *et al.* analyzed in [15] an $O(\log n)$-approximation algorithm for envy-free revenue maximization; the notion of envy-freeness used in their work expresses that every buyer is allocated his utility-wise most preferable allocation under an *item pricing*. Because the algorithm of [15] allocates at most a single item to each buyer and charges him a payment at most equal to his value for the received item, the outcome remains individually rational under the assumption of BC buyers and envy-free according to inequality (2).

4   Identical Valuation Functions on Preference Subsets

In this section we study the cases of BC budgeted buyers with near-identical **SM** valuation functions and, subsequently, with identical **ADD** and **SA** valuation functions. Let us first assume **SM** buyers. We focus on a case of **SM** functions, where every buyer i is equipped with a *preference subset* S_i of items, so that, his value for any subset of items $X \subseteq M$ is defined as $v_i(X) = v(S_i \cap X)$, for some **SM** function $v : 2^M \mapsto \mathbb{R}^+$, which is common across all buyers $i \in N$. Note that every restriction v_i of v, with respect to any preference subset S_i, remains **SM**.

We analyze a straightforward adaptation of the greedy algorithm of Lehmann, Lehmann and Nisan [17], which was originally designed for computing (approximately) social welfare maximizing allocations, for **SM** buyers. Our adaptation of the algorithm, referred to as mLLN, appears in Fig. 1 (for the description of the algorithm, recall the notation $v(A|B) \equiv v(A \cup B) - v(B)$, for any 2 sets A, B). The algorithm chooses iteratively the most valuable buyer-item pair in lines 3.1 and 3.2, by identifying first a buyer i and, subsequently, an item j from the buyer's *currently* eligible preference subset $M_i \subseteq S_i$. If the total value of i after allocating j to him does not exceed his budget b_i, the allocation is performed; otherwise, j is removed from M_i. The algorithm terminates when all eligible preference subsets are empty, and returns the resulting allocation **X**.

The algorithm uses a slightly different selection criterion from [17], in that it always examines the most (marginally) valuable buyer-item pair, instead of scanning the items in arbitrary order and allocating each to the buyer with the highest value for it. It also satisfies the following properties:

1. It *does not allocate* an item if the buyer's total value exceeds his budget.
2. It *does not allocate* to any buyer any item *outside* his preference subset.

It is clear that when we ignore the budgets, mLLN has the same performance as shown in [17], i.e., approximates the optimum social welfare within factor 2.

Fact 1. *If $b_i = +\infty$ for every $i \in N$, mLLN achieves a 2-approximation on the optimal social welfare for buyers with* **SM** *valuation functions.*

The two properties identified above allow us to show:

1. $\mathbf{X} \leftarrow (\emptyset, \ldots, \emptyset)$ // Initialize
2. **for** $i \in N$ **do:** $M_i \leftarrow S_i$ // Eligible items per buyer
3. **while** $(\cup_i M_i) \neq \emptyset$ **do:**   // While eligible items remain
    1. $i := \underset{\ell \in N : M_\ell \neq \emptyset}{\arg\max} \left(\underset{j' \in M_\ell}{\max}\, v_\ell(\{j'\}|X_\ell) \right)$ // Buyer i with most valuable item
    2. $j := \underset{j' \in M_i}{\arg\max}\, v_i(\{j'\}|X_i)$ // Most valuable item of i
    3. **if** $b_i \geq v_i(X_i \cup \{j\})$ **do:**   // If remaining budget suffices
        $X_i \leftarrow X_i \cup \{j\}$;   // Allocate j to i
        for $\ell \in N$ **do:** $M_\ell \leftarrow M_\ell \setminus \{j\}$ // Remove j from eligible subsets
       else do: $M_i \leftarrow M_i \setminus \{j\}$ // Else, j is not eligible for i
4. **return** $\mathbf{X}$

Fig. 1. The modified Lehmann-Lehmann-Nisan (mLLN) algorithm.

Lemma 1. *Let* $\mathbf{X}$ *be an allocation returned by the* mLLN *algorithm for an instance involving BC budgeted buyers with preference-set-restricted identical* **SM** *valuation functions. The outcome* $(\mathbf{X}, \mathbf{p})$ *where* $p_i = v_i(X_i)$ *satisfies individual rationality and envy-freeness.*

We next prove the following lemma:

Lemma 2. *Consider the execution of the* mLLN *algorithm, for BC budgeted buyers with* **SA** *valuation functions. Let* $i \in N$ *and* $j \in M$, *be the buyer and item chosen in the* t-*th iteration. Let* X_i^{t-1} *denote the allocation of* i *in the beginning of the* t-*th iteration and* X_i *be the final allocation of* i. *If* $b_i < v_i(X_i^{t-1} \cup \{j\})$, *then* $v_i(X_i) \geq \frac{1}{2} b_i$.

Using Lemmas 1 and 2, we can establish the following approximation.

Theorem 2. *For a market involving BC budgeted buyers with an identical* **SM** *valuation function on their preference subsets, there exists a polynomial time algorithm that approximates the optimum envy-free revenue within factor 4.*

Proof. We analyze the outcome $(\mathbf{X}, \mathbf{p})$, where $\mathbf{X}$ is the allocation of the mLLN algorithm, and $p_i = v_i(X_i)$. We need to establish that $\mathcal{R}(\mathbf{X}, \mathbf{p}) \geq \frac{1}{4}\mathcal{R}^*$, where $\mathcal{R}^*$ is the optimal envy-free revenue. For this, it suffices to show that $SW(\mathbf{X}) \geq \frac{1}{4}\mathcal{R}^*$. We partition the set of buyers N into two disjoint subsets $\mathcal{V}$ and $\mathcal{B}$, where $\mathcal{V} = N \setminus \mathcal{B}$ and $\mathcal{B}$ is defined to contain buyers for which Lemma 2 can be applied. I.e., *for every buyer* $i \in \mathcal{B}$, *there exists an iteration* t *in the course of the algorithm's execution, during which, the buyer-item pair that is chosen is* $\langle i, j \rangle$, *for some* $j \in S_i$ *(in lines* **3.1, 3.2***), and it occurred that* $b_i < v_i(X_i^{t-1} \cup \{j\})$.

Fix an optimal envy-free outcome, with revenue $\mathcal{R}^*$, and let $\mathcal{R}_\mathcal{B}^*$ (resp. $\mathcal{R}_\mathcal{V}^*$) be the revenue extracted from the buyers of $\mathcal{B}$ (resp. of $\mathcal{V}$) in the optimal outcome. Consider first the case that $\mathcal{R}_\mathcal{B}^* \geq \mathcal{R}_\mathcal{V}^*$. By Lemma 2 we obtain $v_i(X_i) \geq \frac{1}{2} b_i$

for every buyer $i \in \mathcal{B}$. Since b_i is an upper bound on the payment of i in any revenue-optimal envy-free outcome, $SW_{\mathcal{B}}(\mathbf{X}) = \sum_{i \in \mathcal{B}} v_i(X_i) \geq \frac{1}{2}\mathcal{R}^*_{\mathcal{B}}$. Hence:

$$SW(\mathbf{X}) \geq SW_{\mathcal{B}}(\mathbf{X}) \geq \frac{1}{2}\mathcal{R}^*_{\mathcal{B}} \geq \frac{1}{2}\left(\frac{1}{2}\mathcal{R}^*_{\mathcal{B}} + \frac{1}{2}\mathcal{R}^*_{\mathcal{V}}\right) \geq \frac{1}{4}\mathcal{R}^*$$

Consider now the alternative case, where $\mathcal{R}^*_{\mathcal{V}} \geq \mathcal{R}^*_{\mathcal{B}}$. We will argue that $SW(\mathbf{X}) = \sum_i v_i(X_i) \geq \frac{1}{2}\mathcal{R}^*_{\mathcal{V}}$; this, in turn, yields a 4-approximation of the optimal envy-free revenue. The argument is by utilizing Fact 1 on the performance of mLLN for social welfare maximization. We compare the algorithm's execution on two instances of the problem; both concern the same set of buyers, $\mathcal{V}$. For every buyer in $\mathcal{V}$ we ignore his budget constraint for this analysis. Let $M_{\mathcal{V}} = \cup_i X_i$ contain the items allocated to buyers of $\mathcal{V}$ under $\mathbf{X}$. Let the first instance be $I_0 \equiv (\mathcal{V}, M_{\mathcal{V}}, \mathbf{v})$; the second is $I_1 \equiv (\mathcal{V}, M, \mathbf{v})$.

The execution of the mLLN algorithm on I_0 will output an allocation $\mathbf{Y}_0$ for buyers in $\mathcal{V}$, that is identical to their allocation under $\mathbf{X}$; thus, $SW(\mathbf{Y}_0) = SW_{\mathcal{V}}(\mathbf{X})$. Consider now the algorithm's output, $\mathbf{Y}_1$, when executed on I_1. Because the greedy algorithm mLLN has chosen to allocate every item $j \in (\cup_{i \in \mathcal{B}} X_i)$ to some buyer in $\mathcal{B}$ and not to some buyer in $\mathcal{V}$, we know that none of these items can have a higher marginal contribution – in the algorithm's greedy selection order – to $SW_{\mathcal{V}}(\mathbf{Y}_1)$, than their marginal contribution to $SW_{\mathcal{B}}(\mathbf{X})$. A similar argument holds for items $j \in \cup_{i \in \mathcal{V}} X_i$; none of these items may achieve a larger marginal value contribution to $SW_{\mathcal{V}}(\mathbf{Y}_1)$ (in the greedy selection order) than the marginal value it contributed to $SW_{\mathcal{V}}(\mathbf{X}) = SW_{\mathcal{V}}(\mathbf{Y}_0)$. Thus, if $\mathcal{W}^*_1$ denotes the socially optimal welfare value for I_1, we have:

$$SW(\mathbf{X}) = SW_{\mathcal{B}}(\mathbf{X}) + SW_{\mathcal{V}}(\mathbf{X}) = SW_{\mathcal{B}}(\mathbf{X}) + SW_{\mathcal{V}}(\mathbf{Y}_0)$$

$$\geq SW_{\mathcal{V}}(\mathbf{Y}_1) \geq \frac{1}{2}\mathcal{W}^*_1 \geq \frac{1}{2}\mathcal{R}^*_{\mathcal{V}}$$

where the previous to last inequality comes from Fact 1, and the last inequality comes from the fact that $\mathcal{W}^*_1$ is an upper bound on the optimal revenue that can be collected from buyers in $\mathcal{V}$. Hence $SW(X) \geq \frac{1}{4}\mathcal{R}^*$.   □

Next we analyze an algorithm for BC budgeted buyers with fully identical **SA** valuation functions. En route, we prove a factor 2-approximation for buyers with identical **ADD** valuation functions, which improves upon the 4-approximation shown above; this 4-approximation remains valid also for buyers with identical **ADD** valuation functions, because **ADD** $\subset$ **SM**. With respect to the description of the mLLN algorithm in Fig. 1, we assume that the preference subset S_i of each buyer coincides with M, the whole set of items, and that $v_i = v$ for all buyers $i \in N$, where v is **ADD** or **SA**.

Theorem 3. *For a market with BC budgeted buyers and m items, an envy-free outcome can be computed in polynomial time, that approximates the optimum envy-free revenue within factor:*

- 2, *when the buyers have identical additive valuation functions,*
- $O(\sqrt{m})$, *when the buyers have identical subadditive valuation functions.*

No better approximation to the optimum revenue can be achieved in the case of identical subadditive buyers; for high enough budgets, a better revenue approximation factor would result in at least as good an approximation of the optimum social welfare; the latter is excluded by a result from [8] (cf. Theorem 6.1), stating that, for at least $\sqrt{m}$ identical buyers, approximation of the social welfare within less than $O(\sqrt{m})$ requires exponentially many value queries.

5 Additive Buyers

In this section we consider instances with arbitrary additive valuation functions. Recall from Theorem 1 that without the BC assumption, the problem is intractable. Assuming BC buyers, we will analyze a greedy-like algorithm, which yields a constant approximation for a constant number of buyers. For a special case of additive buyers, Colini-Baldeschi *et al.* [6] proved a factor $\Theta(m)$-approximation (cf. [6] Corollary 4), by using the results of [9]. This yields a constant approximation for $m = O(1)$ items. Here, we are able to improve upon this ratio for arbitrary additive valuation functions, under the assumption of budget compatibility. We note that for BC buyers, a factor m-approximation can be obtained straightforwardly, even for **SA** valuation functions. For instances involving 2 buyers only, we can obtain a further improved 4-approximation, again even when the valuation functions of the buyers are subadditive.

Theorem 4. *There exists a polynomial-time 4-approximation algorithm for envy-free revenue maximization in a market wth m items and two budget-compatible buyers having subadditive valuation functions.*

Before we proceed to discuss an algorithm for an arbitrary (constant) number of buyers, we state a useful lemma.

Lemma 3. *Consider any instance $(N, M, \mathbf{v}, \mathbf{b})$ of envy-free revenue maximization, with buyers having additive valuation functions. Let $\mathcal{R}_N^*$ denote the optimum envy-free revenue for this instance. For any subset of buyers $N' \subseteq N$, denote by $\mathcal{R}_{N'}^*$, $\mathcal{R}_{N \setminus N'}^*$ the optimum envy-free revenue values for the sub-instances $(N', M, \mathbf{v}, \mathbf{b})$ and $(N \setminus N', M, \mathbf{v}, \mathbf{b})$ respectively. Then: $\mathcal{R}_{N'}^* + \mathcal{R}_{N \setminus N'}^* \geq \mathcal{R}_N^*$.*

The algorithm we describe allocates items only to a single buyer and charges this buyer his value for the items. For simplicity, we use the notation $[i, X]$ to describe an outcome $(\mathbf{X}, \mathbf{p})$, where $X_i = X$, $p_i = v_i(X_i)$ and $X_{i'} = \emptyset$, $p_{i'} = 0$, for every $i' \neq i$. Throughout this section, we let $k \leq n$ denote the number of distinct budget values among the n buyers; for every budget value b, $N(b) \subseteq N$ is the set of buyers with budget equal to b. Conceivably, k could be much smaller than n, in cases where many buyers can have a similar monetary capacity.

We refer to our algorithm as r-greedy (for *"recursive greedy"*). It is described in pseudo-code on the left of Fig. 2. r-greedy uses a greedy subroutine (right of Fig. 2). The greedy subroutine picks iteratively the pair of a buyer i_0 and a yet unallocated item j_0, with the largest value $v_{i_0}(\{j_0\})$. If $v_{i_0}(X_{i_0} \cup \{j_0\}) \leq b_{i_0}$, it allocates j_0 to i_0; otherwise, it stops and returns the allocation $[i_0, X_{i_0}]$. I.e, it greedily tries to find a buyer who comes close to violating his budget constraint. If no such buyer is found and all items end up being allocated, greedy returns the outcome corresponding to the buyer with the highest value (lines 3 and 4). The performance of greedy is summarized as follows:

Fact 2. *The allocation $[i_0, X_{i_0}]$ returned by greedy is envy-free. Moreover, if it allocates all the items, then $n \cdot v_{i_0}(X_{i_0}) \geq \mathcal{R}_N^*$. Otherwise, $2v_{i_0}(X_{i_0}) \geq b_{i_0}$.*

Envy-freeness follows by the fact that we have additive valuation functions, hence separable, over the set of items. The items that i_0 obtains are not valued higher by the other buyers, thus no envy can be created by charging i_0 his value. The last statement of Fact 2 follows by the budget compatibility assumption and by Lemma 2, shown previously for the algorithm mLLN, which is of similar functionality. Let us now describe r-greedy; first it obtains an allocation $[i_0, X_{i_0}]$ by executing greedy on the whole instance. If all buyers have identical budgets equal to b_{i_0}, the allocation is returned (in line 2). Otherwise, r-greedy calls itself recursively on the sub-instance *not containing* the buyers in $N(b_{i_0})$, and obtains an allocation $[i_1, X_{i_1}]$. Subsequently, it compares the value $v_{i_1}(X_{i_1})$ of the allocation $[i_1, X_{i_1}]$ with the budget b_{i_0} of i_0, so as to determine a final outcome to return. The analysis of the comparison cases is central to the proof of the algorithm's approximation ratio (below). Finally, r-greedy runs in polynomial time, since it calls itself only once on a sub-instance with less buyers.

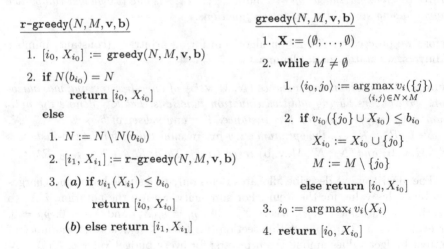

Fig. 2. The $n2^k$-approximation r-greedy algorithm, for n additive buyers with at most $k \leq n$ distinct budget values, accompanied by a greedy subroutine.

Theorem 5. *There exists a polynomial-time $n2^k$-approximation algorithm for envy-free revenue maximization, in a market of m items and n budgeted BC buyers, with additive valuation functions and at most $k \leq n$ distinct budgets.*

Proof. We prove the approximation bound and the outcome's envy-freeness inductively on the number of distinct budgets. Let $[r, X_r]$ be the allocation returned by r-greedy when executed on an instance involving a set N of $n = |N|$ buyers. We will show $(n2^k) \cdot v_r(X_r) \geq \mathcal{R}_N^*$. When $k = 1$, i.e., all buyers have equal budgets, the output of r-greedy is determined in line 2 and coincides with that of greedy. By Fact 2, we have either $v_r(X_r) = \frac{1}{n}\mathcal{R}_N^*$, or $v_r(X_r) \geq \frac{1}{2}b_r \geq \frac{1}{2n}\mathcal{R}_N^*$. Envy-freeness is trivially guaranteed as well. This completes the induction basis.

For the inductive hypothesis, assume that the allocation $[i_1, X_{i_1}]$ of algorithm r-greedy in line 2.2 – for the sub-instance with the set of buyers $N \setminus N(b_{i_0})$ – satisfies envy-freeness among all buyers in $N \setminus N(b_{i_0})$ and:

$$(n - |N(b_{i_0})|)2^{k-1}v_{i_1}(X_{i_1}) \geq \mathcal{R}_{N\setminus N(b_{i_0})}^* \tag{3}$$

We analyze each of the two cases, (a) and (b), in line 2.3 of r-greedy, to assert that the returned allocation $[r, X_r]$ yields an outcome satisfying envy-freeness among all buyers in N and $n2^k \cdot v_r(X_r) \geq \mathcal{R}_N^*$.

Case (a) In this case r-greedy returns $[r, X_r] = [i_0, X_{i_0}]$, which yields an envy-free outcome among the set N of all buyers, with $p_{i_0} = v_{i_0}(X_{i_0})$, by the functionality of greedy. If, additionally, this allocation was determined in line 3 of greedy, then by Fact 2 $v_{i_0}(X_{i_0}) \geq \frac{1}{n}\mathcal{R}_N^*$, thus, is trivially $n2^k$-approximate. Otherwise, the allocation was determined in the "**else**" part of line 2 of greedy, thus: $v_{i_0}(X_{i_0}) \geq \frac{1}{2}b_{i_0} \geq \frac{1}{2|N(b_{i_0})|}\mathcal{R}_{N(b_{i_0})}^*$, where $\mathcal{R}_{N(b_{i_0})}^*$ is the optimum envy-free revenue for the sub-instance involving only the subset $N(b_{i_0})$ of buyers. Also, $p_{i_0} \geq \frac{1}{2}b_{i_0} \geq \frac{1}{2}v_{i_1}(X_{i_1})$, thus, by (3) we have: $(n - |N(b_{i_0})|)2^{k-1}p_{i_0} \geq \frac{1}{2}\mathcal{R}_{N\setminus N(b_{i_0})}^*$. Adding to this inequality $|N(b_{i_0})| \cdot p_{i_0} \geq \frac{1}{2}\mathcal{R}_{N(b_{i_0})}^*$, yields, by Lemma 3:

$$n2^{k-1}p_{i_0} \geq \frac{1}{2}\left(\mathcal{R}_{N(b_{i_0})}^* + \mathcal{R}_{N\setminus N(b_{i_0})}^*\right) \geq \frac{1}{2}\mathcal{R}_N^*$$

which establishes the approximation ratio.

Case (b) is analyzed in a similar manner. $\square$

6 Concluding Remarks and Open Problems

We initiated a systematic study of envy-free revenue maximization for buyers with budgets and asymmetric valuation functions over subsets of items. We established the problem's inapproximability, and introduced the notion of budget compatible buyers so as to alleviate this negative result. Our main positive approximation results include the cases of identical submodular valuation functions, restricted on preference sets, as well as identical subadditive valuation functions. For buyers with arbitrary additive valuation functions we were also able to provide a constant approximation for a constant number of buyers. Several intriguing questions still remain open, particularly with respect to BC buyers. It would be insightful to understand if better approximations are possible for this setting; we believe that the analysis of our algorithm for additive buyers is not tight (we obtained a lower bound of $\Omega(kn)$). For the special case of additive buyers of [6], it is not clear if better results are attainable under the BC assumption. It is also very interesting to obtain hardness results for BC buyers.

References

1. Balcan, M.F., Blum, A.: Approximation algorithms and online mechanisms for item pricing. Theory Comput. **3**(1), 179–195 (2007)
2. Briest, P.: Uniform budgets and the envy-free pricing problem. In: Aceto, L., Damgård, I., Goldberg, L.A., Halldórsson, M.M., Ingólfsdóttir, A., Walukiewicz, I. (eds.) ICALP 2008, Part I. LNCS, vol. 5125, pp. 808–819. Springer, Heidelberg (2008)
3. Briest, P., Krysta, P.: Single-minded unlimited supply pricing on sparse instances. In: Proceedings of the 17th ACM-SIAM Symposium on Discrete Algorithms (SODA), pp. 1093–1102 (2006)
4. Chen, N., Deng, X.: Envy-free pricing in multi-item markets. In: Abramsky, S., Gavoille, C., Kirchner, C., Meyer auf der Heide, F., Spirakis, P.G. (eds.) ICALP 2010. LNCS, vol. 6199, pp. 418–429. Springer, Heidelberg (2010)
5. Cheung, M., Swamy, C.: Approximation algorithms for single-minded envy-free profit-maximization problems with limited supply. In: Proceedings of the 49th IEEE Symposium on Foundations of Computer Science (FOCS), pp. 35–44 (2008)
6. Colini-Baldeschi, R., Leonardi, S., Sankowski, P., Zhang, Q.: Revenue maximizing envy-free fixed-price auctions with budgets. In: Liu, T.-Y., Qi, Q., Ye, Y. (eds.) WINE 2014. LNCS, vol. 8877, pp. 233–246. Springer, Heidelberg (2014)
7. Dobzinski, S., Lavi, R., Nisan, N.: Multi-unit auctions with budget limits. Games econ. behav. **74**(2), 486–503 (2012)
8. Dobzinski, S., Nisan, N., Schapira, M.: Approximation algorithms for combinatorial auctions with complement-free bidders. Math. Oper. Res. **35**(1), 1–13 (2010)
9. Feldman, M., Fiat, A., Leonardi, S., Sankowski, P.: Revenue maximizing envy-free multi-unit auctions with budgets. In: Proceedings of the 13th ACM Conference on Electronic Commerce (EC), pp. 532–549 (2012)
10. Fiat, A., Leonardi, S., Saia, J., Sankowski, P.: Single valued combinatorial auctions with budgets. In: Proceedings of the 12th ACM Conference on Electronic Commerce (EC), pp. 223–232 (2011)
11. Fiat, A., Wingarten, A.: Envy, multi envy, and revenue maximization. In: Leonardi, S. (ed.) WINE 2009. LNCS, vol. 5929, pp. 498–504. Springer, Heidelberg (2009)
12. Foley, D.: Resource allocation and the public sector. Yale Econ. Essays **7**, 45–98 (1967)
13. Goel, G., Mirrokni, V.S., Paes Leme, R.: Polyhedral clinching auctions and the adwords polytope. J. ACM **62**(3), 18 (2015)
14. Gul, F., Stacchetti, E.: Walrasian equilibrium with gross substitutes. J. Econ. Theory **87**(1), 95–124 (1999)
15. Guruswami, V., Hartline, J.D., Karlin, A.R., Kempe, D., Kenyon, C., McSherry, F.: On profit-maximizing envy-free pricing. In: Proceedings of the 16th Annual ACM-SIAM Symposium on Discrete Algorithms (SODA), pp. 1164–1173 (2005)

16. Hartline, J., Yan, Q.: Envy, truth, and profit. In: Proceedings of the 12th ACM Conference on Electronic Commerce (EC), pp. 243–252 (2011)
17. Lehmann, B., Lehmann, D.J., Nisan, N.: Combinatorial auctions with decreasing marginal utilities. Games Econ. Behav. **55**(2), 270–296 (2006)
18. Lehmann, D.J., O'Callaghan, L., Shoham, Y.: Truth revelation in approximately efficient combinatorial auctions. J. ACM **49**(5), 577–602 (2002)
19. Monaco, G., Sankowski, P., Zhang, Q.: Revenue maximization envy-free pricing for homogeneous resources. In: Proceedings of the 24th International Joint Conference on Artificial Intelligence (IJCAI), pp. 90–96 (2015)
20. Varian, H.: Equity, envy and efficiency. J. Econ. Theory **9**, 63–91 (1974)
21. Woeginger, G.H., Yu, Z.: On the equal-subset-sum problem. Inf. Process. Lett. **42**(6), 299–302 (1992)

SBBA: A Strongly-Budget-Balanced Double-Auction Mechanism

Erel Segal-Halevi[✉], Avinatan Hassidim, and Yonatan Aumann

Bar-Ilan University, 5290002 Ramat-Gan, Israel
erelsgl@gmail.com, avinatanh@gmail.com, yaumann@gmail.com

Abstract. In a seminal paper, McAfee (1992) presented the first dominant strategy truthful mechanism for double auction. His mechanism attains nearly optimal gain-from-trade when the market is sufficiently large. However, his mechanism may leave money on the table, since the price paid by the buyers may be higher than the price paid to the sellers. This money is included in the gain-from-trade and in some cases it accounts for almost all the gain-from-trade, leaving almost no gain-from-trade to the traders. We present SBBA: a variant of McAfee's mechanism which is strongly budget-balanced. There is a single price, all money is exchanged between buyers and sellers and no money is left on the table. This means that all gain-from-trade is enjoyed by the traders. We generalize this variant to spatially-distributed markets with transit costs.

Keywords: Mechanism design · Double auction · Budget balance · Social welfare · Gain from trade · Spatially distributed market

1 Introduction

In the simplest *double auction* a single seller has a single item. The seller values the item for s, which is private information to the seller. A single buyer values the item for b, which is private to the buyer. If $b > s$, then trade can increase the utility for both traders; there is a potential *gain-from-trade* of $b - s$. However, there is no truthful, individually rational, budget-balanced mechanism that will perform the trade if-and-only-if it is beneficial to both traders. The reason is that it is impossible to determine a price truthfully. This is easy to see for a deterministic mechanism. If the mechanism chooses a price $p < b$, the seller is incentivized to bid $(p + b)/2$ to force the price up; similarly, if the mechanism chooses a price $p > s$, the buyer is incentivized to force the price down. The impossibility holds even when the valuations are drawn from a known prior distribution and even when the mechanism is allowed to randomize; see the classic papers of [13] and [16].

1.1 McAfee's Trade-Reduction Mechanism

McAfee [12] showed how to circumvent this impossibility result when there are many sellers, seller i having private valuation s_i, and many buyers, buyer i having private valuation b_i. In McAfee's double auction mechanism, each trader is

© Springer-Verlag Berlin Heidelberg 2016
M. Gairing and R. Savani (Eds.): SAGT 2016, LNCS 9928, pp. 260–272, 2016.
DOI: 10.1007/978-3-662-53354-3_21

asked to give his valuation. The sellers are sorted in an ascending order according to their valuations $s_1 \leq s_2 \leq \ldots \leq s_n$, and the buyers are sorted in a descending order $b_1 \geq b_2 \geq \ldots \geq b_n$. Let k be the largest index such that $s_k \leq b_k$. The optimal gain-from-trade is attained by picking any price $p \in [s_k, b_k]$ and performing k deals in that price. But this scheme is not truthful. McAfee attains truthfulness by considering the following two cases:

(a) If there are at least $k + 1$ buyers and $k + 1$ sellers and the price $p_{k+1} :=$ $(b_{k+1} + s_{k+1})/2$ is in the range $[s_k, b_k]$, then p_{k+1} is set as the market price, allowing all k efficient deals to execute in that price.

(b) Otherwise, two prices are used, and $k - 1$ deals are done: all sellers with values $s_1, \ldots, s_{k-1}$ sell their item for s_k, and all buyers with values $b_1, \ldots b_{k-1}$ buy an item for b_k. The mechanism performs a *trade reduction* by canceling a single deal, the deal between b_k and s_k, which is the least efficient of the k efficient deals. Hence, its gain-from-trade is $1 - 1/k$ of the maximum.

Crucially, the gain-from-trade approximated by McAfee's mechanism is the **total-gain-from-trade** - the gain-from-trade including the money left on the table due to the difference between the buyers' price and the sellers' price. Moreover, this money might include almost all the gain, so that the **market-gain-from-trade** - the gain enjoyed by the traders - might be near zero.

Example 1. There are k buyers and k sellers with the following valuations

- $s_i = 0$ and $b_i = B$ for all $i \in \{1, \ldots, k - 1\}$.
- $s_k = \varepsilon$ and $b_k = B - \varepsilon$,   where $0 < \varepsilon \ll B$.

The optimal market-gain-from-trade occurs when all sellers sell and all buyers buy, and it is: $k \cdot B - 2\varepsilon$.

Since there are no $k + 1$-th buyer and seller, McAfee's mechanism sets the buy price at $B - \varepsilon$ and the sell price at ε. The trade includes $k - 1$ buyers and $k - 1$ sellers. The gain-from-trade in each deal is B, so the total-gain-from-trade is $(k - 1) \cdot B$, which is a very good approximation to the optimum for large k.

However, the net gain of each trader is ε, so the market-gain-from-trade is only $(k - 1) \cdot 2\varepsilon$; when $\varepsilon \to 0$, the market-gain-from-trade becomes arbitrarily small, and most gain-from-trade $(k - 1) \cdot (B - 2\varepsilon)$ remains on the table.   □

Money on the table can be desirable in some cases. E.g, the government may want to arrange a double-auction between commercial firms and collect the revenue. However, in other cases it may be considered unfair and drive traders away. For example, if traders in a stock-exchange notice that most gain-from-trade is taken by the operator, they may decide to switch to another operator.

The double-auction literature, e.g. [7], differentiates between mechanisms that are *weakly budget-balanced*, i.e., the auctioneer does not lose but may gain money, and *strongly budget-balanced*, i.e., the auctioneer does not lose nor gain any money. McAfee's mechanism is weakly budget-balanced.

1.2   Our Mechanism

In this paper we introduce SBBA - a Strongly-Budget-Balanced double-Auction mechanism. SBBA attains strong budget-balance by setting a *single* trade price for all traders, in all cases. This may lead to excess supply; to handle the excess supply, a lottery is done between the sellers. At most one seller, selected at random, is excluded from trade. Hence, the expected total-gain-from-trade of SBBA is the same as McAfee's - $1 - 1/k$.

A disadvantage of SBBA is that its approximation holds only in expectation (taken over the randomization of the mechanism), while McAfee's approximation holds in the worst case. An advantage of SBBA is that it is strongly budget-balanced, so the market-gain-from-trade equals the total-gain-from-trade - all gain-from-trade is enjoyed by the traders. Besides these differences, SBBA has all the desirable properties of McAfee's mechanism:

- It is **ex-post individually-rational** - a trader never loses any value from participating in the market (a buyer is never forced to buy an item for more than its declared value; a seller is never forced to sell an item for less than its declared value; a trader who does not participate in the trade pays nothing).
- It is **ex-post dominant-strategy truthful**: for every trader, every vector of declarations by the other traders and every randomization, the trader's net value is always maximized by reporting his true value.
- It is **prior-free** - it does not assume or require any knowledge on the distribution of the traders' valuations. In other words, its approximation ratio is valid even for adversarial (worst-case) valuations.

Below we survey some related literature (Sect. 2). Then, we present the SBBA mechanism in the most basic double auction setting - a single market with single-unit buyers and single-unit sellers (Sect. 3). The idea of SBBA can be used in much more complex settings. To demonstrate its generality, we show how to use it in a *spatially-distributed market* - a collection of markets in different locations, with positive transit costs between markets (Sect. 4).

2 Related Work

2.1 Double Auctions

VCG (Vickrey-Clarke-Grove) is a well-known mechanism that can be used in various settings, including double auction. It is truthful and attains the maximum gain-from-trade. Its main drawback is that it has budget deficit, which means that the auctioneer has to subsidize the market.

McAfee's Trade-Reduction mechanism was extended and generalized in many ways. Some of the extensions are surveyed below.

Babaioff et al. extend McAfee's mechanism to handle spatially-distributed markets with transit costs [3] and supply chains [2,4,5], providing similar welfare guarantees. One variant, **Probabilistic Reduction** [2], achieves *ex-ante* budget balance by randomly selecting between the Vickrey-Clarke-Grove (VCG) mechanism and the Trade-Reduction mechanism. The probability is selected such that the deficit of the VCG exactly balances (in expectation) the surplus of McAfee. However, the probability depends on the distribution of the agents' valuations so the mechanism is not prior-free.

Lately, there has been a surge of interest in a more complicated market, namely double spectrum auctions, in which an auction is used to transfer spectrum from incumbent companies (e.g. TV stations) to modern companies (e.g. cellular operators). [20] adapted the Trade-Reduction mechanism to a double-spectrum-auction by creating groups of non-interfering buyers that can buy the same channel. [17] created on online variant of Trade-Reduction to handle the case in which new buyers arrive over time. [18] adapted Trade-Reduction to enable local markets, in which only some buyer-seller combinations are feasible. [10] adapted it to enable heterogeneous spectra. All these mechanisms are only weakly budget-balanced.

Some recent papers extend McAfee's mechanism to settings with more than one item per trader. These are the **TAHES** mechanism of [10], which is multi-type single-unit; the **Secondary Market** mechanism of [19], which is single-type multi-unit; and the **Combinatorial Reallocation** mechanism of [7], which is multi-type multi-unit. Our mechanism is single-type single-unit; we leave to future work its extension to multi-type and multi-unit settings.

A different approach to double auctions is *random-sampling*. It was introduced by [6] under the assumptions that the traders' valuations are random variables drawn from an unknown bounded-support distribution and there is a single item-type. We recently extended it to a prior-free setting with multiple item-types [15]. The idea is to divide the traders randomly to two half-markets, calculate an optimal price in one half and apply it to the other half, and vice versa. Since there is a single price in each market, the mechanism is strongly-budget-balanced. However, our analysis (for the single-type case) shows that the gain-from-trade is $1 - O(\sqrt{\ln k/k})$. While this still approaches 1 when the market is sufficiently large, the convergence rate is much slower than the $1 - 1/k$ guarantee of SBBA.

Recently, [9] presented a **two-sided sequential posted price mechanism (2SPM)**. Its main objective is to handle matroid constraints on the sets of buyers that can be served simultaneously. Like our mechanism, it is strongly budget-balanced. However, its approximation ratio is multiplicative - the ratio is between 4 to 16, depending on the setting. In particular, the approximation ratio does not approach 1 when the market is large.

The following table compares our work to some typical single-type single-unit double-auction mechanisms. In this table, an asterisk means "in expectation". TGFT means total-gain-from-trade and MGFT means market-gain-from-trade; they are identical for strongly-budget-balanced mechanisms.

2.2   Redistribution Mechanisms

The problem illustrated by Example 1, where the money left on the table eats most of the welfare and leaves little welfare to the agents, happens in other domains besides double auctions. Several authors have suggested a two-step solution: in the first step, the original mechanism is executed and the budget-surplus is collected. In the second step, some of the surplus is re-distributed among the agents. This second step is called a *redistribution mechanism* and it

Mechanism	Prior-free	Budget	TGFT	MGFT
VCG	Yes	Deficit	1	1
Trade reduction (McAfee 1992)[12]	Yes	Surplus	$1 - 1/k$	0
Random sampling (Baliga 2003)[6,15]	Yes	Balance	$1 - O(\sqrt{\ln k}/k)$	
Probabilistic reduction (Babaioff 2009)[3]	No	Balance*	$1 - 1/k$*	
2SPM (Colini 2016)[9]	No	Balance	1/4 to 1/16	
SBBA (this paper)	Yes	Balance	$1 - 1/k$*	

should be carefully designed in order not to harm the truthfulness of the original mechanism. While it is not possible to redistribute the entire surplus in a truthful way, there are some truthful mechanisms that redistribute a large fraction of the surplus [1,8,11]. We take a different approach: we modify the original mechanism such that there is no budget-surplus at all, so no redistribution is needed and all social welfare remains with the agents.

3   The SBBA Mechanism

Order the buyers and sellers as in McAfee's mechanism. In case of ties, impose an arbitrary order, e.g. lexicographic order of name.

Let k be the largest integer for which $s_k \leq b_k$ (or 0 if already $s_1 > b_1$). Call the first k sellers, the "cheap sellers", and the first k buyers, the "expensive buyers". As a convenience, if $k = n$, set $s_{k+1} = \infty$ and $b_{k+1} = 0$. Note that with this notation, we have that $s_i \leq b_i$ for $i \leq k$ and $s_i > b_i$ for $i > k$. The price is:

$$p := \min(s_{k+1}, b_k).$$

There are two cases. We illustrate them below by plotting buyers' valuations as balls and sellers' valuations as squares (in all illustrations, $k = 3$).

Case 1: $s_{k+1} \leq b_k$ (note that by definition of k: $s_{k+1} > b_{k+1}$):

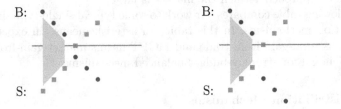

The price is $p = s_{k+1}$. All k expensive buyers and k cheap sellers trade in p.

Case 2: $s_{k+1} > b_k$:

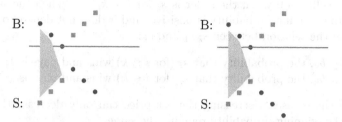

The price is $p = b_k$. From the group of k cheap sellers, select $k-1$ at random and let them trade with the $k-1$ expensive buyers (excluding b_k).

Theorem 1. *The SBBA mechanism is prior-free (PF), individually-rational (IR), strongly-budget-balanced (SBB) and dominant-strategy truthful. Its expected market-gain-from-trade is at least $1 - 1/k$ of the optimum.*

Proof. The mechanism is PF by construction. It is IR since the trade is always between buyers whose value is above p and sellers whose value is below p. It is SBB since there is always a single price and all payments are between buyers and sellers. To analyze the gain-from-trade, note that in case 1, all k efficient deals are carried out and thus the maximum possible gain is achieved. In case 2, a single random deal is canceled, which implies an expected loss of $1/k$. Hence, in expectation, at most a fraction $1/k$ of the gain-from-trade is lost.

To prove truthfulness, we use a characterization of truthful single-parameter mechanisms from [14, Chap. 9]. A mechanism is truthful iff:

(a) the probability of an agent to win, given the bids of other agents, is a weakly monotonically increasing function of the agent's bid, and:

(b) the price paid by a winning agent equals the *critical price* - the lowest value this agent has to bid in order to win, given the other agents' bids.

We prove that SBBA is truthful for the **buyers**. The winning probability of a buyer is either 0 or 1. Hence, it is sufficient to prove that a winning buyer never loses by raising the bid. Consider two cases of a winning buyer:

- The buyer is b_i for $i < k$.
- The buyer is b_k and $b_k \geq s_{k+1}$.

In both cases, if the bid is raised, b_k remains above s_{k+1} and the buyer's index may only decrease, so the buyer still wins.

The critical price when $b_k < s_{k+1}$ is b_k, since a trading buyer (one of the expensive $k-1$ buyers) exits the trade by bidding below b_k and becoming the new b_k. The critical price when $b_k \geq s_{k+1}$ is s_{k+1}, since a trading buyer (one of the expensive k buyers) exits the trade by becoming b_k and bidding below s_{k+1}. In both cases, the price paid by the winning buyers is the critical price.

Finally, we prove that SBBA is truthful for the **sellers**. The sellers' ask-prices represent negative valuations, so monotonicity means that a seller's probability of participation should increase when the ask-price *decreases*. The winning probability of a seller is 0 when the seller is s_i for $i \geq k + 1$. When the seller is s_i for $i \leq k$, the winning probability is positive, and it does not depend on s_i itself but only on the relation between s_{k+1} and b_k:

- If $s_{k+1} \leq b_k$, the probability that s_i (for $i \leq k$) wins and pays is 1;
- If $s_{k+1} > b_k$, the probability that s_i (for $i \leq k$) wins and pays is $1 - 1/k$.

In each of these cases, decreasing the ask-price can only decrease the seller's index, so the winning probability remains the same.

The critical price when $s_{k+1} \leq b_k$ is s_{k+1}, since a trading seller (one of the k cheap sellers) exits the trade by asking above s_{k+1} and becoming the new s_{k+1}. This is indeed the price paid to a winning seller.

The critical price when $s_{k+1} > b_k$ is b_k, since a trading seller (one of the k cheap sellers) exits the trade by asking above b_k, which decreases k by 1 (the seller who increased his ask-price becomes s_k, but now the number of efficient deals is $k - 1$, so s_k is excluded from trade). Indeed, b_k is the price paid to a winning seller. Note that in this case both the winning and the price are realized with probability $1 - 1/k$.

□

3.1 Alternatives

The price set by our mechanism, $\min(b_k, s_{k+1})$, has an interesting economic interpretation: it is the highest price in a price-equilibrium (aka Walrasian equilibrium), i.e. the highest price in which the market can be cleared by balancing supply and demand. If the price is raised above b_k then the demand becomes less than k while the supply is still at least k; if the price is raised above s_{k+1} then the supply becomes at least $k + 1$ while the demand is still at most k; in both cases there is an excess supply.

It is possible to switch the role of buyers and sellers, splitting the cases by whether $b_{k+1} > s_k$. In this case, the price is $\max(s_k, b_{k+1})$ which is the *lowest* price-equilibrium. The alternative mechanism has the same properties of our original mechanism.

There are some other alternatives that come to mind, but are either not truthful or not efficient:

- If in Case 2, instead of using a lottery we select the trading sellers deterministically (e.g. taking the $k - 1$ cheaper sellers), the mechanism will not be truthful, since some agents who want to trade at the market price have an incentive to deviate from their true values in order to enter the trade.
- If we always set the price to s_{k+1}, in some cases this price might be higher than the valuations of all buyers so we might lose all gain-from-trade.
- If we always set the price to b_k, in some cases this price might be higher than the valuations of all sellers, the market will be flooded by inefficient sellers, and the expected gain-from-trade will be low.

4   Spatially Distributed Markets

A *spatially distributed market* is a collection of several markets, each of which is located in a different geographic location. It is possible to transport goods from one market to another for a fixed, positive transit cost, which may be different for each ordered pair of markets. Babaioff et al. [3] extended McAfee's mechanism to handle such markets. Similarly to McAfee's mechanism, their mechanism has a budget surplus. Below we briefly present their mechanism and present a strongly budget-balanced variant of it.

4.1   Create the Market-Flow Graph

Create a network-flow graph representing the market in the following way:

- Create a node for each market. Create a directed edge between each pair of markets, with infinite capacity and with cost equal to the transit cost between the two markets (which may be different in each direction).
- Create an additional *Agents node*, representing the buyers and sellers. For each seller in market i, create an edge FROM the Agents node TO Market i, with unit capacity and cost equal to the seller's ask-price. This edge represents the seller producing an item and sending it to the market. For each buyer in market i, create an edge TO the Agents node FROM Market i, with unit capacity and cost equal to MINUS the buyer's bid. This edge represents the buyer bringing an item from the market.

The following illustration shows a graph representing two markets, with a transit cost of 4 in each direction. Each solid arc represents an infinite-capacity edge. Each dashed arc represents several unit-capacity edges with different costs. The numeric labels are the costs.

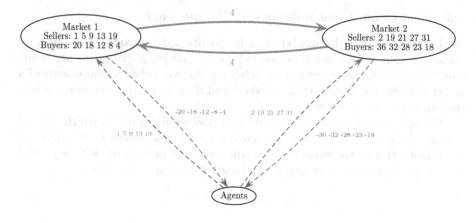

4.2   Calculate a Minimum-Cost Flow

Babaioff et al. use a known polynomial-time algorithm for finding a flow with minimum cost in the market graph. Assuming all data is integral, the flow in every edge is also an integer number. In particular, the flow in every buyer/seller edge (with capacity 1) is either 0 or 1. Hence, a flow in the graph defines an allocation in which each trader trades if and only if the flow in the corresponding edge to/from the Agents node is 1. A minimum-cost flow corresponds to an optimal trade: the (negative) cost of the flow is minus the gain-from-trade. The following illustration shows the minimum-cost flow and the corresponding optimal trade in the above example market (non-trading agents are bracketed):

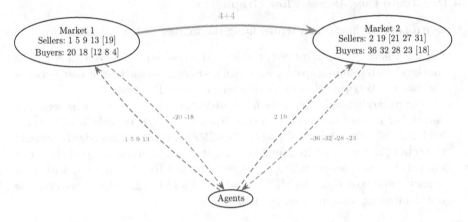

There are 6 efficient deals. The net cost is -100 so the gain-from-trade is 100.

4.3   Find Commercial-Relationship Components

In the optimal flow, the markets can be partitioned to groups, such that all trade is within groups and no trade is between groups. Such groups are called *commercial-relationship components*; they are the connectivity components of a graph in which the nodes are the markets and there is an edge between markets trading in the optimal flow.

The optimal trade can be attained in a price-equilibrium, in which there is a single price in each market. In each component, the prices in the different markets are tied by the equilibrium conditions: if the price in market i is p_i, and there is positive trade from market i to market j, then we must have:

$$p_j = p_i + Cost[i, j] \tag{1}$$

since in equilibrium, the sellers in market i should be indifferent between selling in their local market for a net revenue of p_i, and selling in market j for a net revenue of $p_j - Cost[i, j]$. Therefore, in each component, setting the price in a single market uniquely determines the prices in other markets. Formally, for every two markets i, j in the same component there is a constant $\Delta_{i,j}$ such that, in any price-equilibrium, $p_j = p_i + \Delta_{i,j}$ ($\Delta_{i,j}$ can be calculated, for example, by calculating cheapest paths in the residual graph of the min-cost flow; see [3]).

In our running example, there is a single component. Here, $\Delta_{1,2} = Cost[1, 2] = 4$. The optimal trade can be attained in a price-equilibrium in which the price in Market 1 is p_1 and the price in Market 2 is $p_2 = p_1 + 4$. Any price-vector between $p_1 = 15, p_2 = 19$ and $p_1 = 17, p_2 = 21$ is an equilibrium price-vector.

4.4   Trade Reduction

At this point, Babaioff et al. calculate a *reduced residual graph* of the min-cost flow, remove a single cycle representing the least efficient deal, and determine two prices in each market (buy price and sell price) based on distances in the reduced residual graph. This gives a truthful mechanism with a budget surplus.

Here our mechanism takes a different approach:

- In each commercial-relationship-component, virtually bring all traders to an arbitrary market in that component, e.g., Market i. Adjust their bids according to the price-equilibrium conditions: a bid b in Market j is translated to a bid $b - \Delta_{i,j}$ in Market i.
- Proceed as in the single-market situation (Sect. 3): order the buyers decreasingly and the sellers increasingly and find k - the total number of efficient deals in the component. Set the price in Market i to $p_i := \min(b_k, s_{k+1})$.
- Determine the prices in the other markets of the same component according to the equilibrium conditions: for every market j, set $p_j := p_i + \Delta_{i,j}$. Below we prove that all these prices are non-negative.
- If b_k is the price-setter, then exclude b_k and a random seller in the component from trading. Otherwise, allow all k efficient traders in the component to trade in their market prices.
- Ban any trade between different components.

In the above example, when all traders are brought to Market 1, we have the following valuations (here $\Delta_{1,2} = 4$; the adjusted valuations of traders brought from Market 2 are displayed in slanted digits):

- Sellers: -2, 01, 05, 09, 13, *15, 17,* 19, *23, 27*
- Buyers: *32, 28, 24,* 20, *19,* 18, *14,* 12, 08, 04

Here, the number of efficient deals $k = 6$. We have $b_k = 18$ and $s_{k+1} = 17$, so the price-setter is s_{k+1}, who is originally a seller in Market 2 (where his valuation is 21). The price-vector is $p_1 = 17, p_2 = 21$. All 6 efficient deals are performed.

4.5   Analysis

Similarly to our single-market mechanism, the spatially-distributed-market mechanism is prior-free and strongly-budget-balanced, since there is a single price in each market. The reduction in trade is only at most a single deal *in each component*. Hence, the expected gain-from-trade in each component is $(1 - 1/k)$ of the optimum in that component, where k is the number of efficient deals in the component. When the components are large, this k may be much larger than the number of efficient deals in each market alone.

For truthfulness, we use the monotonicity characterization shown in Theorem 1.

First, we have to prove that a winning trader never loses by increasing/decreasing the bid/ask price. Indeed, if a trader is winning, then the edge from the Agents node to the trader is active in the min-cost flow. When the bid/ask increases/decreases, the cost of that edge decreases. Hence, the cost of the flow decreases, and it remains the min-cost flow. Thus, the partition of the graph to commercial relationship components does not change. Within each component, increasing/decreasing the bid/ask price weakly decreases the index of the trader in the ordering, so a winning buyer/seller is still among the first k buyers/sellers.

Next, we have to prove that each trader pays the critical price. For traders originally from Market i, the critical price is $p_i = \min(b_k, s_{k+1})$; this follows immediately from the proof of Theorem 1. Consider now a winning buyer from Market $j \neq i$ in the same component as Market i. If our buyer bids b, then the bid is translated to Market i as $b - \Delta_{i,j}$. The buyer exits the trade when $b - \Delta_{i,j} < p_i$. Hence, the critical price for our buyer is $p_i + \Delta_{i,j}$, which is exactly the price p_j paid by our buyer. Similar considerations are true for the sellers.

Finally, as promised, we prove that the prices determined by our mechanism are non-negative. Let Market j be a market with a smallest price in some component. Suppose by contradiction that $p_j < 0$. Since we proved that the mechanism is truthful, it implies that there are no active sellers in Market j (since a seller would prefer to lie than to sell in negative price). This means that there is no trade outgoing from Market j. Since there must be active sellers elsewhere in the component, there must be other markets in the component, and this means that there is trade incoming to Market j, say, from Market i. But this means that $p_i = p_j - Cost[i, j]$. Since we assume that transit costs are positive, this contradicts the minimality of p_j. $\qquad\square$

5   Future Work

Besides spatially-distributed markets, there are many other variants of McAfee's mechanism. An interesting line of future work is to survey these variants and see if and how they can be made strongly-budget-balanced.

Acknowledgments. This research was funded in part by the following institutions: The Doctoral Fellowships of Excellence Program at Bar-Ilan University, the Mordechai and Monique Katz Graduate Fellowship Program, and the Israel Science Fund grant 1083/13. We are grateful to Moshe Babaioff for his advice on the spatially-distributed-market mechanism.

References

1. Apt, K.R., Conitzer, V., Guo, M., Markakis, E.: Welfare undominated groves mechanisms. In: Papadimitriou, C., Zhang, S. (eds.) WINE 2008. LNCS, vol. 5385, pp. 426–437. Springer, Heidelberg (2008)
2. Babaioff, M., Nisan, N.: Concurrent auctions across the supply chain. J. Artif. Intell. Res. (JAIR) **21**, 595–629 (2004)
3. Babaioff, M., Nisan, N., Pavlov, E.: Mechanisms for a spatially distributed market. Games Econ. Behav. **66**(2), 660–684 (2009)
4. Babaioff, M., Walsh, W.E.: Incentive-compatible, budget-balanced, yet highly efficient auctions for supply chain formation. Decis. Support Syst. **39**(1), 123–149 (2005)
5. Babaioff, M., Walsh, W.: Incentive compatible supply chain auctions. In: Chaib-draa, B., Müller, J. (eds.) Multiagent Based Supply Chain Management, vol. 28, pp. 315–350. Springer, Heidelberg (2006)
6. Baliga, S., Vohra, R.: Market research and market design. Adv. Theor. Econ. **3**(1), 1059 (2003)
7. Blumrosen, L., Dobzinski, S.: Reallocation mechanisms. In: Proceedings of the Fifteenth ACM Conference on Economics and Computation (EC 2014), NY, USA, p. 617. ACM, New York (2014)
8. Cavallo, R.: Optimal decision-making with minimal waste: strategyproof redistribution of VCG payments. In: Proceedings of the Fifth International Joint Conference on Autonomous Agents and Multiagent Systems (AAMAS 2006), NY, USA, pp. 882–889. ACM, New York (2006)
9. Colini-Baldeschi, R., de Keijzer, B., Leonardi, S., Turchetta, S.: Approximately efficient double auctions with strong budget balance. In: ACM Symposium on Discrete Algorithms, January 2016
10. Feng, X., Chen, Y., Zhang, J., Zhang, Q., Li, B.: TAHES: truthful double auction for heterogeneous spectrums. In: INFOCOM, 2012 Proceedings IEEE, pp. 3076–3080. IEEE, March 2012
11. Guo, M., Conitzer, V.: Optimal-in-expectation redistribution mechanisms. In: Proceedings of the 7th International Joint Conference on Autonomous Agents and Multiagent Systems (AAMAS 2008), vol. 2. pp. 1047–1054. International Foundation for Autonomous Agents and Multiagent Systems, Richland (2008)
12. McAfee, R.P.: A dominant strategy double auction. J. Econ. Theory **56**(2), 434–450 (1992)
13. Myerson, R.B., Satterthwaite, M.A.: Efficient mechanisms for bilateral trading. J. Econ. Theory **29**(2), 265–281 (1983)
14. Nisan, N.: Introduction to mechanism design (for computer scientists). In: Nisan, N., Roughgarden, T., Tardos, E., Vazirani, V. (eds.) Algorithmic Game Theory, pp. 209–241. Cambridge University Press, Cambridge (2007)
15. Segal-Halevi, E., Hassidim, A., Aumann, Y.: MIDA: a multi item-type double-auction mechanism. arXiv preprint (2016). http://arxiv.org/abs/1604.06210
16. Vickrey, W.: Counterspeculation, auctions, and competitive sealed tenders. J. Finance **16**(1), 8–37 (1961)
17. Wang, S., Xu, P., Xu, X., Tang, S., Li, X., Liu, X.: TODA: truthful online double auction for spectrum allocation in wireless networks. In: 2010 IEEE Symposium on New Frontiers in Dynamic Spectrum, pp. 1–10. IEEE, April 2010

18. Wang, W., Li, B., Liang, B.: District: embracing local markets in truthful spectrum double auctions. In: 2011 8th Annual IEEE Communications Society Conference on Sensor, Mesh and Ad Hoc Communications and Networks (SECON), pp. 521–529. IEEE, June 2011
19. Xu, H., Jin, J., Li, B.: A secondary market for spectrum. In: INFOCOM, 2010 Proceedings IEEE, pp. 1–5. IEEE, March 2010
20. Yao, E., Lu, L., Jiang, W.: An efficient truthful double spectrum auction design for dynamic spectrum access. In: 2011 Sixth International ICST Conference on Cognitive Radio Oriented Wireless Networks and Communications (CROWNCOM), pp. 181–185. IEEE, June 2011

Revenue Maximization for Market Intermediation with Correlated Priors

Matthias Gerstgrasser$^{(\boxtimes)}$, Paul W. Goldberg$^{(\boxtimes)}$, and Elias Koutsoupias$^{(\boxtimes)}$

Department of Computer Science, University of Oxford,
Wolfson Building, Parks Road, Oxford, OX1 3QD, UK
{matthias.gerstgrasser,paul.goldberg,elias.koutsoupias}@cs.ox.ac.uk
http://www.cs.ox.ac.uk/

Abstract. We study the computational challenge faced by a intermediary who attempts to profit from trade with a small number of buyers
and sellers of some item. In the version of the problem that we study,
the number of buyers and sellers is constant, but their joint distribution of the item's value may be complicated. We consider discretized
distributions, where the complexity parameter is the support size, or
number of different prices that may occur. We show that maximizing
the expected revenue is computationally tractable (via an LP) if we are
allowed to use randomized mechanisms. For the deterministic case, we
show how an optimal mechanism can be efficiently computed for the one-
seller/one-buyer case, but give a contrasting NP-completeness result for
the one-seller/two-buyer case.

1 Introduction

We consider a double auction scenario from the perspective of a market intermediary, collecting bids from one or more sellers and buyers and determining
payments and allocations. Real-world instances of this are manifold, including
in electronic markets. Companies such as eBay or Amazon match sellers and
buyers, and charge a fee for each successful transaction. Our aim is to maximize
the intermediary's profit in such settings.

There is an extensive literature on this challenge, some of which is discussed
below, but it mostly considers the case of many buyers and/or sellers with independent priors. Our interest here is different, in that we assume only a constant
number of buyers and sellers (in the simplest version, just one of each), and the
complexity arises from their joint probability distribution of valuations for the
item. In the simplest version of this, where there is just one buyer and one seller,
the intermediary can profit from buying the item from the seller and selling at
a higher price to the buyer. We assume their valuations for the item come from
a known joint distribution, which is the input to the problem. We consider two
versions: the "no short selling" version, with the natural constraint that we cannot sell more items than we buy; and the more restrictive "balanced inventory"

M. Gerstgrasser—Recipient of a DOC Fellowship of the Austrian Academy of
Sciences.

© Springer-Verlag Berlin Heidelberg 2016
M. Gairing and R. Savani (Eds.): SAGT 2016, LNCS 9928, pp. 273–285, 2016.
DOI: 10.1007/978-3-662-53354-3_22

version, where in addition we must sell all the items we buy. For multiple buyers and sellers, we assume that the intermediary can buy and sell multiple items (but that each buyer/seller has unit demand/supply).

1.1 Related Work

The problem of optimal mechanisms in a market intermediation setting was first studied by Myerson and Satterthwaite [11]. In addition to an impossibility result for ex-post efficiency in a bilateral trade setting without an intermediary, they show optimal intermediation mechanisms for both social welfare as well as the intermediary's revenue in the case of one buyer and one seller, whose valuations are independent. Their revenue-maximization result is similar to Myerson's seminal single auction result [10] in that it, too, uses virtual valuation functions, for both buyer and seller. Welfare maximization for multiple buyers and sellers has been further studied for instance by McAfee [9] or more recently by [1,4,5].

Our own interest is chiefly in the complexity of computing optimal (revenue-maximizing) or near-optimal mechanisms in the market intermediation setting. Prior work in this area has focused on the case where sellers' and buyers' valuations are independent. Deng et al. [2] show optimal and near-optimal mechanisms that can be computed in polynomial time for several variations of this setting, including continuous or discrete distributions and arbitrary or unlimited supply and demand. Niazadeh et al. [12] as well as Loertscher and Niedermayer [6–8] study a class of mechanisms called respectively *fee-setting mechanisms* or *affine fee schedules* in the independent setting. These are shown by Niazadeh et al. [12] to be able to extract a constant factor of the optimum revenue in the worst case, under certain assumptions on the buyer's and seller's distribution.

Here we are interested in potentially correlated distribution over buyers' and sellers' valuations. The complexity of this has been studied for (non-double) auctions. Papadimitriou and Pierrakos [13] show that for two buyers, an optimal mechanism (for a discrete joint distribution) can be found in polynomial time via a reduction to finding a maximum-weight independent set on a bipartite graph. For continuous distributions they give a FPTAS. For the case of three buyers, in contrast, they show that it is NP-hard to approximate the optimal auction to within a certain constant fraction. Dobzinski et al. [3] show a polynomial-time algorithm for the two-buyer auction through derandomization and give polynomial-time approximation mechanisms for the many-buyers correlated single auction problem, building on previous work by Ronen [14].

2 Preliminaries

2.1 Definitions, Notation

We consider m buyers indexed by j, and k sellers indexed by i (where m, k are constants), each offering (respectively seeking) a single unit of an indivisible good. For fixed m, k, we use "$m \times k$" as shorthand for the m buyers, k sellers case. They cannot trade with each other directly, and can only trade with the intermediary. We assume each seller i has some valuation s_i and each buyer j has some valuation b_j for an item, and that these are drawn from a given joint probability distribution ψ. We focus on discrete distributions. For simplicity, we assume that the support of ψ is a grid of size n^{k+m}, with each player having possible valuations $\{i : 1 \leq i \leq n\}$ (as shown in Fig. 1). The distribution ψ is assumed to be represented as a matrix of the probabilities on each of the grid points.

In this paper we focus on individually rational and incentive compatible mechanisms. Let $\mathbf{b}$ and $\mathbf{s}$ denote the vector of buyers' and sellers' bids received by the mechanism, and let $\mathbf{b}_{-j}$ and $\mathbf{s}_{-i}$ be (respectively) the bids of buyers other than j, and sellers other than i. Analogous to auctions, there are two equivalent ways in which we can define a deterministic, incentive compatible mechanism in this setting. Firstly, we can focus on allocations. For each seller/buyer we define a set S_i/B_j ($\subseteq \mathrm{supp}(\psi)$) of bid vectors in which we buy an item from seller i/sell an item to buyer j. Incentive compatibility means monotonicity of allocations, meaning for each seller i, if $(s, b) \in S_i$, and $s'_i < s_i$, then $(s'_i, s_{-i}, b) \in S_i$. In words, if given everyone else's bids s_{-i}, b, seller i's item would be bough by the intermediary if i bid s_i, then it would also be bought for any lower bid s'_i. For short we will say that S_i is "downward-closed" in the direction of s_i. Similarly for each buyer j, B_j needs to be upward-closed in the direction of b_j.

Equivalently, we may think of a mechanism in terms of critical bids. Myerson tells us that the unique payments that make a monotone allocation rule (as just defined via the regions S_i and B_j) are precisely the critical bids. That is, the lowest (highest) bid for which a buyer (seller) would still be allocated the item (the sale of their item) if everyone else's bids remained fixed. We write $\sigma_i(\mathbf{b}, \mathbf{s}_{-i})$ (respectively, $\beta_j(\mathbf{b}_{-j}, \mathbf{s})$) for these critical bids. (For simplicity sometimes just $\beta_j(\mathbf{b}, \mathbf{s})$ and $\sigma_i(\mathbf{b}, \mathbf{s})$.) If $s_i \leq \sigma_i(\mathbf{b}, \mathbf{s}_{-i})$ we buy an item from seller i (paying $\sigma_i(\mathbf{b}, \mathbf{s}_{-i})$), and similarly if $b_j \geq \beta_j(\mathbf{b}_{-j}, \mathbf{s})$ we sell an item to buyer j (charging $\beta_j(\mathbf{b}_{-j}, \mathbf{s})$). We write $\beta_j(\mathbf{b}_{-j}, \mathbf{s}) = n + 1$ to indicate that a mechanism does not sell to buyer j at all for this combination of others' bids, independently of j's bid. Similarly $\sigma_i(\mathbf{b}, \mathbf{s}_{-i}) = 0$ to indicate not buying from seller i.

It is easy to see that these two yield equivalent definitions. Clearly S_i is simply the region "above" σ_i in the direction of s_i, (the graph of) which in turn is the boundary of S_i. Similarly B_j is the region below β_j in direction b_j. This is a slight generalization of the conceptually simpler picture in auctions. Here we have for each bidder a region B_j where they win the item, and a critical bid function β_j that gives their payment. If there is a single item to be sold, no two of the B_j may overlap. This constraint too generalizes to the market intermediation setting. Consider Fig. 2 in contrast with Fig. 1 to illustrate the difference. As mentioned above, we consider two variants. In the "no short-selling" setting, we must buy at least as many items from sellers as we sell to buyers; in the "balanced inventory" variant we must buy exactly as many as we sell. Formally in terms of critical bids: (Again these can be expressed equivalently in terms of S_i and B_j.)

No Short-Selling:

$$\forall(\mathbf{b},\mathbf{s}), \; |\{j : b_j \geq \beta_j(\mathbf{b},\mathbf{s})\}| \leq |\{i : s_i \leq \sigma_i(\mathbf{b},\mathbf{s})\}| \tag{1a}$$

Balanced Inventory:

$$\forall(\mathbf{b},\mathbf{s}), \; |\{j : b_j \geq \beta_j(\mathbf{b},\mathbf{s})\}| = |\{i : s_i \leq \sigma_i(\mathbf{b},\mathbf{s})\}| \tag{1b}$$

2.2   The Geometry of Deterministic 1×1 Market Intermediation

If there is only a single buyer and a single seller, the constraints simplify significantly, most easily expressed in terms of now simply S and B. In the balanced-inventory case, constraint 1b simplifies to $B = S$. That is, a mechanism in this setting, with only one buyer and seller each, is determined only by a single region of bid-combinations that yield a successful transaction. In the no-short-selling case, constraint 1a simplifies to $B \subseteq S$. That is, S can potentially extend beyond B. However, we can say more, assuming optimality of the mechanism. Recall that by truthfulness, S is down-closed in the seller's direction and B is up-closed in the buyer's direction. If a mechanism is optimal, S must exactly be the down-closure (still in the seller's direction) of B. Firstly, it is easy to see that the down-closure of B must be contained in S: B is contained in S, and S is down-closed. Secondly, if S extended beyond the down-closure of B, we could strictly improve our revenue by removing this protruding part of S. (On the other hand, we may not elect to remove the part of $S - B$ that lies below any

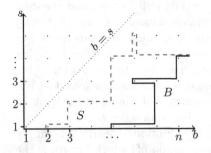

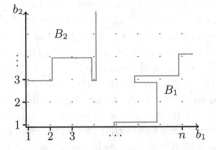

Fig. 1. Example mechanism in the 1×1 case. Note that S contains B, to avoid short-selling. (In a balanced-inventory auction, B and S should coincide.) B and S lie below the diagonal $b = s$: any point above the diagonal is one where the buyer's bid is less than the seller's. The auction shown is suboptimal: in most of the S region, the item is being bought without being sold. Note that we draw the outline of the regions slightly away from the points on the prior support for easier readability.

Fig. 2. Compare this to a two-bidder auction. Here B_1 and B_2 indicate where we sell to each of the two buyers. In the two-bidder auction B_1 and B_2 must be disjoint, as we cannot sell the item twice. In the market intermediation setting, B must be contained in S. Note also that in this setting both B_1 and B_2 are upward-closed in the direction of the respective buyer's bid.

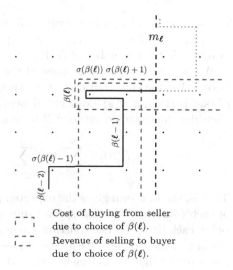

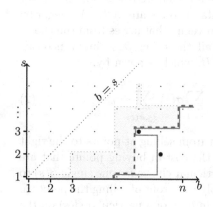

Fig. 3. Removing the indicated area from S as in Fig. 1, the expected revenue of the mechanism cannot decrease. Below red line: Remaining region S, right of black line: Region B. For the remaining part of S that is not also in B, we still buy but not sell the item. This can be optimal, e.g. if there is very high probability weight on the two points indicated. Crucially, if at a point $(b, s) \in S - B$ an optimal mechanism buys but not sells, then there must exist a point $(b, s') \in S \cap B$ with $s' > s$ where it buys and sells. Truthfulness then dictates that it also needs to buy at (b, s). (Color figure online)

Fig. 4. For fixed m_ℓ and $\beta(1), \cdots, \beta(\ell - 1)$, the choice of a particular $\beta(\ell)$ influences the expected revenue in two ways: On the one hand, the revenue from selling to the buyer at all points to the right of $\beta(\ell)$ in row ℓ. On the other hand, the cost of buying from the seller for all points below $\sigma(\beta(\ell)) = \ell$ in rows $\beta(\ell) \leq b < m_\ell$.

point in B due to truthfulness, i.e. down-closedness of S.) Figs. 1 and 3 illustrate this. Note the contrast with a standard 2-bidder auction, where the shape of the region in which we sell to one buyer does not fully determine the region in which we sell to the other. In a way, in the 1×1 market intermediation setting, we have fewer degrees of freedom to consider than in a two-buyer auction setting.

3   The Deterministic One Seller, One Buyer Case

For the case of one seller and one buyer, we show how to compute an optimal deterministic solution using a dynamic programming approach. In the full version of this paper we show how to achieve this via modifications to known 2-bidder auctions in this setting, but the runtime guarantee of that approach, while still polynomial, is substantially worse. We represent a mechanism using values $\beta = (\beta(1), ..., \beta(n))$ for $\beta(s) \in \{1, ..., n+1\}$, where $\beta(s)$ signifies the leftmost point in row s that is a member of B. If $\beta(s) = 1$ then the entire row s is in B. We set $\beta(s) = n+1$ to signify that none of the points in row s are in B. We begin by noting that the contribution to the expected revenue that arises from the choice of one particular $\beta(s)$ does not depend on all the other β_{-s} simultaneously. Consider the expected revenue R for a given B, which is given by:

$$\mathbb{E}[R(\beta)] = \sum_{s=1}^{n} \beta(s) \sum_{b \geq \beta(s)} \psi_{bs} - \sum_{b=1}^{n} \sigma(b) \sum_{s \leq \sigma(b)} \psi_{bs} \tag{2}$$

That is, the first sum gives the expected profit from selling at points to the right of each $(\beta(s), s)$, while the second sum gives the cost of buying points that are below each $(b, \sigma(b))$. The contribution of a particular choice for one single $\beta(s)$ to the first of these sums is easily seen to be simply the profit of selling the points in row s to the right of and including $\beta(s)$. The impact of a particular $\beta(s)$ on the second of the sums is slightly more intricate. There are two ways in which the choice of $\beta(s)$ impacts the cost of buying. Firstly, we may have to buy the item at some points in row s, where we would not buy the item otherwise. Consider the minimum of $\beta(s+1), ..., \beta(n)$, say $\beta(t)$. We know that in row t, we buy and sell at points $(\beta(t), t), ..., (n, t)$. So by truthfulness, we must also buy at points $(\beta(t), s), ..., (n, s)$. This is regardless of our choice of $\beta(s)$. For points to the left of $(\beta(t), s)$, whether we buy the item or not does depend on $\beta(s)$. Secondly, in all those columns in which we buy due to $\beta(s)$, also affect the rows below s. We may increase the buying price from a lower value to s at those points, and (in the no short-selling case) we may have to buy the item (due to truthfulness) at points at which we would not otherwise buy it. The magnitude of this effect depends on all the $\beta(1), ..., \beta(s-1)$. This suggests a bottom-up dynamic programming approach, which we develop in this section.

3.1   Algorithm for the No Short-Selling Case

We next describe our dynamic programming algorithm first for the no short-selling setting. The idea is as follows: because the optimal choice of $\beta(1), ..., \beta(\ell)$ depends only on the minimum of the $\beta(\ell+1), ..., \beta(n)$, we can iteratively compute all the potential optimal values for row 1 given values of $\min\{\beta(2), ..., \beta(n)\}$; then all optimal values of $\beta(1), \beta(2)$ given all possible values of $\min\{\beta(3), ..., \beta(n)\}$. We do not need to consider all n^2 combinations of $\beta(2)$ and $\beta(1)$. Since given $\beta(2)$ and $\min\{\beta(3), ..., \beta(n)\}$, we can immediately look up the best $\beta(1)$ using the information computed in the first step. We then proceed iteratively up the rows until we have computed to optimal values for β.

Let us start by defining $R(n, \beta(\ell), m_\ell)$ to be the expected revenue of the best deterministic mechanism that takes points $(\beta(\ell), \ell)$ and rightward in row ℓ, no points in rows $\ell+1$ and above, and does not have to pay for points in columns m_ℓ to n. We set $R(0, ., .) = 0$. The idea is that we want to capture the best possible revenue extractable from rows 1 to ℓ for a particular choice of $\beta(\ell)$, disregarding the cost of buying in columns m_ℓ to n. We can take $R(n, \beta(\ell), n+1)$ to denote the optimal revenue among mechanisms that have to pay for all rows. More precisely,

$$R(\ell, \beta(\ell), m_\ell) = \max_{\beta(1),\ldots,\beta(\ell-1)} \sum_{s=1}^{\ell} \beta(s) \sum_{b \geq \beta(s)} \psi_{bs} - \sum_{b=1}^{m_\ell - 1} \sigma(b) \sum_{s \leq \sigma(b)} \psi_{bs} \quad (3)$$

It is easy to see that $\max_{\beta(n)} R(n, \beta(n), n+1)$ gives the revenue of the optimal auction. Indeed, by definition this is the maximum expected revenue extractable from all rows, if we have to pay in all columns. We can then show how to recursively compute the values of R, laying the groundwork for our dynamic programming algorithm.

Theorem 1 (Recursion for the no short-selling case). *The $R(\ell, \beta(\ell), m_\ell)$ as defined above satisfy the following recursion:*

$$R(\ell, \beta(\ell), m_\ell) = \max_{\beta(\ell-1)} R\big(\ell - 1, \beta(\ell - 1), \min\{\beta(\ell), m_\ell\}\big) +$$

$$\beta(\ell) \sum_{b \geq \beta(\ell)} \psi_{b\ell} - \ell \sum_{\beta(\ell) \leq b < m_\ell} \sum_{s \leq \ell} \psi_{bs} \quad (4)$$

Proof. We can check this by splitting up the explicit formula for $R(\ell, \beta(\ell), m_\ell)$ into terms for rows below ℓ and row ℓ, and columns to the left of $\min(\beta(\ell), m_\ell)$ and those between the two.

$$R(\ell, \beta(\ell), m_\ell) = \sum_{s=1}^{\ell-1} \beta(s) \sum_{b \geq \beta(s)} \psi_{bs} + \beta(\ell) \sum_{b \geq \beta(\ell)} \psi_{b\ell} -$$

$$\sum_{b=1}^{\min(\beta(\ell),m_\ell)-1} \sigma(b) \sum_{s \leq \sigma(b)} \psi_{bs} - \sum_{b=\beta(\ell)}^{m_\ell - 1} \sigma(b) \sum_{s \leq \sigma(b)} \psi_{bs}$$

Observe that for $b \geq \beta(\ell)$, $\sigma(b)$ will be equal to ℓ (in the $(\ell, \beta(\ell))$-auction), so the last term in the above sum is precisely $\ell \sum_{b=\beta(\ell)}^{m_\ell - 1} \sum_{s \leq \ell} \psi_{bs}$. Similarly, $\min(\beta(\ell), m_\ell)$ is precisely the $m_{\ell-1}$ we used in the recursion, and therefore the first and third term are precisely $R(\ell-1, \beta(\ell - 1), m_{\ell-1})$. Putting these together, we get that:

$$R(\ell, \beta(\ell), m_\ell) = R(\ell - 1, \beta(\ell - 1), m_{\ell-1}) + \beta(\ell) \sum_{b \geq \beta(\ell)} \psi_{b\ell} - \ell \sum_{b=\beta(\ell)}^{m_\ell - 1} \sum_{s \leq \ell} \psi_{bs} \quad (5)$$

i.e. precisely our claimed recursion. (The max follows from optimality of the auction.) The second term on the right hand side is the revenue from selling at points due to the choice of $\beta(\ell)$, while the third term accounts for the cost of buying at points due to this choice. Figure 4 illustrates these two terms. Note that if $\ell = 1$ then the first term vanishes since we defined $R(0, ., .) = 0$, and we are left with the explicit formula for $R(1, ., .)$. $\qquad\square$

We can therefore compute the $R(\ell, \beta(\ell), m_\ell)$ recursively, as claimed. This suggests the following algorithm, listed below as Algorithm 1. This can easily be augmented to keep track of the values used for the $\beta(s)$, and to return the optimal β together with its expected revenue. Therefore, we can compute the optimal region B and thereby the optimal mechanism in the no short-selling setting in time $\mathcal{O}(n^4)$.[1]

Algorithm 1. Optimal revenue in the no short-selling setting

1: **for** $\ell = 1, ..., n$ **do**
2:     **for** $\beta(\ell) = 1, .., n$ **do**
3:         **for** $m_\ell = 1, ..., n+1$ **do**
4:             **if** $\ell = 1$ **then**
5:                 $R(1, \beta(1), m_1) \leftarrow \beta(1) \sum_{b \geq \beta(1)} \psi_{b1} - \sum_{\beta(1) \leq b < m_1} \psi_{b1}$
6:             **else**
7:                 Compute $R(\ell, \beta(\ell), m_\ell)$ using the recursion in theorem 1.
    return $\max_{\beta(n)} R(n, \beta(n), n+1)$.

We can easily modify this algorithm to return the optimal mechanism that satisfies the balanced inventory property. We show the details in the full version of this paper. This modified algorithm runs in time $\mathcal{O}(n^3)$.

4 NP-hardness for the Deterministic Multiple Buyers or Sellers Case

For three or more buyers, it follows from Papadimitriou and Pierrakos [13] that computing the optimal mechanism is NP-hard. We show that this is also true for the 2×1 case (i.e. two buyers, one seller) in the no-short-selling setting by reducing from Maximum Independent Set. The idea here is to place high probability weight on high-revenue points along a diagonal in the $s = 1$ plane for each vertex of a given instance of Independent Set. We then use appropriately placed high-probability points for each of the edges to "force" a higher buying price for (at least) one of any two points corresponding to adjacent vertices. We can do this in a way that ensures that in the optimal mechanism the number of vertex points with a low buying price is maximized and corresponds to the maximum independent set.

Theorem 2 (NP-hardness). *It is NP-hard to compute the optimal mechanism in the 1 seller, 2 buyers setting with no short selling.*

---

[1] Careful analysis of the algorithm presented shows that the last summand in the recursion for $R()$ has $(m_\ell - \beta(\ell)) \cdot \ell$ summands. It is easy to see however that we need not recompute the inner sum from scratch in each iteration. We can thus easily make the computation of the recursion run in linear time, giving the overall running time stated.

Proof. In the following we construct a prior distribution in (b_1, b_2, s)-space. We will "choose" points, and place equal probability mass $\frac{1}{|V|+|E|}$ on all of these chosen points. In the analysis we will omit these weights to simplify the algebra. We place probability 0 on all other points in the prior support. We use K_1 and K_2 as constants whose values we define at the end of the proof.

The Construction. Given a graph (V, E) with $|V| = n$, pick any order of vertices and begin by placing probability weight $\frac{1}{|V|+|E|}$ on point $(K_1 + \lfloor \frac{n}{2} \rfloor - i, K_1 - \lfloor \frac{n}{2} \rfloor + i, 1)$ for each vertex $0 \leq i < n$. Next, enumerate the edges e_j, $0 \leq j < |E|$. We will write each edge as $e_j = (e_{j1}, e_{j2})$, where $e_{j1} < e_{j2}$ in the order of vertices just picked. For each edge put probability weight $\frac{1}{|V|+|E|}$ on point $(K_1 + \lfloor \frac{n}{2} \rfloor - e_{j2}, K_1 - \lfloor \frac{n}{2} \rfloor + e_{j1}, K_2 + j)$. That is, we put probability weight for each edge on a point that has the same b_1-coordinate as the vertex point for its lower-numbered vertex and the same b_2-coordinate as its higher vertex. We choose these edge points with a different s-coordinate each, and all of them with a higher s-coordinate than the vertex points. It is clear that if the mechanism wants to buy and sell at an edge point $(K_1 + \lfloor \frac{n}{2} \rfloor - e_{j2}, K_1 - \lfloor \frac{n}{2} \rfloor + e_{j1}, K_2 + j)$, it will also need to sell (and therefore buy by truthfulness) at one of the points $(K_1 + \lfloor \frac{n}{2} \rfloor - e_{j2}, K_1 - \lfloor \frac{n}{2} \rfloor + e_{j2}, K_2 + j)$ or $(K_1 + \lfloor \frac{n}{2} \rfloor - e_{j1}, K_1 - \lfloor \frac{n}{2} \rfloor + e_{j1}, K_2 + j)$, when it sells to buyer 1 or buyer 2, respectively. But by truthfulness this entails a raised purchase price of $K_2 + j$ at the corresponding vertex points directly below $((K_1 + \lfloor \frac{n}{2} \rfloor - e_{j2}, K_1 - \lfloor \frac{n}{2} \rfloor + e_{j2}, 1)$ or $(K_1 + \lfloor \frac{n}{2} \rfloor - e_{j1}, K_1 - \lfloor \frac{n}{2} \rfloor + e_{j1}, 1))$ where it had otherwise been 1. Figure 5 illustrates this construction.

Reducing from Maximum Independent Set. Now, in order to ensure that the optimal mechanism raises the purchasing price at all vertex points except those that are in an independent set of maximum size, we need to pick constants K_1, K_2 in a way that ensures that:

1. The optimal mechanism always buys and sells at all the edge points.
2. The optimal mechanism raises the purchasing price at as few vertex points as possible.

From condition 1: The worst possible selling price at any edge point is given by $K_1 - \lfloor \frac{n}{2} \rfloor$, and the highest possible purchase price is $K_2 + |E| \leq K_2 + n^2$, for a revenue that is at least $K_1 - \lfloor \frac{n}{2} \rfloor - K_2 - n^2$. On the other hand, buying and selling at an edge point could necessitate a higher purchasing price at a vertex point, raising it by an amount that is bounded above by $K_2 + n^2$ as well. The profit obtained from the edge point must outweigh this. So in order to ensure that the optimal mechanisms buys and sells at all edge points, we need to ensure:

$$K_1 - \lfloor \frac{n}{2} \rfloor - 2K_2 - 2n^2 > 0$$

From condition 2: We need to ensure that if for an edge point $(K_1 + \lfloor \frac{n}{2} \rfloor - e_{j2}, K_1 - \lfloor \frac{n}{2} \rfloor + e_{j1}), K_2 + j)$, only one of the two corresponding vertex points already has a purchase price of at least K_2 due to another edge, but the other is still 1, the optimal mechanism will always prefer to sell to the buyer whose corresponding vertex point already has a high price. In other words, we need to ensure that the potential difference in revenue from selling to one buyer over the other is outweighed by the required raise in the purchase price by (at least) $K_2 - 1$. But the highest difference in selling price is bounded by n, and so we required that $K_2 > n + 1$.

Combining the two we get our desired result: Set $K_2 = 2n$ and $K_1 = 4n^2$ in the above construction for a given instance of Maximum Independent Set. Since the optimal mechanism will buy and sell at all the edge points, it is clear that at most one vertex point corresponding to two adjacent vertices can have a purchase price of 1. On the other hand, in the optimal mechanism the number of vertex points with a raised purchase price will be minimized. Therefore, the vertex points with purchase price 1 in the optimal mechanism correspond to the vertices of the maximum independent set in the graph.   □

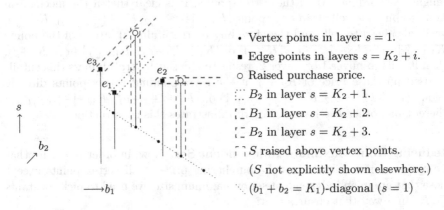

- Vertex points in layer $s = 1$.
- Edge points in layers $s = K_2 + i$.
- Raised purchase price.
- B_2 in layer $s = K_2 + 1$.
- B_1 in layer $s = K_2 + 2$.
- B_2 in layer $s = K_2 + 3$.
- S raised above vertex points.
  (S not explicitly shown elsewhere.)
- $(b_1 + b_2 = K_1)$-diagonal $(s = 1)$

Fig. 5. The construction for the reduction from Maximum Independent Set.

5   Truthful-in-Expectation Mechanisms

While in the preceding section we have shown that we cannot compute an optimal deterministic mechanism for the general case, we can however compute the optimal truthful-in-expectation mechanism for a fixed number of buyers and sellers. In single-item auctions a randomized mechanism is easily described by allocation probabilities $x_i(\mathbf{v})$ and expected payments $p_i(\mathbf{v})$ for all players for each possible bid vector. In the market intermediation setting with multiple buyers and sellers this is not obviously the case. For instance, there are many ways in which to make allocation probabilities of $\frac{1}{2}$ for each of two buyers and two sellers into a randomization over valid outcomes. The mechanism could flip a coin and buy from seller 1 and sell to buyer 1 on heads, seller 2 and buyer 2 on tails.

It could not, however, independently flip four coins if we want to fulfill condition (1a) respectively (1b) ex-post. In the following we will consider the balanced inventory case. Our arguments easily extend to the no short-selling case. We first show that for our purposes, it is indeed sufficient to consider only the marginal allocation probabilities x_i, y_j and expected payments p_i, q_j. First, observe that any two randomized mechanisms that have the same marginal probabilities and expected payments will lead to identical expected utilities for players and expected revenue. It remains to show that any sensible vector of marginal allocation probabilities can be made into a probability distribution over valid outcomes (i.e. allocations which buy exactly as many items as they sell).

Theorem 3. *Let x, y be k-dimensional vectors of probabilities, i.e. $0 \leq x_i, y_i \leq 1$, with $\sum_i x_i = \sum_i y_i$. Then there exists a joint probability distribution over 2 k-dimensional $0/1$ vectors $\{(a,b) \in \{0,1\}^{2k} | \sum_i a_i = \sum_i b_i\}$ which satisfies $\Pr(a_i = 1) = x_i$ and $\Pr(b_i = 1) = y_i$.*

Proof. Let $H_{2k} = [0,1]^{2k}$ be the $2k$-dimensional hypercube, and $H_{2k}^* = \{0,1\}^{2k}$ its vertices. Let $D_{2k} = \{(x,y) \in H_{2k} | \sum x_i = \sum y_i\}$ be the "generalized diagonal" of the hypercube. Let $D_{2k}^* = \{(x,y) \in \{0,1\}^{2k} | \sum x_i = \sum y_i\}$ be the vertices of H_{2k} with as many x-coordinates set to 1 as y-coordinates. That is, this is the set of valid (deterministic) allocation vectors for k buyers and sellers. Then our claim is equivalent to saying that D_{2k} is (in) the convex hull of D_{2k}^*. By the Krein-Milman theorem a convex set S is exactly the convex hull of its extreme points. An extreme point $s \in S$ is any point in S which can not be written as a convex combination of points in $S \setminus s$. Clearly D_{2k} is convex. It remains to show that the extreme points of D_{2k} are precisely D_{2k}^*. Clearly $D_{2k}^* \subseteq D_{2k}$. So let $(\mathbf{x}, \mathbf{y}) \in D_{2k} \setminus D_{2k}^*$ be a point in D_{2k} that does not have all elements equal to 0 or 1. We show that $(\mathbf{x}, \mathbf{y})$ is not an extreme point of D_{2k}.

If there is exactly one x_i with $0 < x_i < 1$, then there must be at least one y_j with $0 < y_j < 1$. (Otherwise $\sum x_i \notin \mathbb{N}$, but $\sum y_j \in \mathbb{N}$, which contradicts the assumption that $\sum x_i = \sum y_j$.) Then for $0 < \epsilon < \min\{x_i, 1-x_i, y_j, 1-y_j\}$, we have that $(x_i + \epsilon, y_j + \epsilon, \mathbf{x}_{-i}, \mathbf{y}_{-j}) \in D_{2k}$, and also $(x_i - \epsilon, y_j - \epsilon, \mathbf{x}_{-i}, \mathbf{y}_{-j}) \in D_{2k}$. Clearly $(\mathbf{x}, \mathbf{y})$ is a convex combination of these two. If there is at least two distinct $0 < x_i, x_\ell < 1, i \neq \ell$, then for $0 < \epsilon < \min\{x_i, 1-x_i, x_\ell, 1-x_\ell\}$, we have that $(x_i + \epsilon, x_\ell - \epsilon, \mathbf{x}_{-i\ell}, \mathbf{y}) \in D_{2k}$, and also $(x_i - \epsilon, x_\ell + \epsilon, \mathbf{x}_{-i\ell}, \mathbf{y}) \in D_{2k}$. Again, clearly $(\mathbf{x}, \mathbf{y})$ is a convex combination of these two. Similarly, if there is no $0 < x_i < 1$ there is at least two such y_j, y_ℓ. So D_{2k} is the convex hull of D_{2k}^*. This shows our claim. $\square$

From this it follows immediately that we need only concern ourselves with the marginal allocation probabilities in computing an optimal randomized mechanism. Therefore we can write this as a LP following the approach of Dobzinski et al. [3] for auctions. We defer the proof of this theorem to the full version.

Theorem 4 (The optimal randomized mechanism as an LP). *For a fixed number of buyers and sellers, we can compute the optimal truthful-in-expectation mechanism using a linear program that is polynomial in the size of the prior.*

6   Discussion and Further Work

One question raised by our results is that of the relation between single seller, single buyer market intermediation and two-bidder auctions. As mentioned, and discussed in the full version the graph algorithm of Papadimitriou and Pierrakos [13] can be used to solve the no short-selling 1×1 market intermediation case, and the derandomization in Dobzinski et al. [3] applies immediately to both this and the 1×1 balanced inventory setting. These give running times of $\mathcal{O}(n^6)$ and $\mathcal{O}(n^7)$ in contrast to a running time of $\mathcal{O}(n^4)$, respectively $\mathcal{O}(n^3)$ in the balanced inventory case, for our approach in the market intermediation setting. It is not clear immediately that the 2-bidder auction design problem could in turn be solved using a modified version of this algorithm, given the additional complexity of two interdependent regions for each seller. We suspect that there might indeed be a gap between the complexity of these two problems. Furthermore, we believe that an optimal 2-bidder reverse auction can be computed using our dynamic program for the balanced inventory case. Thus a gap between auctions and market intermediation would imply an asymmetry between auctions and reverse auctions. An immediate follow-up question is if we can give good approximations in polynomial time. In the full version of this paper we show that no good multiplicative guarantees are possible using prior-independent mechanisms.

References

1. Colini-Baldeschi, R., de Keijzer, B., Leonardi, S., Turchetta, S.: Approximately efficient double auctions with strong budget balance. In: Proceedings of SODA. ACM-SIAM (2016)
2. Deng, X., Goldberg, P., Tang, B., Zhang, J.: Revenue maximization in a bayesian double auction market. Theoret. Comput. Sci. **539**, 1–12 (2014)
3. Dobzinski, S., Fu, H., Kleinberg, R.D.: Optimal auctions with correlated bidders are easy. In: Proceedings of the Forty-third Annual ACM Symposium on Theory of Computing, pp. 129–138. ACM (2011)
4. Dütting, P., Roughgarden, T., Talgam-Cohen, I.: Modularity and greed in double auctions. In: Proceedings of the Fifteenth ACM Conference on Economics and Computation, pp. 241–258. ACM (2014)
5. Feldman, M., Gravin, N., Lucier, B.: Combinatorial auctions via posted prices. In: Proceedings of SODA, pp. 123–135. ACM-SIAM (2015)
6. Loertscher, S., Niedermayer, A.: When is seller price setting with linear fees optimal for intermediaries? Technical report, Discussion Papers, Department of Economics, Universität Bern (2007)
7. Loertscher, S., Niedermayer, A.: Fee setting intermediaries: on real estate agents, stock brokers, and auction houses. Technical report, Discussion paper Center for Mathematical Studies in Economics and Management Science (2008)
8. Loertscher, S., Niedermayer, A.: Fee-setting mechanisms: on optimal pricing by intermediaries and indirect taxation. Technical report, SFB/TR 15 Discussion Paper (2012)
9. McAfee, R.P.: A dominant strategy double auction. J. Econ. Theory **56**(2), 434–450 (1992)

10. Myerson, R.B.: Optimal auction design. Math. Oper. Res. **6**(1), 58–73 (1981)
11. Myerson, R.B., Satterthwaite, M.A.: Efficient mechanisms for bilateral trading. J. Econ. Theory **29**(2), 265–281 (1983)
12. Niazadeh, R., Yuan, Y., Kleinberg, R.D.: Simple and near-optimal mechanisms for market intermediation. arXiv preprint arXiv:1409.2597 (2014)
13. Papadimitriou, C.H., Pierrakos, G.: On optimal single-item auctions. In: Proceedings of the Forty-third Annual ACM Symposium on Theory of Computing, pp. 119–128. ACM (2011)
14. Ronen, A.: On approximating optimal auctions. In: Proceedings of the 3rd ACM Conference on Electronic Commerce, pp. 11–17. ACM (2001)

Mechanism Design

Bribeproof Mechanisms for Two-Values Domains

Matúš Mihalák[1], Paolo Penna[2(✉)], and Peter Widmayer[2]

[1] Department of Knowledge Engineering,
Maastricht University, Maastricht, The Netherlands
[2] Department of Computer Science, ETH Zurich, Zurich, Switzerland
paolo.penna@inf.ethz.ch

Abstract. Schummer [27] introduced the concept of *bribeproof* mechanism which, in a context where monetary transfer between agents is possible, requires that manipulations through bribes are ruled out. Unfortunately, in many domains, the only bribeproof mechanisms are the trivial ones which return a *fixed outcome*.

This work presents one of the few constructions of non-trivial bribeproof mechanisms for this setting. Though the suggested construction applies to rather restricted domains, the results obtained are tight: for several natural problems, the method yields the only possible bribeproof mechanism and no such mechanism is possible on more general domains.

1 Introduction

Strategyproof mechanisms guarantee that the agents never find it convenient to misreport their types, that is, truth-telling is a dominant strategy. Such mechanisms play a key role to cope with selfish behavior, and they received a lot of attention also when considering protocols for optimally allocating resources that necessarily involve selfish entities [21]. One of the critical issues with strategyproof mechanisms is that agents can still manipulate the mechanism, and improve their utilities, by *bribing* one another:

An agent can offer money to another for misreporting her type and in this way the utility of both improves.

The famous second-price auction provides a clear example of such an issue. If two agents are willing to pay 10 and 9 for an item, the one bidding 10 wins and pays 9. However, if before the auction starts the winner offers some money to the other agent for bidding a low value (say 1), then both agents would be better off (now the winner pays only 1 and the other agent gets some money).

The concept of *bribeproof* mechanism [27] strengthens strategyproofness by requiring that bribing another agent is also not beneficial. The appeal of this notion is that it does not consider unreasonably large coalitions.[1] Despite

---

[1] The notion of *coalitional strategyproofness* requires the mechanism to be immune to manipulations by any group of agents. As already observed in [27,29], this notion turns out to be too restrictive as it rules out all but a few unreasonable mechanisms. Moreover, large coalitions would require all members to coordinate their actions.

© Springer-Verlag Berlin Heidelberg 2016
M. Gairing and R. Savani (Eds.): SAGT 2016, LNCS 9928, pp. 289–301, 2016.
DOI: 10.1007/978-3-662-53354-3_23

bribeproofness is apparently adding only a minimal condition, this has a tremendous impact on what the mechanisms can do in general:

- The class of strategyproof mechanisms is extremely rich and, among others, it includes VCG mechanisms which optimize the social welfare;
- In contrast, the class of bribeproof mechanisms consists of only *trivial* mechanisms which output a *fixed outcome* [16,27].

That is to say, while strategyproofness by itself is not an obstacle to optimization, the only way to get bribeproofness is to ignore the agents types, which clashes with most optimization criteria. One example of such VCG mechanisms is for the path auction problem [21] where we want to select the shortest path between two nodes of a given network, every edge is owned by an agent, and the cost of the edges are private. Selecting the shortest path means that we want the solution minimizing the sum of all agents' costs, that is, to optimize the social welfare. Clearly, any trivial mechanism which returns a fixed path has no guarantee to find the shortest path.

1.1 Our Contribution

Because the impossibility results on bribeproof mechanisms hold for unrestricted or for "sufficiently rich" domains [16,27], we are interested in designing bribeproof mechanisms for *restricted domains*. Specifically, we present a novel construction of bribeproof mechanisms for the following class of a *two-values* problems. Every feasible solution corresponds to some *amount of work* allocated to each agent, and every agent has a *private cost* per unit of work which is either L (low) or H (high).[2] Typically the amount of work allocated to the agents cannot be arbitrary, but it is rather determined by the "combinatorial structure" of the problem under consideration. For instance, in the path auction problem [21], the mechanism must select a path in a graph, and each agent owns one edge of the graph (see Fig. 1). Selecting a path means allocating one unit of work to each agent in the path, and no work to all other agents.

In a nutshell our results can be summarized as follows:

- An extremely simple construction yields bribeproof mechanisms if the underlying algorithm satisfies certain monotonicity conditions (Sect. 2).
- One application of the above result, is a class of bribeproof mechanisms optimizing the social welfare for every *binary* allocation problem, that is, whenever each agent is either selected or not selected (Sect. 3).
- These mechanisms actually *characterize* the whole class of bribeproof mechanisms for certain problems, including the path auction one, and the boundary conditions for which such mechanisms exist (Sect. 4).

---

[2] Throughout this work we adopt the terminology used by [1] in the context of procurement auctions, though these domains have been investigated earlier in the context of allocating identical goods, as well as for certain restricted combinatorial auctions. All the results apply to these problems as well (see full version [14]).

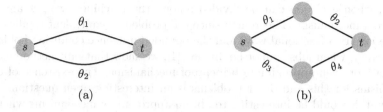

Fig. 1. Two instances of the path auction problem.

- The positive result is more general as it can be applied to non-binary problems and to other optimization criteria (Sect. 5).

More in detail, our mechanisms simply provide all agents the *same* amount of money $\mathcal{M}$ for each unit of work that they get allocated (Definition 1). Such mechanisms are bribeproof if certain monotonicity conditions hold (Theorem 2). Roughly speaking, these conditions relate the "influence" that an agent has on her own allocation to the influence she has on the *others'* allocation. In particular, by taking the special case $\mathcal{M} = \frac{L+H}{2}$ in our construction leads to the following natural sufficient condition (Corollary 1):

> *Bounded influence:* No agent can change the allocation of *another* agent by more than the change caused to *her own* allocation.

For the class of *binary allocations*, where the allocated work is *zero* or *one*, this condition is nothing but *non-bossiness*: no agent can change the allocation of the others, without changing her own allocation (Theorem 3). The main positive result here is that every problem in which one wants to minimize the weighted *sum* of all agents' costs admits an exact strongly bribeproof mechanism (Theorem 4). Interestingly, our general construction provides both characterizations of bribeproof mechanisms as well as the boundary conditions for which such mechanisms exist in several problems:

- For the path auction problem, our mechanism with $\mathcal{M} = \frac{L+H}{2}$ is essentially the only possible, and no mechanism exist on slightly more general domains (with three values, or heterogeneous two values), nor collusion-proof mechanisms for coalitions of three or more agents.
- For the k-items procurement auction, the mechanism with $\mathcal{M} = M$ is bribeproof on *three values* domains L (low), M(medium), H(high). This is the only mechanism for $k = 1$ and no mechanism for *four values* domains exist.

We then turn our attention to problems with different objective function and non-binary allocations. Specifically, we consider minimizing the *maximum* cost among the agents (note that this is different from welfare maximization which would minimize the *sum* of all agents' costs). In the scheduling terminology, we aim at minimizing the *makespan* on related machines [1]. In the fractional version, when each job can be divided among the machines, we get an exact bribeproof mechanism (Theorem 9) since the problem is equivalent to allocating a single job (in a fractional way) and the bounded influence condition holds. On the contrary, when jobs cannot be divided [1], we show that our method cannot give exact or even approximate bribeproof mechanisms. The existence of other mechanisms for this and other problems is an interesting open question. More in general, it would be interesting to obtain approximate mechanisms when the domain does not allow for exact ones.

1.2 Related Work

Schummer [27] introduced the notion of bribeproofness and proved that, on certain domains, the only bribeproof mechanisms are the *trivial* mechanisms which return a *fixed outcome*; [16] proved the same but under weaker assumptions. In simpler domains, bribeproof (or even collusion-proof) mechanisms can be obtained via *take-it-or-leave-it* prices [8,9]: these mechanisms fix a price for each agent, who then wins a copy of the item if bidding above this price, independently of what happens to the other agents. Note that our mechanisms are different from these mechanisms since in our setting we cannot treat agents separately.

Though strategyproofness is much less stringent and quite well understood, restricting the domain is also very common, as unrestricted domains are often unrealistic and impose unnecessary limitations (see e.g. [5,12,25]). In multidimensional domains, minimizing the *makespan* or *min-max fairness* is not possible using strategyproof mechanisms [2,7,11,13,18,21], while for one-parameter domains optimal solutions are possible [1,18] also in polynomial time [4]. Our domains are at the intersection of one-parameter domains in [1] and the two-values domains in [13], and they also appear in study of revenue of take-it-or-leave-it identical items auctions [9]. One-parameter domains have been studied by [20] who characterized strategyproofness and obtained optimal-revenue mechanisms for selling a single item.

The strong limitations imposed by bribeproofness lead to the study of weaker or variants of this notion. *Group strategyproofness* assumes that the members of the coalitions can coordinate their reports but cannot exchange compensations (see e.g. [10,17,19,24]). The restriction to coalitions of size two is called *pairwise strategyproofness* [28], and it corresponds to strong bribeproofness when compensations between agents are not allowed. The class of *deferred acceptance* mechanisms [15] satisfies (weakly) group strategyproofness[3], at the price of significantly worse social welfare even in rather simple settings [6]. Mechanisms with *verification* [23] are based on the assumption that it is possible to partially verify the agents types after the solution is computed. *Collusive dominant-strategy truthfulness* [3] is based on the idea that the mechanism asks the agents to report also their coalitions, and it provides better performance for selling identical items.

---

[3] This condition relaxes group strategyproofness, by requiring that no coalition could deviate from truth-telling in a way that makes *all* of its members strictly better off.

In the so-called *single-peaked* domains, agents receive a variable amount of a divisible item, and they can bribe each other by transferring part of the item (with no money involved). Interestingly enough, [29] characterizes bribeproofness in terms of a *bounded impact* condition which is very similar to our bounded influence, despite the two settings being not equivalent. Finally, while [26] shows that bribeproofness is closely related to Pareto efficiency together with strategyproofness, [22] proved that the latter two requirements cannot be achieved simultaneously in a setting involving finite "small" domains.

1.3 Preliminaries

There is a set $N = \{1, 2, \ldots, n\}$ of $n \geq 2$ agents and a set $\mathcal{A} \subseteq \mathbb{R}^n_+$ of feasible allocations, where each allocation $a \in \mathcal{A}$ is an n-dimensional vector $(a_1, \ldots, a_n)$ with a_i being the amount of work allocated to agent i. For each agent i, her cost for an allocation a is equal to

$$a_i \cdot \theta_i,$$

where $\theta_i \in \mathbb{R}$ is some private number called the *type* of this agent (her cost for a unit of work). Every type θ_i belongs to a publicly-known set Θ_i which is the domain of agent i, and the agent can misreport her type θ_i to any $\hat{\theta}_i \in \Theta_i$. The cross-product $\Theta := \Theta_1 \times \cdots \times \Theta_n$ is the types domain representing the possible type vectors that can be reported by the agents.

A mechanism is a pair (A, p) where $A : \Theta \to \mathcal{A}$ is an algorithm and $p : \Theta \to \mathbb{R}^n$ is a suitable payment function. For any type vector $\hat{\theta} \in \Theta$ reported by the agents, each agent i receives $p_i(\hat{\theta})$ units of money and $A_i(\hat{\theta})$ units of work. A mechanism (A, p) is **bribeproof** if for all $\theta \in \Theta$, all i and j, and all $\hat{\theta}_i \in \Theta_i$

$$
\begin{array}{ccccc}
p_i(\theta) - A_i(\theta) \cdot \theta_i & + & p_j(\theta) - A_j(\theta) \cdot \theta_j & \geq & (1) \\
p_i(\hat{\theta}) - A_i(\hat{\theta}) \cdot \theta_i & + & p_j(\hat{\theta}) - A_j(\hat{\theta}) \cdot \theta_j & &
\end{array}
$$

where $\hat{\theta} = (\hat{\theta}_i, \theta_{-i}) := (\theta_1, \ldots, \theta_{i-1}, \hat{\theta}_i, \theta_{i+1}, \ldots, \theta_n)$ denotes the vector obtained by replacing the ith entry of θ with $\hat{\theta}_i$.

Inequality (1) says that no agent j can bribe another agent i with b units of money to misreport her type so that they both improve. By taking $i = j$ in the definition above, we obtain the (weaker) notion of **strategyproof** mechanism, that is, $p_i(\theta) - A_i(\theta) \cdot \theta_i \geq p_i(\hat{\theta}_i, \theta_{-i}) - A_i(\hat{\theta}_i, \theta_{-i}) \cdot \theta_i$, for all $\theta \in \Theta$, for all i, and for all $\hat{\theta}_i \in \Theta_i$. Strong bribeproofness requires that no two agents can improve even if they *jointly* misreporting their types (see [27, p. 184]). Let $(\hat{\theta}_i, \hat{\theta}_j, \theta_{-ij})$ denote the vector obtained by replacing the ith and the jth entry of θ with $\hat{\theta}_i$ and $\hat{\theta}_j$, respectively. A mechanism (A, p) is **strongly bribeproof** if inequality (1) holds also for all $\hat{\theta} = (\hat{\theta}_i, \hat{\theta}_j, \theta_{-ij})$, with $\theta_i \in \Theta_i$, $\theta_j \in \Theta_j$, and $\theta \in \Theta$.

A domain Θ is a **two-values domain** if there exist two constants L and H with $L < H$ such that $\Theta_i = \{L, H\}$ for all $i \in N$. More generally, for any ordered sequence of reals $w_1 < w_2 < \cdots < w_k$, we denote by $\Theta^{(w_1, w_2, \ldots, w_k)}$ the

k-values domain Θ such that $\Theta_i = \{w_1, w_2, \ldots, w_k\}$ for all $i \in N$. We say that a mechanism is (strongly) bribeproof over a k-values domain if the corresponding condition (1) holds for Θ being a k-values domain.

Example 1 (path auction and perfectly divisible good). In the *path auction problem* instance in Fig. 1, the two feasible allocations are $(1, 1, 0, 0)$ for the "upper path" and $(0, 0, 1, 1)$ for the "lower path". The problem of allocating a single *perfectly divisible* good among the agents corresponds to the set of feasible allocations consists of all vectors $a = (a_1, \ldots, a_n)$ such that $a_i \geq 0$ and $\sum_{i=1}^n a_i = 1$.

In a bribeproof mechanism the payments must depend only on the allocation:

Fact 1. *We say that two type vectors θ' and θ'' are A-equivalent if they differ in exactly one agent's type and algorithm A returns the same allocation in both cases. That is $\theta'' = (\theta_i'', \theta_{-i}')$ and $A(\theta'') = A(\theta')$. In a bribeproof mechanism (A, p) the payment for two A-equivalent type vectors must be the same.*

Proof. Suppose by way of contradiction that $p(\theta') \neq p(\theta'')$ and, without loss of generality, that $p_j(\theta') > p_j(\theta'')$ for some agent j. Since $A(\theta') = A(\theta'')$, this violates bribeproofness (1). $\qquad\square$

2   A Class of Bribeproof Mechanisms

The idea to obtain bribeproof mechanisms is to pay each agent the same fixed amount for each unit of work she gets allocated.

Definition 1 (linear mechanism). *A mechanism (A, p) is a λ-linear mechanism if every agent i receives a fixed payment f_i plus $\lambda L + (1 - \lambda)H$ units of money for each unit of allocated work, where $\lambda \in [0, 1]$. That is,*

$$p_i(\theta) = A_i(\theta) \cdot q^{(\lambda)} + f_i \quad where \quad q^{(\lambda)} = \lambda L + (1 - \lambda)H$$

for all i and for all $\theta \in \Theta$.

Remark 1. Note that, in Definition 1, we limit ourself to $q \in [L, H]$ because otherwise the mechanism would not be even strategyproof in general. Moreover, the constants f_i can be used to rescale the payments without affecting bribeproofness. For instance, one can set each f_i so that truthfully reporting agents are guaranteed a nonnegative utility, i.e., the mechanism satisfies voluntary participation or individual rationality.

In the following we define

$$i\text{-}influence(A_k, \theta) := A_k(L, \theta_{-i}) - A_k(H, \theta_{-i}).$$

The monotonicity condition for strategyproofness [1,20] requires that the allocation of each agent is weakly decreasing in her reported cost, that is, for all $i \in N$ and for all $\theta \in \Theta$

$$i\text{-}influence(A_i, \theta) \geq 0 \qquad \text{(monotonicity)}. \qquad (2)$$

We next show that a stronger condition suffices for bribeproofness.

Theorem 2. *The λ-linear mechanism is bribeproof for a two-values domain if and only if algorithm A satisfies the following conditions: for all $\theta \in \Theta^{(L,H)}$ and for all $i \in N$ condition (2) holds and, for all $\ell \in N$ with $\theta_\ell = L$ and for all $h \in N$ with $\theta_h = H$,*

$$(1 - \lambda) \cdot i\text{-}influence(A_i, \theta) \quad \geq \quad (\lambda - 1) \quad \cdot i\text{-}influence(A_\ell, \theta), \quad (3)$$

$$(1 - \lambda) \cdot i\text{-}influence(A_i, \theta) \quad \geq \quad \lambda \quad \cdot i\text{-}influence(A_h, \theta), \quad (4)$$

$$\lambda \quad \cdot i\text{-}influence(A_i, \theta) \quad \geq \quad (1 - \lambda) \cdot i\text{-}influence(A_\ell, \theta), \quad (5)$$

$$\lambda \quad \cdot i\text{-}influence(A_i, \theta) \quad \geq \quad -\lambda \quad \cdot i\text{-}influence(A_h, \theta). \quad (6)$$

Proof (Sketch). We observe that the utility of a generic agent k is of the form

$$p_k(\hat{\theta}) - A_k(\hat{\theta}) \cdot \theta_k = f_k + A_k(\hat{\theta}) \cdot (H - L) \cdot \begin{cases} 1 - \lambda & \text{if } \theta_k = L; \\ -\lambda & \text{if } \theta_k = H. \end{cases}$$

From this obtain that bribeproofness (1) is equivalent to conditions (3), (4), (5) and (6) when $i \neq j$. In particular, depending on θ_i and θ_j, we have:

$(1) \equiv (3)$ for $\theta_i = L$ and $\theta_j = L$; $(1) \equiv (4)$ for $\theta_i = L$ and $\theta_j = H$

$(1) \equiv (5)$ for $\theta_i = H$ and $\theta_j = L$; $(1) \equiv (6)$ for $\theta_i = H$ and $\theta_j = H$.

$\square$

A simple corollary of the previous theorem is that the following natural condition implies bribeproofness when setting $\lambda = 1/2$.

Definition 2 (bounded influence). *An algorithm A satisfies bounded influence if, for all $\theta \in \Theta$ and for all $i, j \in N$, the following condition holds:*

$$i\text{-}influence(A_i, \theta) \geq |i\text{-}influence(A_j, \theta)|. \quad (7)$$

Corollary 1. *The $\left(\frac{1}{2}\right)$-linear mechanism is bribeproof for two-values domains if and only if its algorithm A satisfies bounded influence.*

3 Binary Allocations

In this section we apply our results to the case of *binary allocations*, that is, the problems in which each agent is allocated either an amount equal 0 or 1 (the path auction and the k-item procurement auction are two examples).

We first observe that bounded influence boils down to the following natural condition called *non-bossiness* (no agent can change the allocation of another agent without changing her own allocation), and for our construction this condition is equivalent to (strong) bribeproofness.

Definition 3 (non-bossiness). *An algorithm A satisfies non-bossiness if, for all i and for all θ, the following implication holds: if i-influence$(A_i, \theta) = 0$ then i-influence$(A_j, \theta) = 0$ for all j.*

Theorem 3. *For binary allocations and two-values domains, the following statements are equivalent:*

1. *The $\left(\frac{1}{2}\right)$-linear mechanism (A, p) is bribeproof.*
2. *Algorithm A satisfies monotonicity and non-bossiness.*
3. *The $\left(\frac{1}{2}\right)$-linear mechanism (A, p) is strongly bribeproof.*

It is not difficult to show that non-bossiness is not a necessary condition in general mechanisms (see full version [14]).

The main application of the previous result is a general construction of exact mechanisms for utilitarian problems (see e.g. [21]), that is, for minimizing the (weighted) *sum* of all agents' costs.

Definition 4 (weighted social cost minimization). *An algorithm A minimizes the weighted social cost if there exist nonnegative constants $\{\alpha_i\}_{i \in N}$ and arbitrary constants $\{\beta_a\}_{a \in \mathcal{A}}$ such that, for all $\theta \in \Theta$, it holds that*

$$A(\theta) \in \arg \min_{a \in \mathcal{A}}\{\operatorname{sum}(a, \theta)\},$$

where sum() *is defined as* $\operatorname{sum}(a, \theta) := \left(\sum_{i \in N} \alpha_i a_i \theta_i\right) + \beta_a$.

Obviously, every mechanism minimizing the sum of all agents' costs correspond to the case $\alpha_i = 1$ and $\beta_a = 0$. Welfare maximization problems correspond to the case in which agents have valuations instead of costs.

Definition 5 (consistent ties). *An algorithm A minimizes the weighted social cost breaking ties consistently if there exists a total order $\preceq$ over the set $\mathcal{A}$ of feasible allocations such that, for all $\theta \in \Theta$ and for all $a' \in \mathcal{A}$, the following implication holds: if* $\operatorname{sum}(A(\theta), \theta) = \operatorname{sum}(a', \theta)$, *then* $A(\theta) \preceq a'$.

Theorem 4. *For binary allocation problems over two-values domains, if algorithm A minimizes the weighted social cost breaking ties consistently, then the corresponding $\left(\frac{1}{2}\right)$-linear mechanism is strongly bribeproof.*

Proof. We show that every algorithm A which minimizes the weighted social cost breaking ties consistently, satisfies non-bossiness and monotonicity (2). The theorem then follows from Theorem 3.

It is convenient to rewrite the weighted social cost into two parts, the contribution of a fixed agent i and the rest:

$$\alpha_i a_i \theta_i + \operatorname{sum}_{-i}(a, \theta_{-i}) \quad \text{where} \quad \operatorname{sum}_{-i}(a, \theta_{-i}) := \left(\sum_{j \in N \setminus \{i\}} \alpha_j a_j \theta_j\right) + \beta_a.$$

For ease of notation, also let $\theta^L := (L, \theta_{-i})$ and $\theta^H := (H, \theta_{-i})$. Observe that since A minimizes the weighted social cost we have the following:

$$\text{sum}(A(\theta^L), \theta^L) \quad = \quad L\alpha_i A_i(\theta^L) + \text{sum}_{-i}(A(\theta^L), \theta_{-i}) \qquad \leq \qquad (8)$$

$$\text{sum}(A(\theta^H), \theta^L) \quad = \quad L\alpha_i A_i(\theta^H) + \text{sum}_{-i}(A(\theta^H), \theta_{-i}), \qquad \text{and}$$

$$\text{sum}(A(\theta^H), \theta^H) \quad = \quad H\alpha_i A_i(\theta^H) + \text{sum}_{-i}(A(\theta^H), \theta_{-i}) \qquad \leq$$

$$\text{sum}(A(\theta^L), \theta^H) \quad = \quad H\alpha_i A_i(\theta^L) + \text{sum}_{-i}(A(\theta^L), \theta_{-i}). \qquad (9)$$

First, we show the following implication:

$$\alpha_i A_i(\theta^H) = \alpha_i A_i(\theta^L) \qquad \Rightarrow \qquad A(\theta^L) = A(\theta^H). \qquad (10)$$

The left-hand side implies that both inequalities (8) and (9) hold with "=". Since ties are broken consistently, we have $A(\theta^L) \preceq A(\theta^H)$ by (8) and $A(\theta^H) \preceq A(\theta^L)$ by (9), thus implying $A(\theta^L) = A(\theta^H)$.

Now observe that (10) implies that A satisfies non-bossiness, and thus it only remains to prove that the monotonicity condition holds. By summing inequalities (8) and (9) we obtain $\alpha_i(H - L)A_i(\theta^H) \leq \alpha_i(H - L)A_i(\theta^L)$. From this the inequality

$$A_i(\theta^H) \leq A_i(\theta^L)$$

follows immediately for the case $\alpha_i > 0$, while for $\alpha_i = 0$ it follows by (10). By definition, the inequality above is equivalent to the monotonicity condition (2).   □

The mechanism for the path auction problem consists in paying each agent in the chosen path an amount equal to $\mathcal{M} = \frac{L+H}{2}$, where the algorithm breaks ties between paths in a fixed order. Similar mechanisms can be obtained for other utilitarian problems like minimum spanning tree [21] or for the k-item procurement auction (see Sect. 4 for the latter).

4 Characterizations for Two Problems

In this section we show that the $(\frac{1}{2})$-linear mechanism is the only bribeproof mechanism for the path auction on general networks (this result applies also to combinatorial auctions with known single minded bidders – see [14] for details). We then obtain analogous characterizations for the k-item procurement auction in terms of our λ-linear mechanisms.

As for the path auction problem, we actually prove a stronger result saying that the $(\frac{1}{2})$-linear mechanism is the only bribeproof for the simple network in Fig. 1b on the following generalization of two-values domains:

Definition 6. *The path auction with ϵ-perturbed domain ($\epsilon \geq 0$) is the path auction problem restricted to the network in Fig. 1b in which the agents domain are as follows: $\Theta_1 = \Theta_2 = \{L - \epsilon, H + \epsilon\}$ and $\Theta_3 = \Theta_4 = \{L, H\}$.*

Clearly the two-values domain corresponds to setting $\epsilon = 0$.

Theorem 5. *A mechanism which is bribeproof for the path auction with ϵ-perturbed domain must be a $\left(\frac{1}{2}\right)$-linear mechanism.*

Proof (Main Ideas). We first show that, no matter how the mechanism breaks ties, the payments must depend only on which path is selected (using Fact 1). This means that the payments are of the form

$$p_i(\theta) = f_i + \begin{cases} q_i & \text{if } i \text{ is selected for types } \theta, \\ 0 & \text{otherwise.} \end{cases}$$

In order to conclude that the mechanism must be a $\left(\frac{1}{2}\right)$-linear mechanism it is enough to prove that $q_i = \frac{L+H}{2}$ for all i. This is the technically involved part, because we have to consider the possible tie breaking rules. At an intermediate step, we show that $q_1 + q_2 = L + H = q_3 + q_4$, for otherwise there exists a coalition which violates bribeproofness. $\square$

By taking $\epsilon = 0$ we obtain a characterization for this problem:

Corollary 2. *The $\left(\frac{1}{2}\right)$-linear mechanism is the only bribeproof mechanism for the path auction on general networks.*

Since in these instances of path-auction problem $\left(\frac{1}{2}\right)$-linear mechanism are *not* bribeproof on three-values domains, we obtain the following result.

Theorem 6. *There is no bribeproof mechanism for the path auction problem on general networks and for three-values domains.*

Theorem 5 implies that we cannot extend the positive result to coalitions of larger size, nor to *heterogeneous* two-values domains in which $\theta_i \in \{L_i, H_i\}$.

Corollary 3. *There is no collusion-proof mechanism for the path auction problem on general networks and two-values domains. The same remains true even if we restrict to coalitions of size three (in which two agents bribe another for misreporting her type).*

Corollary 4. *There is no bribeproof mechanism for the path auction problem on general networks and certain heterogeneous two-values domains.*

We remark that on a simple network consisting of n parallel edges the path auction problem is the same as the 1-item procurement auction. For the k-item procurement auction over three values domains $\Theta^{(L,M,H)}$, we consider the following mechanism:

λ_M-**linear mechanism (normalized to $f_i = 0$):** Select the k agents with smallest types, breaking ties in favor of agents with smaller index; Pay each of the selected agents an amount M, and non-selected agents receive no money.

Note that this is the λ_M-linear mechanism for $\lambda_M := \frac{H-M}{H-L}$.

Theorem 7. *The λ_M-linear mechanism is bribeproof for the k-item procurement auction in the case of three-values domains.*

Also in this problem our construction yields the only mechanism, and results cannot be extended to more complex domains.

Theorem 8. *The λ_M-linear mechanism is the only bribeproof mechanism for the 1-item procurement auction with three-values domains and two agents.*

This implies the impossibility result.

Corollary 5. *There is no bribeproof mechanism for the 1-item procurement auction with two agents and four-values domains.*

5   Min-Max Fairness and Non-binary Problems

In this section we consider problems with *min-max* fairness optimization criteria, and *non-binary* allocations. Thus, the algorithm A should satisfy

$$A(\theta) \in \arg \min_{a \in \mathcal{A}} \max_{i \in N} \{a_i \cdot \theta_i\}. \tag{11}$$

In particular we consider the problem of allocating a perfectly divisible item (Example 1) according to the above min-max fairness criteria (11). In such allocation all agents will get some positive amount so that all costs will be identical.

Theorem 9 (min-max fairness). *There is a strongly bribeproof $\left(\frac{1}{2}\right)$-linear mechanism satisfying min-max fairness for allocating a perfectly divisible item.*

We next consider the problem of scheduling selfish related machines [1]. In this problem, we are given several indivisible items (jobs) each of them with some size. Each item must be assigned to some agent (machine) and the goal is to minimize the maximum cost (the makespan). Note that the allocation of each machine is the sum of the size of the jobs allocated to this machine.

Example 2. Consider three machines and three jobs of size 10, 6, and 6. For $L = 1$ and $H = 2 + \epsilon$, for some small ϵ to be specified (below). The allocation of the jobs minimizing the *makespan* for types $\theta = (L, L, H)$ and $\hat{\theta} = (L, H, H)$, for any $0 < \epsilon < 2/3$, is as follows: $A(L, L, H) = (6+6, 10, 0)$ and $A(L, H, H) = (10, 6, 6)$. This is unique up to a permutation of the allocation of machines with the same type.

Using this example we can show that our construction cannot lead to bribeproof mechanisms for minimizing the makespan in the scheduling problem above, or even to approximate the makespan within some small factor $\alpha > 1$, i.e., returning an allocation whose makespan is at most α times the optimum makespan.

Theorem 10 (selfish related machines). *No bribeproof λ-linear mechanism for the makespan minimization on three agents with two values-domains can approximate the makespan within a factor smaller than $\frac{2}{\sqrt{3}} \approx 1.1547$.*

Impossibility results also apply to *randomized* mechanisms (see [14]).

References

1. Archer, A., Tardos, É.: Truthful mechanisms for one-parameter agents. In: Proceedings of the IEEE FOCS, pp. 482–491 (2001)
2. Ashlagi, I., Dobzinski, S., Lavi, R.: Optimal lower bounds for anonymous scheduling mechanisms. Math. Oper. Res. **37**(2), 244–258 (2012)
3. Chen, J., Micali, S.: Collusive dominant-strategy truthfulness. J. Econ. Theory **147**(3), 1300–1312 (2012)
4. Christodoulou, G., Kovács, A.: A deterministic truthful PTAS for scheduling related machines. SIAM J. Comput. **42**(4), 1572–1595 (2013)
5. Dobzinski, S., Nisan, N.: Multi-unit auctions: beyond roberts. J. Econ. Theory **156**, 14–44 (2015)
6. Dütting, P., Gkatzelis, V., Roughgarden, T.: The performance of deferred-acceptance auctions. In: Proceedings of the 15th ACM EC, pp. 187–204 (2014)
7. Gamzu, I.: Improved lower bounds for non-utilitarian truthfulness. In: Kaklamanis, C., Skutella, M. (eds.) WAOA 2007. LNCS, vol. 4927, pp. 15–26. Springer, Heidelberg (2008)
8. Goldberg, A.V., Hartline, J.D.: Collusion-resistant mechanisms for single-parameter agents. In: Proceedings of the 16th SODA, pp. 620–629 (2005)
9. Goldberg, P.W., Ventre, C.: Using lotteries to approximate the optimal revenue. In: Proceedings of the AAMAS, pp. 643–650 (2013)
10. Juarez, R.: Group strategyproof cost sharing: the role of indifferences. Games Econ. Behav. **82**, 218–239 (2013)
11. Koutsoupias, E., Vidali, A.: A lower bound of $1 + \phi$ for truthful scheduling mechanisms. In: Kučera, L., Kučera, A. (eds.) MFCS 2007. LNCS, vol. 4708, pp. 454–464. Springer, Heidelberg (2007)
12. Lavi, R., Mu'Alem, A., Nisan, N.: Towards a characterization of truthful combinatorial auctions. In: Proceedings of FOCS, pp. 574–583 (2003)
13. Lavi, R., Swamy, C.: Truthful mechanism design for multi-dimensional scheduling via cycle monotonicity. Games Econ. Behav. **67**(1), 99–124 (2009)
14. Mihalák, M., Penna, P., Widmayer, P.: Bribeproof mechanisms for two-values domains. CoRR, abs/1512.04277 (2015)
15. Milgrom, P., Segal, I.: Deferred-acceptance auctions and radio spectrum reallocation. In: Proceedings of the 15th ACM EC, pp. 185–186 (2014)
16. Mizukami, H.: On the constancy of bribe-proof solutions. Econ. Theory **22**(1), 211–217 (2003)
17. Moulin, H.: Incremental cost sharing: characterization by coalition strategy-proofness. Soc. Choice Welfare **16**(2), 279–320 (1999)
18. Mu'alem, A., Schapira, M.: Setting lower bounds on truthfulness. In: Proceedings of ACM SODA, pp. 1143–1152 (2007)
19. Mukherjee, C.: Fair and group strategy-proof good allocation with money. Soc. Choice Welfare **42**(2), 289–311 (2014)
20. Myerson, R.B.: Optimal auction design. Math. Oper. Res. **6**, 58–73 (1981)
21. Nisan, N., Ronen, A.: Algorithmic mechanism design. Games Econ. Behav. **35**, 166–196 (2001)
22. Ohseto, S.: Strategy-proof and efficient allocation of an indivisible good on finitely restricted preference domains. Int J. Game Theory **29**(3), 365–374 (2000)
23. Penna, P., Ventre, C.: Optimal collusion-resistant mechanisms with verification. Games Econ. Behav. **86**, 491–509 (2014)

24. Pountourakis, E., Vidali, A.: A complete characterization of group-strategyproof mechanisms of cost-sharing. Algorithmica **63**(4), 831–860 (2012)
25. Roberts, K.: The characterization of implementable choice rules. In:Aggregation and Revelation of Preferences, pp. 321–348 (1979)
26. Schummer, J.: Eliciting preferences to assign positions and compensation. Games Econ. Behav. **30**(2), 293–318 (2000)
27. Schummer, J.: Manipulation through bribes. J. Econ. Theory **91**(3), 180–198 (2000)
28. Serizawa, S.: Pairwise strategy-proofness and self-enforcing manipulation. Soc. Choice Welfare **26**(2), 305–331 (2006)
29. Wakayama, T.: Bribe-proof division of a private good. SSRN 2349605 (2013). http://dx.doi.org/10.2139/ssrn.2349605

The Anarchy of Scheduling Without Money

Yiannis Giannakopoulos[1], Elias Koutsoupias[2], and Maria Kyropoulou[2(✉)]

[1] University of Liverpool, Liverpool, UK
ygiannak@liverpool.ac.uk
[2] University of Oxford, Oxford, UK
{elias,kyropoul}@cs.ox.ac.uk

Abstract. We consider the scheduling problem on n strategic unrelated machines when no payments are allowed, under the objective of minimizing the makespan. We adopt the model introduced in [Koutsoupias 2014] where a machine is *bound* by her declarations in the sense that if she is assigned a particular job then she will have to execute it for an amount of time at least equal to the one she reported, even if her private, true processing capabilities are actually faster. We provide a (non-truthful) randomized algorithm whose pure *Price of Anarchy* is arbitrarily close to 1 for the case of a single task and close to n if it is applied independently to schedule many tasks. Previous work considers the constraint of truthfulness and proves a tight approximation ratio of $(n+1)/2$ for one task which generalizes to $n(n+1)/2$ for many tasks. Furthermore, we revisit the truthfulness case and reduce the latter approximation ratio for many tasks down to n, asymptotically matching the best known lower bound. This is done via a detour to the relaxed, fractional version of the problem, for which we are also able to provide an optimal approximation ratio of 1. Finally, we mention that all our algorithms achieve optimal ratios of 1 for the social welfare objective.

1 Introduction

We consider a variant of the scheduling problem proposed by Koutsoupias [11] where no payments are allowed and the machines are bound by their declarations. In particular, the goal is to allocate a set of tasks to strategic unrelated machines while minimizing the makespan. The time/cost needed by a machine to execute a task is private information of the machine. Each machine is rational and selfish, and will misreport its costs in an attempt to minimize its own overall running time, under the assumption that if she is allocated a task, she will execute it for at least the declared cost (more specifically, for the maximum among her true and reported execution times). We are interested in designing allocation protocols that do not use payments and the stable outcomes are not far from the non-strategic, centrally enforced optimum makespan.

The field of Mechanism Design [17] focuses on the implementation of desired outcomes. Given the strategic behaviour of the players who provide the input

Supported by ERC Advanced Grant 321171 (ALGAME) and EPSRC grant EP/M008118/1. A full version of this paper can be found in [8].

M. Gairing and R. Savani (Eds.): SAGT 2016, LNCS 9928, pp. 302–314, 2016.
DOI: 10.1007/978-3-662-53354-3_24

and a specific objective function that measures the quality of the outcome, the challenge is to design mechanisms which are able to elicit a desired behaviour from the players, while at the same time optimizing that objective value. A primary designer goal that has been extensively studied is that of *truthfulness*, under the central solution concept of dominant strategies: a player should be able to optimize her own individual utility by reporting truthfully, no matter what strategies the other players follow. However, achieving this is not always compatible with maintaining a good objective value [9,21]. The introduction of *payments* was suggested as a means towards achieving these goals as a carefully designed payment scheme incentivizes the players to make truthful declarations. The goal now becomes to design such algorithms (termed *mechanisms*) which utilize monetary compensations in order to impose truthful behaviour while optimizing the objective function [16].

There are many situations, though, where the use of payments might be considered unethical [17], illegal (e.g. organ donations) or even just impractical. For this reason researchers have started turning their attention to possible ways of achieving truthfulness without the use of payments. In such a setting, in order to circumvent Social Choice impossibility results (e.g. the seminal Gibbard-Satterthwaite [9,21] theorem) domains with richer structure have to be considered. Procaccia and Tennenholtz [20] were the first to consider achieving truthfulness without using payments, by sacrificing the optimality of the solution and settling for just an approximation, in the context of facility location problems. Similar questions have been considered in the context of inter-domain routing [14], in assignment problems [6], and in the setting of allocating items to two players (with the use of a certain artificial currency) [10]. Moreover, (exact, as opposed to approximate) mechanism design without money has a rich history in the social choice literature.

Clearly, truthfulness is a property desired by any mechanism designer; if the mechanism can ensure that no player can benefit from misreporting, the designer knows what kind of player behaviour and outcome to expect. Moreover, the focus on truthful mechanisms has been largely motivated by the Revelation Principle stating that essentially every equilibrium state of a mechanism can be simulated by a truthful mechanism which achieves the same objective. However this is no longer possible in the variant we examine here, due to the fact that the players are bound by their declarations and thus don't have quasi-linear utilities. So, it is no longer without loss of generality if we restrict attention to truthful mechanisms. For mechanisms that are not truthful, *Price of Anarchy* (PoA) [12] analysis is the predominant, powerful tool for quantifying the potential suboptimality of the outcomes/equilibria; it measures the impact the lack of coordination (strategic behaviour) has on the solution quality, by comparing it to the optimal, non-strategic solution.

Scheduling is one of the most influential problems in Algorithmic Game Theory and has been studied extensively. In its most general form, the goal is to schedule m tasks to n parallel machines with arbitrary processing times, in order to minimize the makespan. In the front where payments are allowed and truthfulness comes at no extra cost given the strategic nature of the machines Nisan and Ronen [16] first considered the mechanism design approach of the problem. They prove that the well known VCG mechanism achieves an n-approximation of the optimal makespan, while no truthful deterministic mechanism can achieve approximation ratio better than 2. The currently known best lower bound is 2.61 [13] while Ashlagi et al. [2] prove the tightness of the upper bound for anonymous mechanisms. With respect to randomized (truthful in expectation) mechanisms as well as fractional ones, the best known bounds are $(n + 1)/2$ and $2 - 1/n$ [5,15]. We note that the aforementioned lower bounds disregard computational feasibility and simply rely on the requirement for truthfulness.

In an attempt to get positive results when payments are not allowed in the scheduling context, Koutsoupias [11] first considered the plausible assumption that the machines are bound by their declarations. This was influenced by the notion of impositions that appeared in [7,18] and was applied in facility location as well as digital goods pricing settings. The notion of winner imposition fits within the framework of approximate mechanism design without payments. A more powerful framework that is also very much related to this assumption is the notion of verification that appears in [3,16,19]. The mechanisms in this context are allowed to use payments and simply give or deny payments to machines after they discover their true execution costs. Relevant works include [1,4] where the scheduling problem of selfish tasks is considered again under the assumption that the players who control the tasks are bound by their declarations.

Our Results. In this work we adopt the model of [11]. For the case of scheduling a single task Koutsoupias [11] proved that the approximation ratio of any mechanism is at least $(n+1)/2$ and gave a mechanism matching this bound, where n is the number of machines. When applied to many tasks, this mechanism immediately implies a $n(n+1)/2$ approximation ratio for the makespan objective. In Sect. 3 we provide a (non-truthful) algorithm which performs considerably better than the best truthful mechanism; even the worst pure equilibrium/outcome of our algorithm achieves an optimal makespan, i.e. our algorithm has a pure PoA of 1. If we run this algorithm independently for each job, we get a task-independent and anonymous algorithm, yielding a PoA of n for any number of tasks. Next, revisiting truthfulness, in Sect. 4 we also show that the mechanism inspired by the LP relaxation of the problem is provably truthful and provides an n-approximation ratio when interpreted as a randomized mechanism, while achieving an optimal approximation ratio 1 for the fractional scheduling problem of divisible tasks. This almost matches the lower bound of $(n+1)/2$ for truthful mechanisms known from [11]. Finally, in Sect. 5 we briefly study the more optimistic objective of minimizing the makespan at the best possible equilibrium (instead of the worst one used in the Price of Anarchy metric) and show that the natural greedy algorithm achieves an optimal Price of Stability. Due to lack of space some proofs appear only in the full version [8].

2   Model and Notation

We have a set $N = \{1, 2, \ldots, n\}$ of unrelated parallel machines and m tasks/jobs that need to be scheduled to these machines. Throughout the text we assume that vector $\mathbf{t}$ denotes the true execution times, i.e. $t_{i,j}$ is the time machine i needs to execute task j. This is private knowledge of each machine i. Let $\hat{\mathbf{t}}$ denote the corresponding (not necessarily true) *declarations* of the machines for these costs.

A (randomized) *allocation protocol* takes as input the machines' declarations $\hat{\mathbf{t}}$ and outputs an allocation $\mathbf{A}$ of tasks to machines where A_{ij} is a 0–1 random variable indicating whether or not machine i gets allocated task j and $\mathbf{a}$ is the corresponding probability distribution of allocation, i.e. $a_{i,j} = \Pr[A_{i,j} = 1]$ where of course $\sum_{i=1}^{n} a_{i,j} = 1$ for any task j.

If a machine i is allocated some task j, we assume that the machine will execute the task for time $\max\{t_{i,j}, \hat{t}_{i,j}\}$. So, the expected cost/workload of machine i is defined as

$$C_i(\hat{\mathbf{t}}|\mathbf{t}_i) = \sum_{j=1}^{m} a_{i,j}(\hat{\mathbf{t}}) \max\{\hat{t}_{i,j}, t_{i,j}\}, \qquad (1)$$

while the *makespan* is computed as the average maximum execution time

$$\mathcal{M}(\hat{\mathbf{t}}|\mathbf{t}_i) = \mathbb{E}_{\mathbf{A} \sim \mathbf{a}} \left[\max_{i=1,\ldots,n} \sum_{j=1}^{m} A_{i,j} \max\{\hat{t}_{i,j}, t_{i,j}\} \right].$$

To simplify notation, whenever the true execution times $\mathbf{t}$ are clear from the context we will drop them and simply use $C_i(\hat{\mathbf{t}})$ and $\mathcal{M}(\hat{\mathbf{t}})$.

The allocation protocol is called *truthful*, or truthful mechanism, if it does not give incentives to the machines to misreport their true execution costs. Formally, for every machine i and declarations vector $\hat{\mathbf{t}}$,

$$C_i(\mathbf{t}_i, \hat{\mathbf{t}}_{-i}) \leq C_i(\hat{\mathbf{t}}),$$

where $(\mathbf{x}_i, \mathbf{y}_{-i})$ denotes the vector of declarations where machine i has deviated to $\mathbf{x}_i$ while all other machines report costs as in $\mathbf{y}$. The *approximation ratio* measures the performance of truthful mechanisms and is defined as the maximum ratio, over all instances, of the objective value (makespan) under that mechanism over the optimal objective value achievable by a centralized solution which ignores the truthfulness constraint.

If an allocation protocol is not truthful (we simply refer to it as algorithm), we measure its performance by the quality of its Nash equilibria; the states from which no player has the incentive to unilaterally deviate. The *Price of Anarchy* (PoA) is established as a meaningful benchmark and captures the maximum ratio, over all instances, of the objective value of the worst equilibrium over that of the optimal centralized solution that ignores the machines' incentives. For most part of this paper we restrict attention to pure Nash equilibria where the machines make deterministic reports about their execution costs, and we will from now on refer to them simply as *equilibria*. Then, the corresponding benchmark is called pure PoA. A more optimistic benchmark is the *Price of Stability* (PoS) which compares the objective value of the best equilibrium to the value of the optimal centralized solution.

The makespan objective is inherently different if we consider *divisible tasks*, i.e. fractional allocations. In that case, each machine is allocated a portion of each task by the protocol and the makespan is computed as the maximum of the execution times of the machines, namely

$$\mathcal{M}^f(\mathbf{t}) = \max_{i=1\ldots n} \sum_{j=1}^{m} \alpha_{i,j} t_{i,j}$$

where $\alpha_{i,j} \in [0,1]$ is the *fraction* of task j allocated to machine i. Again, it must be that $\sum_{i=1}^{n} \alpha_{i,j} = 1$ for any task j. Notice here that each fractional algorithm with allocation fractions α naturally gives rise to a corresponding randomized integral algorithm with allocation probabilities $\mathbf{a} = \alpha$, whose makespan is within a factor of n from the fractional one[1], i.e. for any cost matrix $\mathbf{t}$

$$\mathcal{M}^f(\mathbf{t}) \le \mathcal{M}(\mathbf{t}) \le n \cdot \mathcal{M}^f(\mathbf{t}). \tag{2}$$

Except when clearly stated otherwise, in this paper we deal with the integral version of the scheduling problem.

Social Welfare. An alternative objective, very common in the Mechanism Design literature, is that of optimizing *social welfare*, i.e. minimizing the combined costs of all players: $\mathcal{W}(\hat{\mathbf{t}}) = \sum_{i=1}^{n} C_i(\hat{\mathbf{t}})$. It is not difficult to see[3] that the makespan and social welfare objectives are within a factor of n away, whatever the allocation algorithm $\mathbf{a}$ and the input costs $\hat{\mathbf{t}}$ might be:

$$\mathcal{M}(\hat{\mathbf{t}}) \le \mathcal{W}(\hat{\mathbf{t}}) \le n \cdot \mathcal{M}(\hat{\mathbf{t}}). \tag{3}$$

Also notice that for the special case of a single task, since the job is eventually allocated entirely to some machine, the two objectives coincide no matter the number of machines n, i.e. $\mathcal{M}(\hat{\mathbf{t}}) = \mathcal{W}(\hat{\mathbf{t}})$. Because of that and the linearity of the social welfare with respect to the players' costs, it is easy to verify that all algorithms we present in this paper achieve optimal ratios of 1 for that objective, both with respect to equilibrium/PoA and truthfulness analysis (e.g. Theorems 2 and 5). We will not mention that explicitly again in the remaining of the paper and rather focus on the more challenging for our scheduling problem objective of makespan minimization.

3   Price of Anarchy

For clarity of exposition, we first describe our scheduling algorithm in the special case of just $n = 2$ machines (and one task) before presenting the algorithm for

---

[1] This is due to the fact that for any random variables $Y_1, Y_2, \ldots, Y_n$ it is $\mathbb{E}[\max_i Y_i] \le \mathbb{E}[\sum_i Y_i] = \sum_i \mathbb{E}[Y_i] \le n \max_i \mathbb{E}[Y_i]$, and also $\max_i \mathbb{E}[Y_i] \le \mathbb{E}[\max_i Y_i]$ due to the convexity of the max function.

the general case of $n \geq 1$. Since we treat the case of only one task in this section, we use $\hat{t}_i$ and t_i to denote the declared and the true execution time of machine i, respectively, and use a_i to denote i's allocation probability.

3.1 Warm Up: The Case of Two Machines

To simplify notation, throughout this section we will assume without loss of generality that $\hat{t}_1 \leq \hat{t}_2$, i.e. the input to our algorithm is sorted in nondecreasing order. Notice that the true bids $\mathbf{t} = (t_1, t_2)$ do not have to preserve this ordering, since the highest biding machine might very well in reality have the fastest execution capabilities.

Our algorithm for the case of two machines, parametrized by two constants $L > 2$, $c > 1$, and denoted by $\mathcal{A}^{(2)}_{L,c}$ is defined by the allocation probabilities in Fig. 1. Whenever parameter c is insignificant in a particular context[2], we will just use $\mathcal{A}^{(2)}_{L}$.

	a_1	a_2
if $\hat{t}_1 = \hat{t}_2$	$\frac{1}{2}$	$\frac{1}{2}$
if $\hat{t}_1 < \hat{t}_2 < c \cdot \hat{t}_1$	$\frac{1}{L}$	$1 - \frac{1}{L}$
if $c \cdot \hat{t}_1 \leq \hat{t}_2$	$1 - \frac{1}{L}\frac{\hat{t}_1}{\hat{t}_2}$	$\frac{1}{L}\frac{\hat{t}_1}{\hat{t}_2}$

Fig. 1. Algorithm $\mathcal{A}^{(2)}_{L,c}$ for scheduling a single task to two machines, parametrized by $L > 2$ and $c > 1$. The probability that machine $i = 1, 2$ gets the task is denoted by a_i, and $\hat{t}_1, \hat{t}_2$ are the reported execution times by the machines.

The main result of this section is the following theorem, showing that by choosing parameter L arbitrarily high, the above algorithm can achieve an optimal Price of Anarchy:

Theorem 1. *For the case of one task and two machines, algorithm $\mathcal{A}^{(2)}_{L}$ has a (pure) Price of Anarchy of $1 + \frac{1}{L}$ (for any $L > 2$).*

We break down the proof of Theorem 1 in distinct claims.

Claim 1. *At any equilibrium $\hat{\mathbf{t}}$ the ratio of the two bids must be at least c, i.e. $\hat{t}_2 \geq c \cdot \hat{t}_1$.*

---

[2] In such case, as it is for example in the statement of Theorem 1, one can simply pick e.g. $c = 1 + \frac{1}{L}$.

Proof. Without loss assume $\hat{t}_1 \neq 0$, since otherwise the claim is trivially true. First, assume for a contradiction that $\hat{t}_1 < \hat{t}_2 < c \cdot \hat{t}_1$. Then the machine with largest report would have an incentive to deviate to bid $t'_2 = \max\{c\hat{t}_1, t_2\}$:

$$C_2(\hat{t}) = \left(1 - \frac{1}{L}\right) \max\{\hat{t}_2, t_2\} > \frac{1}{L}\hat{t}_1 = \frac{\hat{t}_1}{Lt'_2} \max\{t'_2, t_2\} = C_2(\hat{t}_1, t'_2)$$

where the inequality holds since $L > 2$ and the final two equalities hold because for the deviating bid it is $t'_2 \geq t_2, c\hat{t}_1$. Thus $\hat{t} = (\hat{t}_1, \hat{t}_2)$ could not have been an equilibrium under the assumption that $\hat{t}_1 < \hat{t}_2 < c \cdot \hat{t}_1$.

A similar contradiction can be obtained for the remaining case of $\hat{t}_1 = \hat{t}_2$. In this case, both machines have an incentive to deviate to a bid $t'_1 = \frac{\hat{t}_1}{c} < \hat{t}_1$, since

$$C_1(\hat{t}) = \frac{1}{2} \max\{\hat{t}_1, t_1\} \geq \frac{1}{2} \max\{t'_1, t_1\} > \frac{1}{L} \max\{t'_1, t_1\} = C_1(t'_1, \hat{t}_2).$$

Claim 2. *At any equilibrium $\hat{t}$ the machine with the largest report will never have underbid, i.e. $\hat{t}_2 \geq t_2$.*

Proof. Assume for a contradiction that $\hat{t}_2 < t_2$. Then

$$C_2(\hat{t}) = \frac{\hat{t}_1}{L\hat{t}_2} \max\{\hat{t}_2, t_2\} = \frac{\hat{t}_1}{L\hat{t}_2} t_2 > \frac{\hat{t}_1}{Lt_2} t_2 = C_2(\hat{t}_1, t_2),$$

the first equality holding due to Claim 1 and the last one because $t_2 > \hat{t}_2 \geq \hat{t}_1$.

Claim 3. *At any equilibrium $\hat{t}$ the smallest bid is given by $\hat{t}_1 = \min\{t_1, \frac{t_2}{c}\}$.*

Proof. Assume for a contradiction that $\hat{t}_1 \neq t'_1 = \min\{t_1, \frac{t_2}{c}\}$. Then, we will show that the lowest bidding machine would have an incentive to deviate from $\hat{t}_1$ to t'_1.

Indeed, first consider the case when $\hat{t}_1 < t'_1$. Then

$$C_1(\hat{t}) = \left(1 - \frac{\hat{t}_1}{L\hat{t}_2}\right) \max\{\hat{t}_1, t_1\} > \left(1 - \frac{t'_1}{L\hat{t}_2}\right) \max\{t'_1, t_1\} = C_1(t_1, \hat{t}_2).$$

In the remaining case of $\hat{t}_1 > t'_1 = \min\{t_1, \frac{t_2}{c}\}$, because of Claim 1 it must be that $t'_1 = t_1 < \hat{t}_1 \leq \frac{t_2}{c}$, thus

$$C_1(\hat{t}) = \left(1 - \frac{\hat{t}_1}{L\hat{t}_2}\right) \max\{\hat{t}_1, t_1\} = \left(1 - \frac{\hat{t}_1}{L\hat{t}_2}\right) \hat{t}_1 > \left(1 - \frac{t_1}{L\hat{t}_2}\right) t_1 = C_1(t_1, \hat{t}_2).$$

where the inequality holds since $x \mapsto \left(1 - \frac{x}{y}\right) x$ is a strictly increasing function for $x \in [0, \frac{y}{2}]$, and indeed $t_1 < \hat{t}_1 < \hat{t}_2 < \frac{L\hat{t}_2}{2}$.

Claim 4. *At any equilibrium $\hat{t}$ bidding must preserve the relative order of the true execution times, i.e. $t_1 \leq t_2$.*

Proof. For a contradiction assume that $t_2 < t_1$, and first consider the case when $t_2 < \hat{t}_1$. If we pick $t_2' \in \left(\frac{\hat{t}_1}{c}, \hat{t}_1\right)$, we have

$$C_2(\hat{\mathbf{t}}) = \frac{\hat{t}_1}{L\hat{t}_2} \max\{\hat{t}_2, t_2\} = \frac{\hat{t}_1}{L} > \frac{1}{L} \max\{t_2', t_2\} = C_2(\hat{t}_1, t_2').$$

For the remaining case of $\hat{t}_1 \le t_2 < t_1$, first note that if $t_1 \le \frac{\hat{t}_2}{c}$ then by Claim 3 we would immediately derive that $\hat{t}_1 = t_1$, which is a contradiction. Hence, we can assume that $\hat{t}_1 = \frac{\hat{t}_2}{c} < t_1$. Then, if $\hat{t}_2 > t_1$ we have that

$$C_1(\hat{\mathbf{t}}) = \left(1 - \frac{\hat{t}_1}{L\hat{t}_2}\right) \max\{\hat{t}_1, t_1\} = \left(1 - \frac{\hat{t}_1}{L\hat{t}_2}\right) t_1 > \frac{1}{L} t_1 = C_1(t_1, \hat{t}_2),$$

the inequality holding because $\frac{\hat{t}_1}{\hat{t}_2} \frac{1}{L} \le \frac{1}{L} < \frac{1}{2}$, and if $\hat{t}_2 \le t_1$ then, in the same way, for $t_1' = \max\{t_1, c\hat{t}_2\}$

$$C_1(\hat{\mathbf{t}}) > \frac{1}{L} t_1 \ge \frac{\hat{t}_2}{L} = \frac{\hat{t}_2}{L\hat{t}_1} \max\{t_1', t_1\} = C_1(t_1, \hat{t}_2).$$

Proof (Proof of Theorem 1). Claims 1–4 imply that the makespan (and thus also the social cost since we have a single task) of any allocation at equilibrium can be bounded by

$$\mathcal{M}(\hat{\mathbf{t}}) = \left(1 - \frac{\hat{t}_1}{L\hat{t}_2}\right) \max\{\hat{t}_1, t_1\} + \frac{\hat{t}_1}{L\hat{t}_2} \max\{\hat{t}_2, t_2\}$$

$$\le \max\{\hat{t}_1, t_1\} + \frac{1}{L}\hat{t}_1$$

$$\le \left(1 + \frac{1}{L}\right) t_1,$$

where t_1 is the optimal makespan.

Also, it is important to mention that it can be verified that there exists at least one (pure Nash) equilibrium, e.g. reporting $\hat{t}_1 = t_1$ and $\hat{t}_2 = \max\{Lc{\cdot}t_1, t_2\}$.

3.2   The General Case

The algorithm for two machines (and a single task) can be naturally generalized to the case of any number of machines $n \ge 2$. Due to lack of space we only give the definition of the algorithm here and the proof can be found in the full version of the paper [8]. We note that the essence of the techniques and the core ideas we presented in Sect. 3.1 carry over to the general case.

To present our algorithm $\mathcal{A}_{L,c}$ we first need to add some notation. We use $\hat{t}_{min}$ and $\hat{t}_{sec}$ to denote the smallest and second smallest declarations in $\hat{\mathbf{t}}$, and $N_{\min}, N_{\sec}$ the corresponding sets of machine indices that make these declarations. (If $N = N_{\min}$, i.e. all machines make the same declaration we define $\hat{t}_{sec} = \hat{t}_{min}$). Also let $n_{\min} = |N_{\min}|$ and $n_{\sec} = |N_{\sec}|$.

	$i \in N_{\min}$	$i \in N_{\sec}$	$i \in N\setminus(N_{\min}\cup N_{\sec})$
if $\hat{t}_{min} = \hat{t}_{sec}$	$\frac{1}{n}$	$\frac{1}{n}$	$\frac{1}{n}$
if $\hat{t}_{min} < \hat{t}_{sec} < c\cdot\hat{t}_{min}$	$\left(\frac{1}{L}\right)/n_{min}$	$\left(1-\frac{1}{L}\right)/n_{sec}$	0
if $\hat{t}_{sec} \geq c\cdot\hat{t}_{min}$	$\left(1 - \displaystyle\sum_{k\in N\setminus N_{min}} \frac{\hat{t}_{min}}{L\cdot\hat{t}_k}\right)/n_{min}$	$\frac{\hat{t}_{min}}{L\cdot t_i}$	$\frac{\hat{t}_{min}}{L\cdot t_i}$

Fig. 2. Algorithm $\mathcal{A}_{L,c}$ for scheduling a single task to $n \geq 2$ machines, parametrized by $L > 2(n-1)$ and $c > 1$. The first and second highest reported execution times by the machines are denoted by $\hat{t}_{min}$ and $\hat{t}_{sec}$ respectively, while $N_{\min}$, $N_{\sec}$ denote the corresponding sets of machine indices, and $n_{\min}$, $n_{\sec}$ their cardinalities.

Our main algorithm $\mathcal{A}_{L,c}$ for the case of one task and n machines, parametrized by $L > 2(n-1)$, $c > 1$, is defined by the allocation probabilities a_i for each machine $i \in N$ given in Fig. 2.

As the following theorem suggests, by picking a high enough value for L the above algorithm can achieve an optimal performance under equilibrium:

Theorem 2. *For the problem of scheduling one task without payments to $n \geq 2$ machines, algorithm $\mathcal{A}_L$ has a (pure) Price of Anarchy of $1 + \frac{n}{L}$ (for any $L > 2(n-1)$).*

Multiple Tasks. It is not difficult to extend our single-task algorithm and the result of Theorem 2 to get a task-independent, anonymous algorithm with a pure PoA of n for any number of tasks $m \geq 1$: simply run $\mathcal{A}_L$ *independently* for each job. Then, the equilibria of the extended setting correspond exactly to players not having an incentive to deviate for any task/round, and the approximation ratio of $1 + \frac{n}{L}$ with respect to the minimum cost $\min_i t_{i,j}$ at every such round $j = 1, \ldots, m$, guarantees optimality with respect to the social welfare and thus provides indeed a worst-case n-approximation for the makespan objective (see Eq. 3).

4   Truthful Mechanisms

In this section we turn our attention to truthful algorithms for many tasks and provide a mechanism that achieves approximation ratio n, almost matching the $(n + 1)/2$ known lower bound on truthfulness [11]. The best known ratio before our work was $n(n+1)/2$, achieved by running the algorithm of Koutsoupias [11] independently for each task. Unfortunately this guarantee turns out to be tight for the particular algorithm (see the full version [8] for a bad instance), thus here we have to devise more involved, non task-independent mechanisms.

4.1   The LP Mechanism

It is a known fact that the LP relaxation of a problem can be a useful tool for designing mechanisms (both randomized and fractional). We recall that the LP relaxation for the scheduling problem is as shown in Fig. 3.

$$\text{minimize } \mu$$

$$\forall j : \qquad \sum_{i=1}^{n} \alpha_{i,j} = 1 \qquad \text{(each task is allocated entirely)}$$

$$\forall i : \qquad \mu - \sum_{j=1}^{m} \alpha_{i,j} t_{i,j} \geq 0 \qquad \text{(the cost of each machine does not exceed makespan)}$$

$$\forall i, j : \qquad \alpha_{i,j} \geq 0 \qquad \text{(the allocation probabilities are positive)}$$

Fig. 3. The LP relaxation for the scheduling problem. Our LP mechanism is defined by an optimal solution $\alpha_{i,j}^{\mathrm{LP}}(\mathbf{t})$ to this program.

We denote an *optimal* solution[3] to the above LP by $\alpha^{\mathrm{LP}}(\mathbf{t}), \mu_{\mathbf{t}}^{\mathrm{LP}}$ (dropping the LP superscript whenever this is clear from the context). The vector $\alpha^{\mathrm{LP}}(\mathbf{t})$ can be straightforwardly interpreted as allocation probabilities or allocation fractions giving rise to a randomized and a fractional mechanism, respectively. We refer to the corresponding mechanisms as the LP randomized and the LP fractional mechanism. In Theorem 3 we show that both mechanisms are truthful, hence, we can think of $\mu_{\mathbf{t}}^{\mathrm{LP}}$ as corresponding to the maximum (expected) cost/workload perceived by any machine.

It is a simple observation that in an optimal solution the workload must be fully balanced among all machines and that μ^{LP} can only increase when all execution times increase, i.e. $\mu_{\mathbf{t}}^{\mathrm{LP}} \leq \mu_{\mathbf{t}'}^{\mathrm{LP}}$ for $\mathbf{t} \leq \mathbf{t}'$ (pointwise).

Note that the proof of Theorem 3 is identical in both cases where the α correspond to fractions or allocation probabilities. Hence, the result holds for both the LP randomized and the LP fractional mechanism.

Theorem 3. *Under the LP (fractional or randomized) mechanism, truthfully reporting the execution times is a (weakly) dominant strategy for every machine.*

Proof. Recall that $\mathbf{t}$ and $\hat{\mathbf{t}}$ denote the true and (some) declared execution times for all the machines. Fix some machine i and define vector $\mathbf{t}^{\max_i}$ as follows: row i, $\mathbf{t}_i^{\max_i}$, is the vector of point-wise maxima between true and declared times for machine i, that is

$$\mathbf{t}_i^{\max_i} = (\max\{\hat{t}_{i,1}, t_{i,1}\}, \max\{\hat{t}_{i,2}, t_{i,2}\}, \dots, \max\{\hat{t}_{i,m}, t_{i,m}\}),$$

---

[3] Notice that although $\mu_{\mathbf{t}}^{\mathrm{LP}}$ is unique, there might be various allocation fractions $\alpha_{i,j}$ that give rise to the optimal makespan $\mu_{\mathbf{t}}^{\mathrm{LP}}$, in which case we can choose an arbitrary one for $\alpha_{\mathbf{t}}^{\mathrm{LP}}$, e.g. take the lexicographically smaller.

while every other row $k \neq i$ is $t_k^{\mathrm{max}_i} = \hat{t}_k$, i.e. $\mathbf{t}^{\mathrm{max}_i} = (\mathbf{t}_i^{\mathrm{max}_i}, \hat{\mathbf{t}}_{-i})$. Seen as a vector of declarations, $\mathbf{t}^{\mathrm{max}_i}$ corresponds to machine i's deviation from $\hat{\mathbf{t}}$ to $\mathbf{t}_i^{\mathrm{max}_i}$. Then we can derive the following:

$$\sum_{j=1}^{m} \alpha_{k,j}(\hat{\mathbf{t}}) t_{k,j}^{\mathrm{max}_i} = \sum_{j=1}^{m} \alpha_{k,j}(\hat{\mathbf{t}}) \hat{t}_{k,j} = \mu_{\hat{\mathbf{t}}} = \sum_{j=1}^{m} \alpha_{i,j}(\hat{\mathbf{t}}) \hat{t}_{i,j} \leq \sum_{j=1}^{m} \alpha_{i,j}(\hat{\mathbf{t}}) t_{i,j}^{\mathrm{max}_i}$$

and thus from the optimality of the LP solutions it must be that

$$\mu_{\mathbf{t}^{\mathrm{max}_i}} \leq \max_{l=1,\ldots,n} \left\{ \sum_{j=1}^{m} \alpha_{l,j}(\hat{\mathbf{t}}) t_{l,j}^{\mathrm{max}_i} \right\} = \sum_{j=1}^{m} \alpha_{i,j}(\hat{\mathbf{t}}) t_{i,j}^{\mathrm{max}_i}.$$

Bringing everything together and taking into consideration that $(\mathbf{t}_i, \hat{\mathbf{t}}_{-i}) \leq \mathbf{t}^{\mathrm{max}_i}$ we get

$$C_i(\hat{\mathbf{t}}) = \sum_{j=1}^{m} \alpha_{i,j}(\hat{\mathbf{t}}) \max\{\hat{t}_{i,j}, t_{i,j}\} \geq \mu_{\mathbf{t}^{\mathrm{max}_i}} \geq \mu_{(\mathbf{t}_i, \hat{\mathbf{t}}_{-i})} = \sum_{j=1}^{m} \alpha_{i,j}(\mathbf{t}_i, \hat{\mathbf{t}}_{-i}) t_{i,j} = C_i(\mathbf{t}_i, \hat{\mathbf{t}}_{-i}),$$

which shows that indeed, whatever the declarations of the other machines $\hat{\mathbf{t}}_{-i}$, machine i is always (weakly) better of by truthfully reporting $\mathbf{t}_i$.

Theorem 3 gives rise to the following two results.

Theorem 4. *The LP fractional mechanism has approximation ratio 1 for the fractional scheduling problem without money, for any number of machines and tasks.*

As discussed in Sect. 2, by 2 we know that the above performance guarantee can deteriorate at most by a factor of n when we use the fractions as allocation probabilities for the integral case:

Theorem 5. *The LP randomized mechanism has approximation ratio n for (integrally) scheduling any number of tasks to n machines without money.*

4.2   The Proportional Mechanism

In this section we briefly consider the proportional mechanism which allocates to each machine i a $t_i^{-1} / \sum_{k=1}^{n} t_k^{-1}$ fraction of the task or probability of getting the task, respectively, depending on whether we consider the randomized or the fractional variant. In [11] it was shown that this algorithm is truthful and that its approximation ratio for randomized allocations of a single task is n. With the following theorem we wish to stress the difference between fractional and (randomized) integral allocations. The theorem is about the fractional case and proves the optimality of the proportional mechanism for scheduling one task without payments.

Theorem 6. *The proportional mechanism has an optimal approximation ratio of 1 for the fractional scheduling problem of a single task. For m tasks the approximation ratio increases to at least m.*

5   Price of Stability and Mixed Equilibria

In this section we attempt a more optimistic approach regarding the problem of scheduling without payments. We consider the benchmark of the *best* (mixed Nash) equilibrium and prove that the following, most natural greedy algorithm can achieve optimality: allocates each task independently to the machine declaring the minimum cost (breaking ties arbitrarily).

Theorem 7. *The Price of Stability of the Greedy algorithm is 1 for scheduling without money any number of tasks to any number of machines.*

References

1. Angel, E., Bampis, E., Pascual, F., Tchetgnia, A.-A.: On truthfulness and approximation for scheduling selfish tasks. J. Sched. **12**(5), 437–445 (2009)
2. Ashlagi, I., Dobzinski, S., Lavi, R.: Optimal lower bounds for anonymous scheduling mechanisms. Math. Oper. Res. **37**(2), 244–258 (2012)
3. Auletta, V., De Prisco, R., Penna, P., Persiano, G.: The power of verification for one-parameter agents. In: Díaz, J., Karhumäki, J., Lepistö, A., Sannella, D. (eds.) ICALP 2004. LNCS, vol. 3142, pp. 171–182. Springer, Heidelberg (2004)
4. Christodoulou, G., Gourvès, L., Pascual, F.: Scheduling selfish tasks: about the performance of truthful algorithms. In: Lin, G. (ed.) COCOON 2007. LNCS, vol. 4598, pp. 187–197. Springer, Heidelberg (2007)
5. Christodoulou, G., Koutsoupias, E., Kovács, A.: Mechanism design for fractional scheduling on unrelated machines. ACM Trans. Algorithms **6**(2), 38: 1–38: 18 (2010)
6. Dughmi, S., Ghosh, A.: Truthful assignment without money. In: EC, pp. 325–334 (2010)
7. Fotakis, D., Tzamos, C.: Winner-imposing strategyproof mechanisms for multiple facility location games. In: Saberi, A. (ed.) WINE 2010. LNCS, vol. 6484, pp. 234–245. Springer, Heidelberg (2010)
8. Giannakopoulos, Y., Koutsoupias, E., Kyropoulou, M.: The anarchy of scheduling without money. CoRR, abs/1607.03688 (2016). http://arxiv.org/abs/1607.03688
9. Gibbard, A.: Manipulation of voting schemes: a general result. Econometrica **41**(4), 587–601 (1973)
10. Guo, M., Conitzer, V.: Strategy-proof allocation of multiple items between two agents without payments or priors. In: AAMAS, pp. 881–888 (2010)
11. Koutsoupias, E.: Scheduling without payments. Theory Comput. Syst. **54**(3), 375–387 (2014)
12. Koutsoupias, E., Papadimitriou, C.: Worst-case equilibria. Comput. Sci. Rev. **3**(2), 65–69 (2009)
13. Koutsoupias, E., Vidali, A.: A lower bound of $1+\varphi$ for truthful scheduling mechanisms. Algorithmica **66**(1), 211–223 (2013)
14. Levin, H., Schapira, M., Zohar, A.: Interdomain routing and games. In: STOC, pp. 57–66 (2008)
15. Mu'alem, A., Schapira, M.: Setting lower bounds on truthfulness: extended abstract. In: SODA, pp. 1143–1152 (2007)
16. Nisan, N., Ronen, A.: Algorithmic mechanism design. Games Econ. Behav. **35**(1/2), 166–196 (2001)

17. Nisan, N., Roughgarden, T., Tardos, E., Vazirani, V.V.: Algorithmic Game Theory. Cambridge University Press, New York (2007)
18. Nissim, K., Smorodinsky, R., Tennenholtz, M.: Approximately optimal mechanism design via differential privacy. In: ITCS, pp. 203–213 (2012)
19. Penna, P., Ventre, C.: Optimal collusion-resistant mechanisms with verification. Games Econ. Behav. **86**, 491–509 (2014)
20. Procaccia, A.D., Tennenholtz, M.: Approximate mechanism design without money. In: EC, pp. 177–186 (2009)
21. Satterthwaite, M.A.: Strategy-proofness and Arrow's conditions: existence and correspondence theorems for voting procedures and social welfare functions. J. Econ. Theory **10**(2), 187–217 (1975)

An Almost Ideal Coordination Mechanism for Unrelated Machine Scheduling

Ioannis Caragiannis[1]($\boxtimes$) and Angelo Fanelli[2]

[1] CTI "Diophantus" and University of Patras, Rion, Greece
caragian@ceid.upatras.gr
[2] CNRS (UMR-6211), Caen, France
angelo.fanelli@unicaen.fr

Abstract. Coordination mechanisms aim to mitigate the impact of selfishness when scheduling jobs to different machines. Such a mechanism defines a scheduling policy within each machine and naturally induces a game among the selfish job owners. The desirable properties of a coordination mechanism includes simplicity in its definition and efficiency of the outcomes of the induced game. We present a broad class of coordination mechanisms for unrelated machine scheduling that are simple to define and we identify one of its members (mechanism DCOORD) that is superior to all known mechanisms. DCOORD induces potential games with logarithmic price of anarchy and only constant price of stability. Both bounds are almost optimal.

1 Introduction

We consider a selfish scheduling setting where each job owner acts as a non-cooperative player and aims to assign her job to one of the available machines so that the completion time of the job is as low as possible. An algorithmic tool that can be utilized by the designer of such a system is a *coordination mechanism* [8]. The coordination mechanism uses a *scheduling policy* within each machine that aims to mitigate the impact of selfishness to performance.

We focus on *unrelated machine scheduling*. There are m available machines and n players, each controlling a distinct job. The job (owned by player) u has a (possibly infinite) positive processing time (or load) $w_{u,j}$ when processed by machine j. A scheduling policy defines the way jobs are scheduled within a machine. In its simplest form, such a policy is *non-preemptive* and processes jobs uninterruptedly according to some order. *Preemptive* scheduling policies (which is our focus here) do not necessarily have this feature (e.g., they may process jobs in parallel) and may even introduce some idle time.

Naturally, a coordination mechanism induces a game with the job owners as players. Each player has all machines as possible *strategies*. The term *assignment*

This work was partially supported by the Caratheodory grant E.114 from the University of Patras and the project ANR-14-CE24-0007-01 *"CoCoRICo-CoDec"*. Part of the work was done while the second author was visiting the Institute for Mathematical Sciences, National University of Singapore in 2015.

M. Gairing and R. Savani (Eds.): SAGT 2016, LNCS 9928, pp. 315–326, 2016.
DOI: 10.1007/978-3-662-53354-3_25

is used for a snapshot of the game, where each player has selected a strategy, i.e., it has selected a particular machine to process her job. Given an assignment, the cost a player experiences is the completion time of her job on the machine she has selected. This is well-defined by the scheduling policy of the machine and typically depends on the characteristics of all jobs assigned to the machine.

Assignments in which no player has any incentive to change her strategy are called *pure Nash equilibria* (or, simply, *equilibria*). When studying a coordination mechanism, we are interested in bounding the inefficiency of equilibria of the game induced by the mechanism. We use the *maximum completion time* among all jobs to measure the social cost. A related quantity is the load of a machine which is defined as the total processing time of the jobs assigned to the machine. The *makespan* of an assignment is the maximum load over all machines. Clearly, the makespan of an assignment is a lower bound on the maximum completion time. The *price of anarchy* (respectively, *price of stability*) of the game induced by a coordination mechanism is defined as the worst (respectively, best) ratio of the maximum completion time over all equilibria over the optimal makespan.

We prefer mechanisms that induce games that always have equilibria. Furthermore, we would like these equilibria to be easy to find. A highly desirable property that ensures that equilibria can be reached by the players themselves (with best-response play) is the existence of a *potential function*. A potential function is defined over all possible assignments and has the property that, in any two assignments differing in the strategy of a single player, the difference of the two values of the potential and the difference of the completion time of the deviating player have the same sign.

Coordination mechanisms for scheduling were introduced by Christodoulou et al. [8]. Immorlica et al. [11] were the first to consider coordination mechanisms in the unrelated machine setting and studied several intuitive mechanisms, including ShortestFirst and Makespan. In ShortestFirst, the jobs in each machine are scheduled non-preemptively, in monotone non-decreasing order of their processing time. Since ties are possible, the mechanism has to distinguish between jobs with identical processing times, e.g., using distinct IDs for the jobs. This is necessary for every deterministic non-preemptive coordination mechanism in order to be well-defined. In contrast, in Makespan, each machine processes the jobs assigned to it "in parallel" so that they all have the same completion time. So, no ID information is required by Makespan. We use the term *anonymous* to refer to coordination mechanisms having this property. These two coordination mechanisms are *strongly local* in the sense that the only information that is required to compute the schedule of jobs within a machine is the processing time of the jobs on that machine only. A *local* coordination mechanism may use all parameters of the jobs that are assigned to a machine (e.g., the whole load vector of each job).

Azar et al. [4] prove lower bounds of $\Omega(m)$ and $\Omega(\log m)$ on the price of anarchy for any strongly local and local non-preemptive coordination mechanism, respectively. On the positive side, they presented two local coordination mechanisms with price of anarchy $o(m)$. Their first coordination mechanism

Table 1. A comparison between DCOORD and other local coordination mechanisms from the literature.

Coordination mechanism	PoA	PoS	PNE	Pot.	IDs.	Preempt	Reference
AFJMS-1	$\Theta(\log m)$	-	No	No	Yes	No	[4]
AFJMS-2	$O(\log^2 m)$	-	Yes	Yes	Yes	Yes	[4]
ACOORD	$O(\log m)$	$\Theta(\log m)$	Yes	Yes	Yes	Yes	[7]
Balance	$O(\log m)$	$\Theta(\log m)$	Yes	Yes	Yes	Yes	[9]
BCOORD	$\Theta(\frac{\log m}{\log\log m})$	-	?	No	No	Yes	[7]
CCOORD	$O(\log^2 m)$	$O(\log m)$	Yes	Yes	No	Yes	[7]
DCOORD	$O(\log m)$	$O(1)$	Yes	Yes	No	Yes	This paper

(henceforth called AFJMS-1) is non-preemptive and may induce game without equilibria. When the induced game has equilibria, the price of anarchy is at most $O(\log m)$. Their second coordination mechanism (henceforth called AFJMS-2) is preemptive, induces potential games, and has price of anarchy $O(\log^2 m)$. Both mechanisms are not anonymous.

Caragiannis [7] presents three more coordination mechanisms. The mechanism ACOORD, induces potential games with price of anarchy $O(\log m)$. The mechanism uses the distinct IDs of the jobs to ensure that the equilibria of the game are essentially assignments that are reached by a greedy-like online algorithm for minimizing the p-norm of machine loads. [3] and [6] study this online scheduling problem; the results therein imply that the price of stability of mechanism ACOORD is $\Omega(\log m)$ as well. A different coordination mechanism with similar characteristics (called Balance) is presented in [9]. The coordination mechanism BCOORD (defined also in [7]) has even better price of anarchy $O\left(\frac{\log m}{\log\log m}\right)$ (matching a lower bound due to Abed and Huang [2] for all deterministic coordination mechanisms) but the induced games are not potential ones and may not even have equilibria. However, the price of anarchy bound for BCOORD indicates that preemption may be useful in order to beat the $\Omega(\log m)$ lower bound for non-preemptive mechanisms from [4]. Interestingly, this mechanism is anonymous. The third mechanism CCOORD is anonymous as well, induces potential games, and has price of anarchy and price of stability $O(\log^2 m)$ and $O(\log m)$, respectively. To the best of our knowledge, this is the only anonymous mechanism that induces potential games and has polylogarithmic price of anarchy.[1] Table 1 summarizes the known local coordination mechanisms.

In the discussion above, we have focused on papers that define the social cost as the maximum completion time (among all players). An alternative social cost that has received much attention is the *weighted average completion time*; see

---

[1] Even though their mechanism Balance heavily uses job IDs, Cohen et al. [9] claim that it is anonymous. This is certainly false according to our terminology since anonymity imposes that two jobs with identical load vectors should be indistinguishable.

[1,5,10] for some recent related results. Interestingly, the design principles that lead to efficient mechanisms in their case are considerably different.

Our contribution is as follows. We introduce a quite broad class (called $\mathcal{M}(d)$) of local anonymous coordination mechanisms that induce potential games. The class contains the coordination mechanism CCOORD as well as the novel coordination mechanism DCOORD, which has additional *almost ideal* properties. In particular, we prove that it has logarithmic price of anarchy and only constant price of stability. A (qualitative and quantitative) comparison of DCOORD to other known local coordination mechanisms is depicted in Table 1.

The rest of the paper is structured as follows. We begin with preliminary definitions in Sect. 2. Section 3 is devoted to the definition of the class of mechanisms $\mathcal{M}(d)$ and to the proof that all mechanisms in this class induce potential games. Then, the novel mechanism DCOORD from this class is defined in Sect. 4; its feasibility as well as preliminary statements that are useful for the analysis are also presented there. Finally, in Sect. 5, we prove the bounds on the price of anarchy and stability.

2   Definitions and Preliminaries

Throughout the paper, we denote the number of machines by m. The index j always refers to a machine; the sum $\sum_j$ runs over all available machines. An assignment is a partition $N = (N_1, ..., N_m)$ of the players to the m machines. So, N_j is the set of players assigned to machine j under N. We use the notation $L_j(N_j)$ to refer to the load of machine j, i.e., $L_j(N_j) = \sum_{u \in N_j} w_{u,j}$.

A coordination mechanism uses a scheduling policy per machine. For every set of jobs assigned to machine j, the scheduling policy of the machine defines a detailed schedule of the jobs in the machine, i.e., it defines which job is executed in each point in time, whether more than one jobs are executed in parallel, or whether a machine stays idle for particular time intervals. Instead of defining coordination mechanisms at this level of detail, it suffices to focus on the definition of the completion time $\mathcal{P}(u, N_j)$ for the job of each player $u \in N_j$. This definition should correspond to some feasible detailed scheduling of jobs in the machine. A sufficient condition that guarantees *feasibility* is to define completion times that are never smaller than the machine load.

Like the coordination mechanisms in [4,7,9], our coordination mechanisms are local. The completion time $\mathcal{P}(u, N_j)$ of the job belonging to player u in machine j depends on the processing times the jobs in N_j have on machine j, as well as on the minimum processing time $w_u = \min_j w_{u,j}$ of job u over all machines.

Our proofs exploit simple facts about Euclidean norms of machine loads. Recall that, for $p \geq 1$, the p-norm of the vector of machine loads $L(N) = (L_1(N_1), L_2(N_2), ..., L_m(N_m))$ under an assignment N is $\|L(N)\|_p = \left(\sum_j L_j(N_j)^p\right)^{1/p}$. By convention, we denote the makespan $\max_j L_j(N_j)$ as $\|L(N)\|_\infty$. The following property follows easily by the definition of norms; we use it extensively in the following.

Lemma 1. *For any $p \geq 1$ and any assignment N, $\|L(N)\|_\infty \leq \|L(N)\|_p \leq m^{1/p}\|L(N)\|_\infty$.*

We also use the well-known Minkowski inequality (or triangle inequality for the p-norm). For machine loads, it reads as follows:

Lemma 2 (Minkowski inequality). *For every $p \geq 1$ and two assignments N and N', $\|L(N) + L(N')\|_p \leq \|L(N)\|_p + \|L(N')\|_p$.*

The notation $L(N)+L(N')$ denotes the m-entry vector with $L_j(N_j)+L_j(N'_j)$ at the j-th entry. Another necessary technical lemma follows by the convexity properties of polynomials; see [7] for a proof.

Lemma 3. *For $r \geq 1$, $t \geq 0$, positive integer p, and $a_i \geq 0$ for $i = 1, ..., p$, it holds*

$$\sum_{i=1}^{p}((t + a_i)^r - t^r) \leq \left(t + \sum_{i=1}^{p} a_i\right)^r - t^r.$$

3 A Broad Class of Coordination Mechanisms

In this section, we show that the coordination mechanism CCOORD from [7] can be thought of as belonging to a broad class of coordination mechanisms, which we call $\mathcal{M}(d)$. This class contains also our novel coordination mechanism DCOORD, which will be presented in Sect. 4.

The definition of CCOORD uses a positive integer $d \geq 2$ and the functions Ψ_j that map sets of players to the reals as follows. For any machine j, $\Psi_j(\emptyset) = 0$ and for any non-empty set of players $U = \{u_1, u_2, ..., u_\ell\}$,

$$\Psi_j(U) = d! \sum_{t_1+t_2+...+t_\ell=d} \prod_{k=1}^{\ell} w_{u_k,j}^{t_k}.$$

The sum runs over all multi-sets of non-negative integers $\{t_1, t_2, ..., t_\ell\}$ that satisfy $t_1 + t_2 + ... + t_\ell = d$. So, $\Psi_j(U)$ is the sum of all possible degree-d monomials of the processing times of the jobs belonging to players from U on machine j, with each term in the sum having a coefficient of $d!$. CCOORD schedules the job of player u_i on machine j in an assignment N so that its completion time is

$$\mathcal{P}(u_i, N_j) = \left(\frac{w_{u_i,j}\Psi_j(N_j)}{w_{u_i}}\right)^{1/d}.$$

We will extend CCOORD to define a broad class of coordination mechanisms; we use $\mathcal{M}(d)$ to refer to this class, where $d \geq 2$ is a positive integer. Each member of $\mathcal{M}(d)$ is identified by a *coefficient function* γ. The coefficient functions are defined over multi-sets of non-negative integers that have sum equal to $d+1$ and take non-negative values. An important property of the coefficient functions is that they are *invariant to zeros* that requires that for a multi-set A of integers that sum up to $d+1$, $\gamma(A) = \gamma(A \cup \{0\})$. Hence, the value returned by γ depends only on the non-zero elements in the multiset it takes as argument.

The definition of a coordination mechanism in $\mathcal{M}(d)$ uses the quantity $\Lambda_{u_i,j}(U)$, which is defined as follows for a machine j and a job u_i from a subset of jobs $U = \{u_1, u_2, ..., u_\ell\}$:

$$\Lambda_{u_i,j}(U) = \sum_{\substack{t_1+t_2+...+t_\ell=d+1 \\ t_i \geq 1}} \gamma(\{t_1, t_2, ..., t_\ell\}) \prod_{k=1}^{\ell} w_{u_k,j}^{t_k}. \tag{1}$$

The sum runs over all multi-sets of non-negative integers, with each integer corresponding to a distinct player of U, so that the integer t_i corresponding to player u_i is strictly positive. Notice that γ is defined over (unordered) multi-sets; this implies that symmetric monomials have the same coefficient. For example, for the set of players $U = \{u_1, u_2\}$ and a machine j,

$$\Lambda_{u_1,j}(U) = \gamma(\{3,0\})w_{u_1,j}^3 + \gamma(\{2,1\})w_{u_1,j}^2 w_{u_2,j} + \gamma(\{1,2\})w_{u_1,j}w_{u_2,j}^2.$$

Clearly, $\{2,1\}$ and $\{1,2\}$ denote the same multi-set and, hence, the coefficients of the (symmetric) second and third monomial are identical.

A coordination mechanism of $\mathcal{M}(d)$ sets the completion time of player u_i to

$$\mathcal{P}(u_i, N_j) = \left(\frac{\Lambda_{u_i,j}(N_j)}{w_{u_i}} \right)^{1/d}. \tag{2}$$

when its job is scheduled on machine j under assignment N.

By simply setting $\gamma(A) = d!$ for every multi-set A of non-negative integers summing up to $d+1$, we obtain CCOORD. Indeed, it is easy to see that $\Lambda_{u_i,j}(U) = w_{u_i,j}\Psi_j(U)$ in this case.

The definition of $\mathcal{M}(d)$ guarantees that all its members satisfy two important properties. First, every coordination mechanism in $\mathcal{M}(d)$ is anonymous. This is due to the fact that the definition of the completion time in (2) does not depend on the identity of a player and the jobs of two different players u and u' that have equal processing times $w_{u,j} = w_{u',j}$ at machine j and the same minimum processing time (over all machines) will enjoy identical completion times therein, when each is scheduled together with a set U of other players (i.e., $\mathcal{P}(u, U \cup \{u\}) = \mathcal{P}(u', U \cup \{u'\})$) or when the set of players N_j assigned to machine j contains both u and u' ($\mathcal{P}(u, N_j) = \mathcal{P}(u', N_j)$ in this case).

Another important property of the coordination mechanisms in $\mathcal{M}(d)$ is that they always induce potential games. We will prove this in a while, after defining the function $\Lambda_j(U)$, again for a machine j and a set of players $U = \{u_1, u_2, ..., u_\ell\}$, as follows:

$$\Lambda_j(U) = \sum_{t_1+t_2+...+t_\ell=d+1} \gamma(\{t_1, t_2, ..., t_\ell\}) \prod_{k=1}^{\ell} w_{u_k,j}^{t_k}. \tag{3}$$

Compared to the definition of $\Lambda_{u_i,j}(U)$ in (1), the sum in (3) runs just over all multi-sets of non-negative integers (corresponding to players in U) that sum up to $d + 1$, without any additional constraint.

We will sometimes use the informal term Λ-functions to refer to the functions defined in both (1) and (3). We can now state (in Lemma 4) the following property of Λ-functions that we will use several times in our analysis below; its proof is omitted due to lack of space. For example, it will be particularly useful in order to prove that mechanisms of $\mathcal{M}(d)$ induce potential games (in Theorem 1).

Lemma 4. *Consider a machine j and a set of players $U = \{u_1, u_2, ..., u_\ell\}$. Then, for every player $u_i \in U$,*

$$\Lambda_j(U) = \Lambda_{u_i,j}(U) + \Lambda_j(U \setminus \{u_i\}).$$

Theorem 1. *The non-negative function Φ, which is defined over assignments of players to machines as $\Phi(N) = \sum_j \Lambda_j(N_j)$, is a potential function for the game induced by any coordination mechanism in $\mathcal{M}(d)$.*

Proof. Consider two assignments N and N' that differ in the assignment of a single player u. Assume that player u is assigned to machine j_1 and j_2 in the assignments N and N', respectively. Using the definition of function Φ and Lemma 4, we have

$$\Phi(N) - \Phi(N') = \sum_j \Lambda_j(N_j) - \sum_j \Lambda_j(N_j)$$
$$= \Lambda_{j_1}(N_{j_1}) + \Lambda_{j_2}(N_{j_2}) - \Lambda_{j_1}(N'_{j_1}) - \Lambda_{j_2}(N'_{j_2})$$
$$= \Lambda_{u,j_1}(N_{j_1}) + \Lambda_{j_1}(N_{j_1} \setminus \{u\}) + \Lambda_{j_2}(N_{j_2})$$
$$\quad - \Lambda_{j_1}(N'_{j_1}) - \Lambda_{u,j_2}(N'_{j_2}) - \Lambda_{j_2}(N'_{j_2} \setminus \{u\}).$$

Now observe that $N_{j_1} \setminus \{u\} = N'_{j_1}$ and $N'_{j_2} \setminus \{u\} = N_{j_2}$. Hence, using this observation and the definition of the completion time for u in assignments N and N', the above derivation becomes

$$\Phi(N) - \Phi(N') = \Lambda_{u,j_1}(N_{j_1}) - \Lambda_{u,j_2}(N'_{j_2})$$
$$= w_u \left(\mathcal{P}(u, N_{j_1})^d - \mathcal{P}(u, N'_{j_2})^d \right),$$

which implies that the difference in the potentials and the difference $\mathcal{P}(u, N_{j_1}) - \mathcal{P}(u, N'_{j_2})$ in the completion time of the deviating player u in the two assignments have the same sign as desired. $\square$

4   The Coordination Mechanism DCOORD

Like CCOORD, our new coordination mechanism DCOORD belongs to class $\mathcal{M}(d)$. It uses the coefficient function defined as

$$\gamma(\{t_1, t_2, ..., t_\ell\}) = \begin{cases} 1 & \text{if } \exists i \text{ such that } t_i = d + 1 \\ \frac{d!d}{t_1! t_2! ... t_\ell!} & \text{otherwise} \end{cases}$$

for every multi-set of integers $\{t_1, t_2, ..., t_\ell\}$ such that $t_1 + t_2 + ... + t_\ell = d + 1$.

Observe that $\gamma(\{t_1, t_2, ..., t_\ell\})$ is very similar (but not identical) to the multinomial coefficient defined as $\binom{d+1}{t_1, t_2, ..., t_\ell} = \frac{(d+1)!}{t_1! ... t_\ell!}$. This is exploited in the proof of the next statement.

Lemma 5. *Consider a machine j and a subset of players $U = \{u_1, u_2, ..., u_\ell\}$. Then,*

$$\Lambda_j(U) = \frac{d}{d+1} L_j(U)^{d+1} + \frac{1}{d+1} \sum_{u \in U} w_{u,j}^{d+1}.$$

Proof. By the definition of $\Lambda_j(U)$ and the coefficient function γ, we have

$$\Lambda_j(U) = \sum_{t_1+t_2+...+t_\ell=d+1} \gamma(\{t_1, t_2, ..., t_\ell\}) \prod_{k=1}^{\ell} w_{u_k,j}^{t_k}$$

$$= \frac{d}{d+1} \sum_{t_1+t_2+...+t_\ell=d+1} \binom{d+1}{t_1, t_2, ..., t_\ell} \prod_{k=1}^{\ell} w_{u_k,j}^{t_k} + \frac{1}{d+1} \sum_{u \in U} w_{u,j}^{d+1}$$

$$= \frac{d}{d+1} L_j(U)^{d+1} + \frac{1}{d+1} \sum_{u \in U} w_{u,j}^{d+1}$$

as desired. $\qquad\qquad\square$

We proceed with two properties which relate Λ-functions to machine loads. The first one (Corollary 1) follows as a trivial corollary of Lemma 5 after observing that $\sum_{u \in U} w_{u,j}^{d+1} \leq L_j(U)^{d+1}$.

Corollary 1. *Consider a machine j and a set of players U. Then,*

$$\frac{d}{d+1} L_j(U)^{d+1} \leq \Lambda_j(U) \leq L_j(U)^{d+1}.$$

The second one (Lemma 6) will be very useful in proving that DCOORD is feasible and in bounding its price of anarchy. Its proof is omitted due to lack of space.

Lemma 6. *Let $U = \{u_1, ..., u_\ell\}$ be a set of players. For every player $u_i \in U$ and every machine j, it holds that*

$$w_{u_i,j} L_j(U)^d \leq \Lambda_{u_i,j}(U) \leq d \cdot w_{u_i,j} L_j(U)^d.$$

Feasibility follows easily now.

Theorem 2. *DCOORD produces feasible schedules.*

Proof. Consider player u_1 and any assignment N which assigns it to machine j together with $\ell - 1$ other players $u_2, u_3, ..., u_\ell$. By the leftmost inequality of Lemma 6, we have that

$$P(u_1, N_j) = \left(\frac{\Lambda_{u_1,j}(N_j)}{w_{u_1}}\right)^{1/d} \geq \left(\frac{w_{u_1,j}}{w_{u_1}}\right)^{1/d} L_j(N_j) \geq L_j(N_j),$$

as desired. The inequality holds since, by definition, $w_{u_1,j} \geq w_{u_1}$. □

5 Bounding the Price of Anarchy and Stability

For proving the price of anarchy bound, we will need the following lemma which relates the load of any machine at an equilibrium with the optimal makespan.

Lemma 7. *Let N be an equilibrium and N^* an assignment of optimal makespan. Then, for every machine j, it holds that*

$$L_j(N) \leq m^{\frac{1}{d+1}} \frac{d+1}{\ln 2} \|L(N^*)\|_\infty.$$

Proof. Consider a player u that is assigned to machine j in the equilibrium assignment N and to machine j' in the assignment $\tilde{N}$ that minimizes the l_{d+1}-norm of the machine loads. First, consider the case where $j \neq j'$. In the equilibrium assignment N, player u has no incentive to deviate from machine j to machine j' and, hence, $P(u, N_j) \leq P(u, N_{j'} \cup \{u\})$. By the definition of DCOORD, we obtain that $\Lambda_{u,j}(N_j) \leq \Lambda_{u,j'}(N_{j'} \cup \{u\})$. Using this observation, Lemmas 4, and 5, we get

$$\Lambda_{u,j}(N_j) \leq \Lambda_{u,j'}(N_{j'} \cup \{u\}) = \Lambda_{j'}(N_{j'} \cup \{u\}) - \Lambda_{j'}(N_{j'})$$

$$= \frac{d}{d+1} L_{j'}(N_{j'} \cup \{u\})^{d+1} - \frac{d}{d+1} L_{j'}(N_{j'})^{d+1} + \frac{1}{d+1} w_{u,j'}^{d+1}$$

$$= \frac{d}{d+1}(L_{j'}(N_{j'}) + w_{u,j'})^{d+1} - \frac{d}{d+1} L_{j'}(N_{j'})^{d+1} + \frac{1}{d+1} w_{u,j'}^{d+1}$$

We will now prove that the same inequality holds when $j = j'$. In this case, together with Lemmas 4 and 5, we need to use a different argument that exploits a convexity property of polynomials. We have

$$\Lambda_{u,j}(N_j) = \Lambda_{u,j'}(N_{j'}) = \Lambda_{j'}(N_{j'}) - \Lambda_{j'}(N_{j'} \setminus \{u\})$$

$$= \frac{d}{d+1} L_{j'}(N_{j'})^{d+1} - \frac{d}{d+1} L_{j'}(N_{j'} \setminus \{u\})^{d+1} + \frac{1}{d+1} w_{u,j'}^{d+1}$$

$$= \frac{d}{d+1} L_{j'}(N_{j'})^{d+1} - \frac{d}{d+1}(L_{j'}(N_{j'}) - w_{u,j'})^{d+1} + \frac{1}{d+1} w_{u,j'}^{d+1}$$

$$\leq \frac{d}{d+1}(L_{j'}(N_{j'}) + w_{u,j'})^{d+1} - \frac{d}{d+1} L_{j'}(N_{j'})^{d+1} + \frac{1}{d+1} w_{u,j'}^{d+1}.$$

The last inequality follows since $z^{d+1} - (z-\alpha)^{d+1} \leq (z+\alpha)^{d+1} - z^{d+1}$ for every $z \geq \alpha \geq 0$, due to the convexity of the polynomial function z^{d+1}.

Let us sum the above inequality over all players. We obtain

$$\sum_j \sum_{u \in N_j} \Lambda_{u,j}(N_j)$$

$$\leq \frac{d}{d+1} \sum_j \sum_{u \in \tilde{N}_j} \left((L_j(N_j) + w_{u,j})^{d+1} - L_j(N_j)^{d+1} \right) + \frac{1}{d+1} \sum_j \sum_{u \in \tilde{N}_j} w_{u,j}^{d+1}$$

$$\leq \frac{d}{d+1} \sum_j \left(\left(L_j(N_j) + \sum_{u \in \tilde{N}_j} w_{u,j} \right)^{d+1} - L_j(N_j)^{d+1} \right) + \frac{1}{d+1} \sum_j L_j(\tilde{N}_j)^{d+1}$$

$$= \frac{d}{d+1} \|L(N) + L(\tilde{N})\|_{d+1}^{d+1} - \frac{d}{d+1} \|L(N)\|_{d+1}^{d+1} + \frac{1}{d+1} \|L(\tilde{N})\|_{d+1}^{d+1}$$

$$\leq \frac{d}{d+1} \left(\|L(N)\|_{d+1} + \|L(\tilde{N})\|_{d+1} \right)^{d+1} - \frac{d-1}{d+1} \|L(N)\|_{d+1}^{d+1}. \tag{4}$$

The second inequality follows by Lemma 3 and since $\sum_{u \in \tilde{N}_j} w_{u,j}^{d+1} \leq L_j(\tilde{N})^{d+1}$. The equality follows by the definition of l_{d+1}-norms and the last inequality follows by Minkowski inequality (Lemma 2) and by the fact that $\|L(\tilde{N})\| \leq \|L(N)\|$.

Using the definition of norms and Lemma 6, we also have

$$\|L(N)\|_{d+1}^{d+1} = \sum_j L_j(N_j)^{d+1} = \sum_j \sum_{u \in N_j} w_{u,j} L_j(N_j)^d \leq \sum_j \sum_{u \in N_j} \Lambda_{u,j}. \tag{5}$$

By combining (4) and (5), we have

$$2\|L(N)\|_{d+1}^{d+1} \leq \left(\|L(N)\|_{d+1} + \|L(\tilde{N})\|_{d+1} \right)^{d+1}$$

and, equivalently,

$$\|L(N)\|_{d+1} \leq \frac{1}{2^{\frac{1}{d+1}} - 1} \|L(\tilde{N})\|_{d+1} \leq \frac{d+1}{\ln 2} \|L(N^*)\|_{d+1}$$

$$\leq m^{\frac{1}{d+1}} \frac{d+1}{\ln 2} \|L(N^*)\|_\infty.$$

The second inequality follows since, by definition, $\|L(\tilde{N})\|_{d+1} \leq \|L(N^*)\|_{d+1}$ and by the inequality $e^z \geq z + 1$. The third inequality follows by Lemma 1. Since $\|L(N)\|_{d+1} \geq L_j(N_j)$ for every machine j, the lemma follows.   □

For the price of stability bound, we will use a qualitatively similar (to Lemma 7) relation between machine loads at a particular equilibrium and the optimal makespan.

Lemma 8. *Let N be the equilibrium that minimizes the potential function and N^* an assignment of optimal makespan. Then, for every machine j, it holds that*

$$L_j(N_j) \leq \left(\frac{d+1}{d} m \right)^{\frac{1}{d+1}} \|L(N^*)\|_\infty.$$

Proof. Observe that $\Phi(N) \leq \Phi(N^*)$ since every equilibrium that is reached when players repeatedly best-respond starting from assignment N^* has potential at most $\Phi(N^*)$. Using this observation, the definition of norms, Corollary 1, and the definition of the potential function (see the statement of Theorem 1), we have

$$\|L(N)\|_{d+1}^{d+1} = \sum_j L_j(N_j)^{d+1} \leq \frac{d+1}{d} \sum_j \Lambda_j(N_j) = \frac{d+1}{d}\Phi(N)$$

$$\leq \frac{d+1}{d}\Phi(N^*) = \frac{d+1}{d}\sum_j \Lambda_j(N_j^*)^{d+1} \leq \frac{d+1}{d}\sum_j L_j(N_j^*)^{d+1}$$

$$= \frac{d+1}{d}\|L(N^*)\|_{d+1}^{d+1}.$$

Hence, for every machine j, by exploiting Lemma 1, we have $L_j(N) \leq \|L(N)\|_{d+1} \leq \left(\frac{d+1}{d}\right)^{\frac{1}{d+1}} \|L(N^*)\|_{d+1} \leq \left(\frac{d+1}{d}m\right)^{\frac{1}{d+1}} \|L(N^*)\|_\infty$ as desired. $\qquad \square$

We are now ready to complete the price of anarchy/stability proofs. We will do so by comparing the completion time of any player to the optimal makespan $\|L(N^*)\|_\infty$.

Theorem 3. *By setting $d = O(\log m)$, DCOORD has price of anarchy $O(\log m)$ and price of stability $O(1)$.*

Proof. Consider a player u that is assigned to machine j at some equilibrium N and satisfies $w_u = w_{u,j^*}$ for some machine j^*. We will use the fact that player u (is either already at or) has not incentive to deviate to machine j^* at equilibrium to bound its completion time as follows:

$$\mathcal{P}(u, N_j) \leq \mathcal{P}(u, N_{j^*} \cup \{u\}) = \left(\frac{\Lambda_{u,j^*}(N_{j^*} \cup \{u\})}{w_u}\right)^{1/d}$$

$$\leq \left(\frac{dw_{u,j^*} L_{j^*}(N_{j^*} \cup \{u\})^d}{w_u}\right)^{1/d} \leq d^{1/d}(L_{j^*}(N_{j^*}) + w_u).$$

The equality follows by the definition of DCOORD, and the second inequality follows by Lemma 6. The third inequality follows since $w_u = w_{u,j^*}$ and by observing that $L_{j^*}(N_{j^*} \cup \{u\}) = L(N_{j^*}) + w_u$ if $u \notin N_{j^*}$ (i.e., $j \neq j^*$) and $L_{j^*}(N_{j^*} \cup \{u\}) = L(N_{j^*})$ otherwise.

Now, using Lemma 7 to bound $L_{j^*}(N_{j^*})$, we obtain that

$$\mathcal{P}(u, N_j) \leq d^{1/d}\left(m^{\frac{1}{d+1}}\frac{d+1}{\ln 2} + 1\right)\|L(N^*)\|_\infty.$$

If the equilibrium N is a potential-minimizing assignment, Lemma 8 can be further used to obtain the better guarantee

$$\mathcal{P}(u, N_j) \leq d^{1/d}\left(\left(\frac{d+1}{d}m\right)^{\frac{1}{d+1}} + 1\right)\|L(N^*)\|_\infty.$$

The theorem follows since, by setting $d = \Theta(\log m)$, the factors (ignoring $\|L(N^*)\|_\infty$ in the rightmost expressions become $O(\log m)$ and $O(1)$, respectively. So, in general, we have that the completion time of any player at equilibrium is at most $O(\log m)$ times the optimal makespan (hence, the price of anarchy bound) while there exists a particular equilibrium where the completion time of any player is at most $O(1)$ times the optimal makespan (hence, the price of stability bound). □

References

1. Abed, F., Correa, J.R., Huang, C.-C.: Optimal coordination mechanisms for multi-job scheduling games. In: Schulz, A.S., Wagner, D. (eds.) ESA 2014. LNCS, vol. 8737, pp. 13–24. Springer, Heidelberg (2014)
2. Abed, F., Huang, C.-C.: Preemptive coordination mechanisms for unrelated machines. In: Epstein, L., Ferragina, P. (eds.) ESA 2012. LNCS, vol. 7501, pp. 12–23. Springer, Heidelberg (2012)
3. Awerbuch, B., Azar, Y., Grove, E.F., Kao, M.-Y., Krishnan, P., Vitter, J.S.: Load balancing in the L_p norm. In: Proceedings of the 36th Annual IEEE Symposium on Foundations of Computer Science (FOCS), pp. 383–391 (1995)
4. Azar, Y., Fleischer, L., Jain, K., Mirrokni, V.S., Svitkina, Z.: Optimal coordination mechanisms for unrelated machine scheduling. Oper. Res. **63**(3), 489–500 (2015)
5. Bhattacharya, S., Im, S., Kulkarni, J., Munagala, K.: Coordination mechanisms from (almost) all scheduling policies. In: Proceedings of the 5th Conference on Innovations in Theoretical Computer Science (ITCS), pp. 121–134 (2014)
6. Caragiannis, I.: Better bounds for online load balancing on unrelated machines. In: Proceedings of the 19th Annual ACM/SIAM Symposium on Discrete Algorithms (SODA), pp. 972–981 (2008)
7. Caragiannis, I.: Efficient coordination mechanisms for unrelated machine scheduling. Algorithmica **66**, 512–540 (2013)
8. Christodoulou, G., Koutsoupias, E., Nanavati, A.: Coordination mechanisms. Theor. Comput. Sci. **410**(36), 3327–3336 (2009)
9. Cohen, J., Dürr, C., Thang, N.K.: Smooth inequalities and equilibrium inefficiency in scheduling games. In: Goldberg, P.W. (ed.) WINE 2012. LNCS, vol. 7695, pp. 350–363. Springer, Heidelberg (2012)
10. Cole, R., Correa, J.R., Gkatzelis, V.S., Gkatzelis, N.: Decentralized utilitarian mechanisms for scheduling games. Games Econ. Behav. **92**, 306–326 (2015)
11. Immorlica, N., Li, L., Mirrokni, V.S., Schulz, A.S.: Coordination mechanisms for selfish scheduling. Theor. Comput. Sci. **410**(17), 1589–1598 (2009)

Designing Cost-Sharing Methods
for Bayesian Games

George Christodoulou[1], Stefano Leonardi[2], and Alkmini Sgouritsa[1]([✉])

[1] Computer Science Department, University of Liverpool, Liverpool, UK
{gchristo,salkmini}@liv.ac.uk
[2] Department of Computer, Control and Management Engineering Antonio Ruberti,
Sapienza University of Rome, Rome, Italy
leonardi@dis.uniroma1.it

Abstract. We study the design of cost-sharing protocols for two fundamental resource allocation problems, the *Set Cover* and the *Steiner Tree Problem*, under environments of incomplete information (Bayesian model). Our objective is to design protocols where the worst-case Bayesian Nash equilibria, have low cost, i.e. *the Bayesian Price of Anarchy (PoA)* is minimized. Although budget balance is a very natural requirement, it puts considerable restrictions on the design space, resulting in high PoA. We propose an alternative, relaxed requirement called *budget balance in the equilibrium (BBiE)*. We show an interesting connection between algorithms for *Oblivious Stochastic* optimization problems and cost-sharing design with low PoA. We exploit this connection for both problems and we enforce approximate solutions of the stochastic problem, as Bayesian Nash equilibria, with the same guarantees on the PoA. More interestingly, we show how to obtain the same bounds on the PoA, by using *anonymous* posted prices which are desirable because they are easy to implement and, as we show, induce *dominant strategies* for the players.

Keywords: Price of Anarchy · Bayesian games · Network design

1 Introduction

A *cost-sharing game*, is an abstract setting that describes interactions of selfish players in environments where the cost of the produced solution needs to be shared among the participants. A *cost-sharing protocol* prescribes how the incurred cost is split among the users. This defines a game that is played by

Part of this work was done while all authors were visiting the Economics and Computation program of the Simons Institute for the Theory of Computing.

G. Christodoulou—This work was supported by EP/M008118/1, EP/K01000X/1 and LT140046.

S. Leonardi—The work of S. Leonardi was partially supported by Google Focused award "Algorithms for Large-scale Data analysis" and EU FET project no. 317532 "Multiplex".

M. Gairing and R. Savani (Eds.): SAGT 2016, LNCS 9928, pp. 327–339, 2016.
DOI: 10.1007/978-3-662-53354-3_26

the participants, who try to select outcomes that incur low personal costs. Chen, Roughgarden and Valiant [6] initiated the *design* aspect, seeking for protocols that induce approximately efficient equilibria, *with low Price of Anarchy (PoA)* [25]. Similarly, we study the design of cost-sharing protocols, for two well-studied and very general resource allocation problems with numerous applications, the *Set Cover* and the *Steiner tree (multicast)* problem.

Set Cover Game. In the (weighted) set cover problem, there is a universe of n elements, $U = \{1, \ldots, n\}$, and a family of subsets of U, $\mathcal{F} = \{F_1, \ldots, F_m\}$, with weights/costs $c_{F_1}, \ldots, c_{F_m}$. A subset of elements, $X \subseteq U$, needs to be covered by the $F_i's$ so that the total cost is *minimized*. We are interested in a game theoretic version, where there are $|X|$ players and $|U|$ possible *types*; X corresponds to the set of players and each player's type associates her with a specific element of U. Multiple players may have the same type. A player's action is to chose a subset from $\mathcal{F}$ that covers her element, and pay some cost-share for using it. A cost-sharing method prescribes how the subsets' costs are split among players.

Multicast Game. In a multicast game, there is a rooted (connected) undirected graph $G = (V, E, t)$, where each edge e carries a nonnegative weight c_e and t is a designated root. There are k players and $|V| = n$ possible types; each player's type associates him with a specific vertex of V which needs to establish connectivity with t. The players' strategies are all the paths that connect their terminal with t. A cost-sharing method defines the cost-shares of the players.

Cost-Sharing Under Uncertainty. There are two different possible sources of uncertainty that may need to be considered in the above scenarios. Firstly, the designer needs to specify the cost-sharing protocol, having only partial information about the players' types. Moreover, the players themselves, when they select their actions, may have incomplete knowledge about the types of the other players. We approach the former by using a stochastic model similar to [10], and the latter, as a *Bayesian game*, introduced by [21], which is an elegant way of modelling selfishness in partial-information settings. In a Bayesian game, players do not know the private types of the other players, but only have *beliefs*, expressed by probability distributions over the possible realizations of the types.

The order of events is as follows; first, the designer specifies the cost-sharing methods, using the product probability distribution over the players' types, then the players interact in the induced Bayesian game, and end up in a Bayesian Nash Equilibrium. We are interested in the design of protocols, where *all* equilibria have low cost i.e., the (Bayesian) PoA of the induced game is *low*.

Budget-Balance in the Equilibrium (BBiE). One of the axioms that [6] required in their design space, that every cost-sharing protocol should satisfy, is *budget balance* i.e., that the players' cost-shares cover *exactly* the cost of *any* solution. Although budget balance is a very natural requirement, it puts considerable restrictions on the design space. However, since we expect that the players will end up in a Nash equilibrium, it is not clear why one should be interested to impose budget balance in non-equilibrium states; the players are going to deviate from such states anyways. We propose an alternative, relaxed requirement that we call *budget balance in the equilibrium (BBiE)*. A BBiE cost-sharing protocol satisfies budget balance in *all equilibria*; for any non-equilibrium profile we do not impose this requirement. This natural relaxation, enlarges the design space but maintains the desired property of balancing the cost in the equilibrium. More importantly, this amplification of the design space, allows us to design protocols that dramatically outperform the best possible PoA bounds obtained by budget-balanced protocols. Indeed, by restricting to budget-balanced protocols, a lower bound of $\Omega(n)$ exists, for the complete information set cover game [6]; we extend this lower bound for the Bayesian setting. We further show a lower bound of $\Omega(\sqrt{n})$, for the multicast Bayesian game. We demonstrate that, by designing BBiE protocols, we can enforce better solutions, that dramatically improve the PoA. For the set cover game, we improve the PoA to $O(n/\log n)$ (or $O(\log n)$ if $m = poly(n)$). Regarding the multicast game, we improve the PoA to $O(1)$.

Posted Prices. It is a very common practice, especially in large markets and double auctions, for sellers to use posted prices. More closely to cost-sharing games is the model proposed by Kelly [23] regarding *bandwidth allocation*. Kelly's mechanism processes players' willingness to pay and posts a price for the whole bandwidth. Then each player pays a price proportional to the bandwidth she uses. This can be seen as pricing an infinitesimal quantity of bandwidth and the players, acting as price-takers, choose some number of quantities to buy. It turns out that it is in the best interest of the players to buy the whole bandwidth.

The use of posted prices, to serve as cost-sharing mechanism, is highly desirable, but not always possible to achieve; a price is posted for each resource and then the players behave as price takers, picking up the cheapest possible resources that satisfy their requirements. Such a mechanism is desirable because it is extremely easy to implement and also induces *dominant strategies*. We stress that our main results can be implemented by *anonymous* posted prices.

1.1 Results and Discussion

We study the design of cost-sharing protocols for two fundamental resource allocation problems, the *Set Cover* and the *Steiner tree problem*. We are interested in environments of incomplete information where both the designer and the players have partial information, described by prior probability distributions over types. Our objective is to design cost-sharing protocols that are *BBiE* and the worst-case equilibria have low cost, i.e. *the Bayesian PoA* is minimized.

We show an interesting connection between algorithms for *Oblivious Stochastic* optimization problems and cost-sharing design with low PoA. We exploit this for both problems and we are able to enforce approximate solutions of the stochastic problem, as Bayesian Nash equilibria, with the same guarantees on the PoA. Although this connection is quite simple, it results in significant improvement on the PoA comparing to budget-balanced protocols. More precisely, we map each player to a *single* specific strategy and charge very high costs for any alternative strategy. In this way, their mapped strategy becomes a (strongly) *dominant strategy*. For the set cover game, we enforce the oblivious solution given by [19]. They apriori map each player i to some subset $F_i \in \mathcal{F}$; then, if i is sampled, F_i should be in the induced solution. For the multicast game, the algorithm of [17], for the online Steiner tree problem, provides an oblivious solution.

Budget-Balanced Protocols (Sect. 3). First, we provide lower bounds for the PoA of budget-balanced protocols. It is not hard to see that there exists a set cover game that reduces to the lower bound of Chen, Roughgarden and Valiant [6] for the multicast directed network games, resulting in PoA= $\Omega(n)$ in the complete information case; we refer the reader to the full version of the paper for this reduction. For the stochastic or Bayesian setting, where players are i.i.d., we show that the same lower bound holds, but a further analysis is needed. We refer the reader to the full version of the paper for this reduction. Regarding the multicast game, the PoA is $O(1)$ for the complete information case [6] and the stochastic case [10,17]. However, we show that for the Bayesian setting there is a lower bound of $\Omega(\sqrt{n})$ (see Table 1 for a summary of the main results).

BBiE Protocols (Sect. 4). For the Bayesian (and stochastic) set cover game there exists an *ex-post*[1] BBiE protocol (determined in polynomial time) with PoA $O(\log n)$, if $m = poly(n)$, and $O\left(\frac{\log m}{\log\log m - \log\log n}\right)$, if $m \gg n$. An *ex-post* BBiE protocol also exists for the Bayesian multicast game resulting in constant PoA.

Posted Prices (Sect. 5). For the Bayesian (and stochastic) settings, ex-post BBiE cannot be obtained by anonymous prices. Hence, we examine prices that are *ex-ante* BBiE. In the full version of the paper, we discuss limitations of other concepts, such as BBiE with "high" probability or bounded possible excess and deficit. In Sect. 5 we present anonymous prices with the same upper bounds as the BBiE protocols, for the unweighted set cover and for the multicast games, respectively. We stress that oblivious solutions may not be sufficient to guarantee low PoA for anonymous posted prices, in contrast to the BBiE protocols. This is because it is not clear anymore how to enforce players to choose desirable strategies, since *anonymous* prices are available to anyone. The reason that they exist here is due to the specific properties of the oblivious solution.

Regarding the weighted set cover game, we can only provide *semi-anonymous* prices with the same bounds; by semi-anonymous we mean that the prices for each player do not depend on her identity, but only on her type. We leave the case of anonymous prices as an open question. We remark that in all cases, posted prices induce *dominant strategies* for the players. At last, for the poly-time determinable prices, we give tight lower bounds.

Prior-Independent Mechanisms. Clearly, the above BBiE protocols and posted prices depend on the prior distribution. Prior-independent mechanisms are also of high interest and in Sect. 6 we discuss their limitations.

---

[1] In ex-post budget-balance we require budget-balance in every realization of the game. If the *expected* excess and deficit are zero, the budget balance is called ex-ante.

Table 1. PoA of budget-balanced and BBiE protocols.

	BB protocols		BBiE protocols/posted prices	
	Set cover	Undirected	Set cover	Undirected
Full information	$\Theta(n)$ [6]	$O(1)$ [6]	1	1
Bayesian	$\Omega(n)$	$\Omega(\sqrt{n})$	$O(n/\log n)$	$O(1)$

In the full version of the paper we further study the complete information setting (see Table 1). Due to lack of space, we refer the reader to the full version of the paper for all the missing proofs.

1.2 Related Work

There is a vast amount of research in cost-sharing games and so, we only mention some of the most related. Moulin and Shenker [27] studied cost-sharing games under mechanism design context. In similar context, other papers considered (group)strategy proof and efficient mechanisms and relaxed the budget-balanced constraint; Devanur, Mihail and Vazirani [12] and Immorlica, Mahdian and Mirrokni [22] studied the set cover game under this context showing positive and negative bounds on the fraction of the cost that is covered.

Regarding the network design games, there is a long line of works mainly focusing on fair cost allocation originated by Anshelevich et al. [2]. They showed a tight $\Theta(\log k)$ bound on the PoS for directed networks, while for undirected networks the exact value of PoS still remains an open problem. For multicast games, Li [26] proved an upper bound of $O(\log k/\log\log k)$, while for broadcast games, a constant upper bound is known due to Bilò, Flammini and Moscardelli [4]. Chen, Roughgarden and Valiant [6] were the first to study the design aspects for this game, identifying the best protocol with respect to the PoA and PoS in various cases, followed by [10,13,18]. The Bayesian Price of anarchy was first studied in auctions by [8]; see also [28] for routing games, and [30] for the PoS of Shapley protocol in cost-sharing games.

Close in spirit to our work is the notion of Coordination Mechanisms [7] which provide a way to improve the PoA in cases of incomplete information. Similar to our context, the designer has to decide in advance game-specific policies, without knowing the exact input. Such mechanisms have been used for scheduling and simple routing games, see [1,3,9] and the papers cited therein.

Posted prices have been used for pricing in large markets. Kelso and Crawford [24] and Gul and Stacchetti [20] proved the existence of prices, for gross substitute valuations, that clear the market efficiently. Pricing bundles for combinatorial Walrasian equilibria was introduced by Feldman, Gravin and Lucier [15], who showed that half of the social welfare can be achieved. In a follow-up work [16], they considered Bayesian combinatorial auctions and they could guarantee half of the optimum welfare, by using anonymous posted prices.

The underlying problems that we consider here, the set cover and the minimum Steiner tree problems, are well studied NP-complete problems. The best known approximations are $O(\log(k))$ [11] (by using a simple greedy algorithm) and 1.39 [5]; in fact, for the set cover problem, Feige [14] showed that no improvement by a constant factor is likely. Research has been done regarding the stochastic model, Grandoni et al. [19] showed a roughly $O(\log nm)$ tight bound for the set cover problem and Garg et al. [17] gave bounds on the approximation of the stochastic online Steiner tree problem. A slightly different distribution is the independent activations; [10,29] demonstrated constant approximation algorithms or the universal TSP problem and the multicast game, respectively.

2   Model

Cost-Sharing Protocol. In the cost-sharing games, we consider that there are k players who are interested in a set of resources, $R = \{r_1, \ldots, r_m\}$. Each resource r carries a cost c_r. Whenever a subset of players uses a resource r, they are charged some cost-share, defined by a cost-sharing (resource-specific) method ξ. A cost-sharing protocol Ξ decides a cost-sharing method for each resource. In accordance with previous works, [6,10,13], the following are some natural properties that Ξ needs to satisfy:

- *Stability*: The induced game has always a *pure* Nash equilibrium.
- *Separability*: The cost shares of each resource r are completely determined by the set of players that choose it.
- *BBiE*: In any pure (Bayes) Nash equilibrium profile, the cost shares of the players choosing r should cover exactly the cost of r.

For the rest of the paper, by k we denote the number of players and by n the number of different types of the players, i.e. in the set cover game, $|U| = n$, and in the multicast game, $|V| = n$.

Information Models. We study several information models, from the point of view of the designer and of other players, regarding the knowledge of players' type. A player's type is some resource: in the set cover game, it is some element from U that needs to be covered, and in the multicast game, it is some vertex of G, on which the player's terminal lies, and requires connectivity with the root t. The parameters of the game is known to both the protocol designer and the participants. To be more specific, the tuple $(U, \mathcal{F}, c)$ in the set cover game and the underlying (weighted) graph in the multicast game are commonly known.

The information models that we consider are the following:

- *Complete Information*: The types of the players are common knowledge, i.e. they are known to all players and to the designer.
- *Stochastic/A priori*: The players' types are drawn from some product distribution D defined over the type set (U for set cover and V for multicast). The actual types are unknown to the designer, who is only aware of D. However, the players decide their strategies by knowing other players' types.

– *Bayesian*: The players' types are drawn from some product distribution D defined over the type sets. Both the designer and the players know only D. The players now decide their strategies by knowing only D and not the actual types. A natural assumption is that every player knows her own type.

We assume that the players' types are distributed i.i.d. ($D = \pi^k$) and the type of each player is drawn independently from some probability distribution $\pi : R \to [0,1]$, with $\sum_{r \in R} \pi(r) = 1$; R is either U in the set cover or V in the multicast. For simplicity we write π_r instead of $\pi(r)$.

Price of Anarchy (PoA). Let $opt(\mathbf{t})$ be the optimum solution given the players' types $\mathbf{t}$, and $NE(\mathbf{t})$ and BNE be the set of pure Nash equilibria and pure Bayesian Nash equilibria, respectively. We denote the cost of any solution A as $c(A)$. Then, the *Price of Anarchy* (PoA) for the complete information, stochastic and Bayesian settings is defined, respectively, as:

$$PoA = \max_{\substack{\mathbf{t} \\ \mathbf{s} \in NE(\mathbf{t})}} \frac{c(\mathbf{s})}{c(opt(\mathbf{t}))}; \qquad PoA = \max_D \frac{\mathbb{E}_{\mathbf{t} \sim D}[\max_{\mathbf{s} \in NE(\mathbf{t})} c(\mathbf{s})]}{\mathbb{E}_{\mathbf{t} \sim D}[c(opt(\mathbf{t}))]};$$

$$PoA = \max_{D, \mathbf{s} \in BNE} \frac{\mathbb{E}_{\mathbf{t} \sim D, \mathbf{s}(\mathbf{t})}[c(\mathbf{s}(\mathbf{t}))]}{\mathbb{E}_{\mathbf{t} \sim D}[c(opt(\mathbf{t}))]}.$$

3   Lower Bounds for Budget-Balanced Protocols

Theorem 1. *The Bayesian or stochastic PoA of any budget-balanced protocol, for the unweighted set cover game, is $\Omega(n)$.*

Proof. Consider n players and n elements/types $U = (1, \ldots, n)$ and the family of sets $\mathcal{F} = \{F_1 = \{1\}, F_2 = \{2\}, \ldots F_n = \{n\}, F_{all} = U\}$ with unit costs. Suppose that π is the uniform distribution over U. Then the probability that element i is drawn as the type of at least one player is $q_i = 1 - \left(1 - \frac{1}{n}\right)^n \geq 1 - \frac{1}{e}$. By using any budget-balanced protocol, it is a (Bayes) Nash equilibrium if each player of type i chose set F_i. Her cost-share does not exceed 1, while by deviating to F_{all} her cost-share becomes 1. The expected cost of that equilibrium is $nq_i = \Omega(n)$, whereas the optimum solution (all players choose the set F_{all}) has cost 1.   □

Theorem 2. *The Bayesian PoA of any budget-balanced protocol, for the multicast game, is $\Omega(\sqrt{n})$.*

Proof. Consider the graph of Fig. 1. We set $p = 1 - \left(1 - \frac{1}{\sqrt{n}}\right)^{\frac{1}{n}}$, such that the probability that vertex v_i is drawn as the type of at least one player is $q_i = 1 - (1 - p)^n = \frac{1}{\sqrt{n}}$. We claim that, for any budget-balanced protocol, it is a Bayes-Nash equilibrium if any player with type v_i uses the direct edges (v_i, t). Indeed, if player i uses any other path (v_i, v, v_j, t) her cost-share will be at least $\frac{2}{\sqrt{n}} + (1 - q_j) = 1 + \frac{1}{\sqrt{n}}$, which is greater than her current cost-share of at most 1. The expected social cost and optimum are: $\mathbb{E}[SC] = \sum_i q_i = \sqrt{n}$ and $\mathbb{E}[Opt] \leq \sum_i q_i \cdot \frac{1}{\sqrt{n}} + 1 = n\frac{1}{n} + 1 = 2$. So, the Bayes PoA is at least $\frac{1}{2}\sqrt{n}$.   □

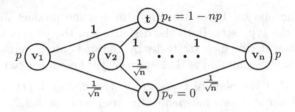

Fig. 1. Lower bound on the PoA of any budget-balanced protocol.

4 BBiE Protocols

In this section we drop the requirement of budget balance and instead we consider a more general class of cost-sharing protocols $\mathcal{C}$, where the requirement is to preserve the budget balance in the equilibrium. For the rest of the paper, by h we denote a very high value with respect to the parameters of the game. h should be larger than the total cost-share of any player by using any budget-balanced protocol. It is safe to assume that $h > \sum_{r \in R} c_r$. For the set cover game it is sufficient that $h > \max_j c_{F_j}$. To show our results we will use known oblivious algorithms of the corresponding optimization problems and we will enforce their solution by applying appropriate cost-sharing protocols (or posted prices in Sect. 5); e.g. choices, not consistent with this solution, are highly expensive.

The types of the players correspond to the input components of the problem, and the set of the resources are the domain of players action space. An oblivious algorithm assigns an action for each input component, based on the prior distribution, and *independently of the realization* of all other input components. Take as an example, the multicast game, where the actions of an input (source) corresponds to the paths connecting the source to the root. An oblivious solution, maps each vertex to some path that connects it to the root, and is used in *any realization* of the input that contains this source.

Theorem 3. *Let G be any cost-sharing game and Π the underlying optimization resource allocation problem. Given any oblivious algorithm of Π with approximation ratio ρ, there exists a cost-sharing protocol $\Xi \in \mathcal{C}$ for G with $PoA = O(\rho)$.*

The following corollaries hold for both the Bayesian and the stochastic setting.

Set Cover Game. Grandoni et al. [19] studied the stochastic problem, and they showed two mapping algorithms for the oblivious set cover problem (one for the *unweighted* problem which is *length-oblivious* and one for the *unweighted* problem which is *length-oblivious*), which are almost $O(\log mn)$-competitive. Theorem 3 implies the following corollary.

Corollary 4. *In the unweighted and weighted set cover game, there exist length-oblivious protocol $\Xi_1 \in \mathcal{C}$ and length-aware protocol $\Xi_2 \in \mathcal{C}$, respectively, both computed in polynomial time, and with PoA of $O(\log n)$, if $m = poly(n)$, and $O\left(\frac{\log m}{\log \log m - \log \log n}\right)$, if $m \gg n$.*

Multicast Game. Garg et al. [17] showed a constant approximation on the online Steiner tree problem. The idea is the following: sample a set S from the distribution π^k over the vertices and construct a minimum Steiner tree (or a constant approximation). Then connect each other vertex with its nearest vertex from S via shortest path. That way we end up with a spanning tree T (standard derandomization techniques can apply [10,29,31]). T defines a single path from each vertex to the root and this is an oblivious strategy for each players' type. By using Theorem 3 and any constant approximation of the minimum Steiner tree (the best known is by [5]), the following corollary holds.

Corollary 5. *In the multicast game, there exists $\Xi \in \mathcal{C}$ with $PoA = O(1)$.*

5   Posted Prices

In this section, we show how to set *anonymous* or *semi-anonymous* prices for the resources. Ex-post BBiE cannot be obtained by using anonymous posted prices. Instead, we require *ex-ante* BBiE. For the rest of the section we define k_A to be the expected number of players having type in A and k_A^1 to be the expected number of players having type in A, given there exists at least one such player:

$$k_A = \mathbb{E}_\mathbf{t}[|i : t_i \in A|] = k \sum_{i \in A} \pi_i;$$

$$k_A^1 = \mathbb{E}_\mathbf{t}[|i : t_i \in A| \text{ given } |i : t_i \in A| \geq 1] = \frac{k \sum_{i \in A} \pi_i}{1 - \left(1 - \sum_{i \in A} \pi_i\right)^k}. \qquad (1)$$

Set Cover Game. To determine anonymous prices for the unweighted set cover game, we first state Lemma 6 to be used in stability arguments.

Lemma 6. *For any $a > b > 0$ and integer $k \geq 2$, $\frac{a}{1-(1-a)^k} > \frac{b}{1-(1-b)^k}$.*

Proposition 7. *In the unweighted set cover game, there exist length-oblivious and anonymous prices (computed in polynomial time) with PoA $O(\log n)$, if $m = poly(n)$, and $O\left(\frac{\log m}{\log\log m - \log\log n}\right)$, if $m \gg n$.*

Proof. In order to set the prices, we run the greedy algorithm of [11] and at each step we set the price for the selected set. Algorithm 1 describes this procedure.

We first argue that there exists a unique Bayes-Nash equilibrium, where each player i chooses the set picked earlier by Algorithm 1 and covers her. For that it is sufficient to show that for any two sets A and B, such that $\sum_{i \in A} \pi_i > \sum_{i \in B} \pi_i$, $k_A^1 > k_B^1$. From (1), we need to show that $\frac{k \sum_{i \in A} \pi_i}{1-\left(1-\sum_{i \in A} \pi_i\right)^k} > \frac{k \sum_{i \in B} \pi_i}{1-\left(1-\sum_{i \in B} \pi_i\right)^k}$, which is true due to Lemma 6, by setting $a = \sum_{i \in A} \pi_i$ and $b = \sum_{i \in B} \pi_i$; note that for $k = 1$, there exists only one player and this is a trivial case.

Next notice that, given that a set F is chosen by some player, the expected number of players paying for it is k_F^1, resulting in ex-ante BBiE. As for the PoA,

---

ALGORITHM 1. Bayesian posted prices.

Input: $(U, \mathcal{F})$.
while $U \neq \emptyset$ **do**
    let $F \leftarrow$ set in $\mathcal{F}$ maximizing $\sum_{i \in F \cap U} \pi_i$;
    set the price for F to $\frac{1}{k_{F \cap U}^1}$; Let $U \leftarrow U \setminus F$.
end
Set the price of all other sets to h.

Grandoni et al. [19] analyzed the performance of Algorithm 1, for the stochastic problem. They didn't consider any prices, instead they mapped each player to the first set considered by the algorithm and they used the mapping in order to form a set cover. Their cover though coincide with the equilibrium solution and therefore their results immediately provide bounds on the PoA. □

Proposition 8. *In the* weighted *set cover game, there exist* length-aware *and* semi-anonymous *prices (computed in polynomial time) with PoA* $O(\log n)$, *if* $m = poly(n)$, *and* $O\left(\frac{\log nm}{\log\log m - \log\log n}\right)$, *if* $m \gg n$.

Proposition 9. *For* $k = \Omega(n)$, *there are* no *anonymous prices for the* unweighted set cover, *or* semi-anonymous *prices for the* weighted set cover, *with* $PoA = o\left(\frac{\log m}{\log\log m - \log\log n}\right)$, *for* $m \gg n$. *Moreover, there are* no *such prices computed in poly-time, with* $PoA = o(\log n)$ *for* $m = poly(n)$, *unless* $NP \subseteq DTIME(n^{O(\log\log n)})$.

Multicast Game. We construct a spanning tree T in the same way as in Sect. 4 and we use it to set the posted prices (computed in polynomial time).

Proposition 10. *In the multicast game, there exist prices with* $PoA = O(1)$.

Proof. For each edge $e \in E(T)$, let $V(e)$ be the set of vertices that are disconnected from the root t in $T \setminus \{e\}$. We set the price for each $e \in E(T)$ as $c_e / k_{V(e)}^1$. For each $e \notin E(T)$, the price is set to h. In the equilibrium each player chooses the path that connects her terminal with t via T. The constant PoA follows by [17] and the approximation of [5]. The expected total prices for $e \in E(T)$ is $k_{V(e)}^1 c_e / k_{V(e)}^1 = c_e$, if e is used, and 0 otherwise, resulting in ex-ante BBiE. □

6 Prior-Independent Mechanisms

The design of prior-independent mechanisms is a more difficult task, as the objective now is to identify a single mechanism that always has good performance, under any distributional assumption. In this section, we show limitations of prior-independent mechanisms even for the restricted class of i.i.d. prior distributions.

BBiE Protocols. Satisfying BBiE with prior-independent protocols highly restricts the class of cost-sharing protocols and seems hard for natural classes of distribution, e.g. i.i.d., to find ex-post BBiE protocols with low PoA.

Proposition 11. *In the* weighted *set cover game, any prior-independent, ex-post BBiE protocol* $\Xi \in \mathcal{C}$ *has PoA* $= \Omega(\sqrt{n})$.

Proof. Consider n players, $n+1$ elements/types $U = \{0, 1, \ldots, n\}$ and the family of sets $\mathcal{F} = \{F_0, F_1, \ldots F_n, F_{all}\}$, with $F_j = \{j\}$, $c_{F_j} = 1$ for all j, and $F_{all} = \{1, \ldots, n\}$, $c_{F_{all}} = \sqrt{n}$. Note that 0 is covered only by F_0, serving as dummy set.

Given a BBiE, prior-independent protocol Ξ, suppose that there exists some F_j, $j \neq 0$, where Ξ is not budget-balanced, i.e. there exists a set of players S, such that if only S chooses F_j, the sum of their cost-shares are different from 1. Consider the prior distribution $D_1 = \pi^n$ with $\pi(0) = \pi(j) = 1/2$ and $\pi(j') = 0$ for any $j' \notin \{0, j\}$. With positive probability, $1/2^n$, all player of S have type j and all other players have type 0. If all players of S choose F_j in any *pure* Bayes-Nash equilibrium, ex-post BBiE is violated. So, there exists a player choosing F_{all} (and this happens with probability $1/2$) which results in PoA $= \Omega(\sqrt{n})$.

Suppose now that Ξ is budget-balanced for any F_j, where $j \neq 0$. Let I be the set of players such that whenever $i \in I$ is the only player choosing F_{all}, Ξ doesn't charge $\sqrt{n}$ to i. Consider the prior distribution $D_2 = \pi^n$ with $\pi(0) = 1/2$ and $\pi(j) = 1/2n$ for all other j. With positive probability, $1/(2^n n)$, player i's type is some $j \neq 0$ and all other players' type is 0. If for any type $j \neq 0$ player i chooses F_{all} in any Bayes-Nash equilibrium, ex-post BBiE is violated.

We claim that the strategy profile, where any player i with type t_i chooses F_{t_i} is a Bayes-Nash equilibrium. For any player $i \in I$ there is no other valid strategy. For each player $i \notin I$, whenever $t_i \neq 0$, player i always pays at most 1 (due to budget balanced in F_{t_i}), whereas if she deviates to F_{all} she pays $\sqrt{n}$.

Each element $j \neq 0$ is a type of a player with probability $1 - \left(1 - \frac{1}{2n}\right)^n \geq 1 - \frac{2}{e}$, giving an expected cost of $\Omega(n)$ in the equilibrium. The expected optimum is at most $1 + \sqrt{n}$ by using only F_0 and F_{all} and so PoA $= \Omega(\sqrt{n})$. $\square$

Posted Prices. Setting posted prices in the adversarial model cannot guarantee any budget-balance in equilibrium, even ex-ante. Consider the set cover game (similar example exists for the multicast game) with n players, n elements and two subsets of unit costs, one containing element 1 and the other containing the rest. Suppose now that we post a price q for the first subset. If $q \leq 1/\sqrt{n}$, for the uniform prior distribution, the expected number of players with type 1, given that there exists at least one, is $\frac{n \cdot 1/n}{1 - (1 - 1/n)^n} \leq \frac{e}{e-1}$. The expected cost shares for the first set are $O(1/\sqrt{n})$, meaning that its cost is undercovered by a factor of $\Omega(\sqrt{n})$. If $q > 1/\sqrt{n}$, consider the prior $D = \pi^n$, where $\pi(1) = 1$ and $\pi(j) = 0$ for all $j \neq 1$. All players choose the first set and their total shares are $n \cdot 1/\sqrt{n} = \sqrt{n}$ which exceeds the set's cost by a factor of $\sqrt{n}$. So, there is no way to avoid an over/under-charge of a resource by a factor better than $\Theta(\sqrt{n})$.

References

1. Abed, F., Correa, J.R., Huang, C.-C.: Optimal coordination mechanisms for multi-job scheduling games. In: Schulz, A.S., Wagner, D. (eds.) ESA 2014. LNCS, vol. 8737, pp. 13–24. Springer, Heidelberg (2014)

2. Anshelevich, E., Dasgupta, A., Kleinberg, J.M., Tardos, É., Wexler, T., Roughgarden, T.: The price of stability for network design with fair cost allocation. SIAM J. Comput. **38**(4), 1602–1623 (2008)
3. Bhattacharya, S., Kulkarni, J., Mirrokni, V.: Coordination mechanisms for selfish routing over time on a tree. In: Esparza, J., Fraigniaud, P., Husfeldt, T., Koutsoupias, E. (eds.) ICALP 2014. LNCS, vol. 8572, pp. 186–197. Springer, Heidelberg (2014)
4. Bilò, V., Flammini, M., Moscardelli, L.: The price of stability for undirected broadcast network design with fair cost allocation is constant. In: FOCS, pp. 638–647. IEEE (2013)
5. Byrka, J., Grandoni, F., Rothvoß, T., Sanità, L.: An improved LP-based approximation for steiner tree. In: STOC, pp. 583–592. ACM (2010)
6. Chen, H., Roughgarden, T., Valiant, G.: Designing network protocols for good equilibria. SIAM J. Comput. **39**(5), 1799–1832 (2010)
7. Christodoulou, G., Koutsoupias, E., Nanavati, A.: Coordination mechanisms. Theor. Comput. Sci. **410**(36), 3327–3336 (2009)
8. Christodoulou, G., Kovács, A., Schapira, M.: Bayesian combinatorial auctions. In: Aceto, L., Damgård, I., Goldberg, L.A., Halldórsson, M.M., Ingólfsdóttir, A., Walukiewicz, I. (eds.) ICALP 2008, Part I. LNCS, vol. 5125, pp. 820–832. Springer, Heidelberg (2008)
9. Christodoulou, G., Mehlhorn, K., Pyrga, E.: Improving the price of anarchy for selfish routing via coordination mechanisms. Algorithmica **69**(3), 619–640 (2014)
10. Christodoulou, G., Sgouritsa, A.: Designing networks with good equilibria under uncertainty. In: SODA, pp. 72–89. SIAM (2016)
11. Chvatal, V.: A greedy heuristic for the set-covering problem. Math. Oper. Res. **4**(3), 233–235 (1979)
12. Devanur, N.R., Mihail, M., Vazirani, V.V.: Strategyproof cost-sharing mechanisms for set cover and facility location games. Decis. Support Syst. **39**(1), 11–22 (2005)
13. von Falkenhausen, P., Harks, T.: Optimal cost sharing for resource selection games. Math. Oper. Res. **38**(1), 184–208 (2013)
14. Feige, U.: A threshold of ln n for approximating set cover. J. ACM **45**(4), 634–652 (1998)
15. Feldman, M., Gravin, N., Lucier, B.: Combinatorial walrasian equilibrium. In: STOC, pp. 61–70. ACM (2013)
16. Feldman, M., Gravin, N., Lucier, B.: Combinatorial auctions via posted prices. In: SODA, pp. 123–135. SIAM (2015)
17. Garg, N., Gupta, A., Leonardi, S., Sankowski, P.: Stochastic analyses for online combinatorial optimization problems. In: SODA, pp. 942–951. SIAM (2008)
18. Gkatzelis, V., Kollias, K., Roughgarden, T.: Optimal cost-sharing in weighted congestion games. In: Liu, T.-Y., Qi, Q., Ye, Y. (eds.) WINE 2014. LNCS, vol. 8877, pp. 72–88. Springer, Heidelberg (2014)
19. Grandoni, F., Gupta, A., Leonardi, S., Miettinen, P., Sankowski, P., Singh, M.: Set covering with our eyes closed. SIAM J. Comput. **42**(3), 808–830 (2013)
20. Gul, F., Stacchetti, E.: Walrasian equilibrium with gross substitutes. J. Econ. Theory **87**(1), 95–124 (1999)
21. Harsanyi, J.C.: Games with incomplete information played by "Bayesian" players, i–iii. part ii. Bayesian equilibrium points. Manage. Sci. **14**(5), 320–334 (1968)
22. Immorlica, N., Mahdian, M., Mirrokni, V.S.: Limitations of cross-monotonic cost-sharing schemes. ACM Trans. Algorithms **4**(2), 24 (2008)
23. Kelly, F.: Charging and rate control for elastic traffic. Eur. Trans. Telecomm. **8**(1), 33–37 (1997)

24. Kelso, A.S., Crawford, V.P.: Job matching, coalition formation, and gross substitutes. Econometrica **50**(6), 1483–1504 (1982)
25. Koutsoupias, E., Papadimitriou, C.: Worst-case equilibria. In: Meinel, C., Tison, S. (eds.) STACS 1999. LNCS, vol. 1563, pp. 404–413. Springer, Heidelberg (1999)
26. Li, J.: An o(log(n)/log(log(n))) upper bound on the price of stability for undirected shapley network design games. Inf. Process. Lett. **109**(15), 876–878 (2009)
27. Moulin, H., Shenker, S.: Strategyproof sharing of submodular costs: budget balance versus efficiency. Econ. Theory **18**(3), 511–533 (2001)
28. Roughgarden, T.: The price of anarchy in games of incomplete information. ACM Trans. Econ. Comput. **3**(1), 6 (2015)
29. Shmoys, D., Talwar, K.: A constant approximation algorithm for the *a priori* traveling salesman problem. In: Lodi, A., Panconesi, A., Rinaldi, G. (eds.) IPCO 2008. LNCS, vol. 5035, pp. 331–343. Springer, Heidelberg (2008)
30. Syrgkanis, V.: Price of stability in games of incomplete information. CoRR (2015)
31. Williamson, D.P., van Zuylen, A.: A simpler and better derandomization of an approximation algorithm for single source rent-or-buy. Oper. Res. Lett. **35**(6), 707–712 (2007)

Abstracts

Essential μ-Compatible Subgames for Obtaining a von Neumann-Morgenstern Stable Set in an Assignment Game

Keisuke Bando[1,2]($\boxtimes$) and Yakuma Furusawa[1,2]

[1] Department of Business Economics, School of Management,
Tokyo University of Science, Tokyo, Japan
k.bando@rs.tus.ac.jp
[2] Department of Social Engineering,
Graduate School of Decision Science and Technology,
Tokyo Institute of Technology, Tokyo, Japan

Abstract. We study von Neumann-Morgenstern (vNM) stable sets in an assignment game. Previous research has shown that for any given optimal matching μ, the union of the extended cores of all μ-compatible subgames is a vNM stable set. Typically, the set of all μ-compatible subgames includes many elements, most of which are inessential for obtaining the vNM stable set. We introduce the notion of essential μ-compatible subgames, without which one cannot obtain the vNM stable set. We provide an algorithm to find a collection of essential μ-compatible subgames for obtaining the vNM stable set under a mild assumption for the valuation matrix. Our simulation result reveals that the average number of essential μ-compatible subgames is significantly lower than that of all μ-compatible subgames.

Keywords: Two-sided matching · Assignment game · Stable set · Essential μ-compatible subgames

The full version of this paper is available at http://www.soc.titech.ac.jp/info/docs/ DP2016-5.pdf. Keisuke Bando acknowledges the Japan Society for the Promotion of Science for financial support through the Grant-in-Aid for Young Scientists (WAKATE B, No. 16K17079).

© Springer-Verlag Berlin Heidelberg 2016
M. Gairing and R. Savani (Eds.): SAGT 2016, LNCS 9928, p. 343, 2016.
DOI: 10.1007/978-3-662-53354-3

Repeated Multimarket Contact
with Observation Errors

Atsushi Iwasaki[1], Tadashi Sekiguchi[2], Shun Yamamoto[3],
and Makoto Yokoo[3]

[1] University of Electro-Communications, Chofu, Japan
iwasaki@is.uec.ac.jp
[2] Kyoto University, Kyoto, Japan
sekiguchi@kier.kyoto-u.ac.jp
[3] Kyushu University, Fukuoka, Japan
syamamoto@agent.inf.kyushu-u.ac.jp, yokoo@inf.kyushu-u.ac.jp

This paper analyzes repeated multimarket contact with observation errors where two players operate in multiple markets simultaneously. A firm, e.g., Uber, provides its taxi service in multiple distinct markets (areas) and determines its price or allocation in each area, facing an oligopolistic competition, which is often modeled as a prisoners' dilemma. To improve profits, it is inevitably helpful to realize how the firm's rival should behave in an equilibrium. Alternatively, it is pointed out that tacit collusion among firms is likely to occur. It is also desirable for a regulatory agency to theoretically understand the extent of the profits firms earn by collusion. However, despite vast empirical studies that have examined whether multimarket contact fosters cooperation or collusion, little is theoretically known as to how players behave in an equilibrium when each player receives a noisy observation or signal of other firms' actions.

This paper considers a different, but realistic noisy situation where players do not share common information on each other's past history, i.e., *private* monitoring where each player may observe a different signal. For example, although a firm cannot directly observe its rival's action, e.g., prices, it can observe a noisy signal, e.g., its rival's sales amounts. Though analytical studies on this class of games have not been very successful, a *belief-free* approach has established a general characterization where an equilibrium strategy is constructed. However, it is not obvious whether this approach is helpful in examining the effects of multimarket contact. Its tractability may be lost if we deal with any number of markets, so that the number of available actions exponentially increases.

The goal of this paper is to answer the following question: under multimarket contact with private monitoring, can we find a particular class of strategies which can sustain a better outcome than an equilibrium strategy for a single market? To the best of our knowledge, we are the first to construct a strategy designed for multiple markets whose per-market equilibrium payoffs exceed one for a single market. First, we construct an entirely novel strategy whose behavior is specified

---

A full version of the paper can be found at http://arxiv.org/abs/1607.03583. Supported in part by KAKENHI 26280081 and 24220003.

M. Gairing and R. Savani (Eds.): SAGT 2016, LNCS 9928, pp. 344–345, 2016.
DOI: 10.1007/978-3-662-53354-3

by a nonlinear function of the signal configurations. Precisely, a player chooses her action at a period according to in which markets she receives bad signals at the previous period. We call this class of the strategies *nonlinear transition, partial defection*. Then, we show that the per-market equilibrium payoff improves when the number of markets is sufficiently large via the theoretical and numerical analysis.

Author Index

Printed in the United States
By Bookmasters